Analytical Applications of Immobilized Microbial Cells

Analytical Applications of Immobilized Microbial Cells

Contributors

Xia Wang, Zhonghui Gai et al.

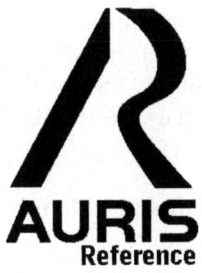

AURIS
Reference

www.aurisreference.com

Analytical Applications of Immobilized Microbial Cells

Contributors: Xia Wang, Zhonghui Gai et al.

Published by Auris Reference Limited

www.aurisreference.com

United Kingdom

Copyright 2016

Printed in 2017 for Sale in the Indian Subcontinent

Notice

Contributors, whose names have been given on the book cover, are not associated with the Publisher. The editors and the Publisher have attempted to trace the copyright holders of all material reproduced in this publication and apologise to copyright holders if permission has not been obtained. If any copyright holder has not been acknowledged, please write to us so we may rectify.

Reasonable efforts have been made to publish reliable data. The views articulated in the chapters are those of the individual contributors, and not necessarily those of the editors or the Publisher. Editors and/or the Publisher are not responsible for the accuracy of the information in the published chapters or consequences from their use. The Publisher accepts no responsibility for any damage or grievance to individual(s) or property arising out of the use of any material(s), instruction(s), methods or thoughts in the book.

Analytical Applications of Immobilized Microbial Cells

ISBN: 978-1-78154-844-8

British Library Cataloguing in Publication Data
A CIP record for this book is available from the British Library

Printed in the United Kingdom

Exclusively distributed by CBS Publishers & Distributors Pvt. Ltd.

Sales & Distribution Rights only for India, Pakistan, Bangladesh, Sri Lanka, Nepal and Bhutan.This book is not to be sold outside these territories.

Contents

List of Abbreviations

AIST	Advanced Industrial Science and Technology
ANOVA	Analysis of Variance
BE	Bioelectrochemical Systems
BOD	Biological Oxygen Demand
COD	Chemical Oxygen Demand
DC	Distal Colon
DGGE	Denaturing Gradient Gel Electrophoresis
DMRT	Duncan's Multiple Range Test
DMSO	Dimethyl Sulfoxide
DOM	Dissolved Organic Matter
DON	Deoxynivalenol
FTIR	Fourier Transform Infrared Spectroscopy
GEO	Gene Expression Omnibus
HPLC	High Pressure Liquid Chromatography
JSPS	Japan Society for the Promotion of Science
LED	Light-Emitting Diode
MBAS	Methylene Blue Active Substances
ME	Microbial Electrolysis Cells
MF	Microbial Fuel Cells
MSM	Mineral Salts Medium
NCWRP	North City Water Reclamation Plant
ORF	Open Reading Frame
OUT	Operational Taxonomy Unit
PA	Polyacrylamide
PA	Polycyclic Aromatic Hydrocarbons
PCD	Programmed Cell Death
PCR	Polymerase Chain Reaction
PDC	Pyruvate Decarboxylase
PPFD	Photosynthetic Photon Flux Density
PUF	Polyurethane Foam
RLU	Relative Light Units
SA	Sodium Alginate
SCFA	Short Chain Fatty Acids
SD	Standard Deviations
SDH	Succinate Dehydrogenase
SDS	Sodium Dodecyl Sulfate
SEM	Scanning Electron Microscope
SHE	Standard Hydrogen Electrode
TAP	Tris-Acetate-Phosphate
TLC	Thin Layer Chromatography

List of Contributors

Xia Wang
State Key Laboratory of Microbial Technology, Shandong University, Jinan 250100, People's Republic of China

Zhonghui Gai
Key Laboratory of Microbial Metabolism, Ministry of Education, College of Life Science and Biotechnology, Shanghai Jiao Tong University, Shanghai 200240, People's Republic of China

Bo Yu
Institute of Microbiology, Chinese Academy of Sciences, Beijing 100080, People's Republic of China

Jinhui Feng
State Key Laboratory of Microbial Technology, Shandong University, Jinan 250100, People's Republic of China

Changyong Xu
State Key Laboratory of Microbial Technology, Shandong University, Jinan 250100, People's Republic of China

Yong Yuan
State Key Laboratory of Microbial Technology, Shandong University, Jinan 250100, People's Republic of China

Zhixin Lin
Key Laboratory of Microbial Metabolism, Ministry of Education, College of Life Science and Biotechnology, Shanghai Jiao Tong University, Shanghai 200240, People's Republic of China

Ping Xu
State Key Laboratory of Microbial Technology, Shandong University, Jinan 250100, People's Republic of China
Key Laboratory of Microbial Metabolism, Ministry of Education, College of Life Science and Biotechnology, Shanghai Jiao Tong University, Shanghai 200240, People's Republic of China
Institute of Microbiology, Chinese Academy of Sciences, Beijing 100080, People's Republic of China

Tadahiro Suzuki
Applied Microbiology Division, National Food Research Institute, 2-1-12 Kannondai, Tsukuba, Ibaraki 305-8642, Japan

Yumiko Iwahashi
Applied Microbiology Division, National Food Research Institute, 2-1-12 Kannon-dai, Tsukuba, Ibaraki 305-8642, Japan

Veena S. More
Department of Biochemistry, MSRCASC, Mattikere, MSRIT Post, Bangalore 560054, India

Preeti N. Tallur
Department of Chemistry, Government Arts and Science College, Karwar 581301, India

Francois N. Niyonzima
Department of Maths, Science and PE, College of Education, University of Rwanda, Kigali 5039, India

Sunil S. More
Department of Biochemistry, Center for Post Graduate Studies, Jain University, Bangalore 560011, India

Joanna Berlowska
Institute of Fermentation Technology and Microbiology, Technical University of Lodz, ul. Wolczanska 171/173, 90-924 Lodz, Poland

Dorota Kregiel
Institute of Fermentation Technology and Microbiology, Technical University of Lodz, ul. Wolczanska 171/173, 90-924 Lodz, Poland

Wojciech Ambroziak
Institute of Fermentation Technology and Microbiology, Technical University of Lodz, ul. Wolczanska 171/173, 90-924 Lodz, Poland

Andreas Blecha
Institut für Genetik, Technische Universität Dresden, D-01062 Dresden, Germany

Kristof Zarschler
Institut für Genetik, Technische Universität Dresden, D-01062 Dresden, Germany

Klaas A Sjollema
Eukaryotic Microbiology, Groningen Biomolecular Sciences and Biotechnology Institute (GBB), University of Groningen, PO Box 14, NL-9750 AA Haren, The Netherlands.

Marten Veenhuis
Eukaryotic Microbiology, Groningen Biomolecular Sciences and Biotechnology Institute (GBB), University of Groningen, PO Box 14, NL-9750 AA Haren, The Netherlands.

Gerhard Rödel
Institut für Genetik, Technische Universität Dresden, D-01062 Dresden, Germany

Valeria Mozzetti
Laboratory of Food Biotechnology, Institute of Food Science and Nutrition, Schmelzbergstrasse 7, ETH-Zurich, 8092 Zürich, Switzerland

Franck Grattepanche
Laboratory of Food Biotechnology, Institute of Food Science and Nutrition, Schmelzbergstrasse 7, ETH-Zurich, 8092 Zürich, Switzerland

Déborah Moine
Nestlé Research Center, Vers-chez-les-Blanc, 1000 Lausanne 26, Switzerland

Bernard Berger
Nestlé Research Center, Vers-chez-les-Blanc, 1000 Lausanne 26, Switzerland

Enea Rezzonico
Nestlé Research Center, Vers-chez-les-Blanc, 1000 Lausanne 26, Switzerland

Leo Meile
Laboratory of Food Biotechnology, Institute of Food Science and Nutrition, Schmelzbergstrasse 7, ETH-Zurich, 8092 Zürich, Switzerland

Fabrizio Arigoni
Nestlé Research Center, Vers-chez-les-Blanc, 1000 Lausanne 26, Switzerland

Christophe Lacroix
Laboratory of Food Biotechnology, Institute of Food Science and Nutrition, Schmelzbergstrasse 7, ETH-Zurich, 8092 Zürich, Switzerland

Meerza Abdul Razak
Natco Pharma Limited, Natco House, Road No. 2, Banjara Hills, Hyderabad 500 034, India

Buddolla Viswanath
Department of Virology, Sri Venkateswara University, Tirupati 517502, A. P, India

Rachel C Wagner
Department of Civil and Environmental Engineering, 212 Sackett Building, The Pennsylvania State University, University Park, PA 16802, USA

Sikandar Porter-Gill
Department of Civil and Environmental Engineering, 212 Sackett Building, The Pennsylvania State University, University Park, PA 16802, USA

Bruce E Logan
Department of Civil and Environmental Engineering, 212 Sackett Building, The Pennsylvania State University, University Park, PA 16802, USA

Sophie Fehlbaum
Laboratory of Food Biotechnology, Institute of Food, Nutrition and Health, ETH Zurich, Zurich, Switzerland

Christophe Chassard
Laboratory of Food Biotechnology, Institute of Food, Nutrition and Health, ETH Zurich, Zurich, Switzerland

Martina C. Haug
Laboratory of Food Biotechnology, Institute of Food, Nutrition and Health, ETH Zurich, Zurich, Switzerland

Candice Fourmestraux
Danone Nutricia Research, Palaiseau, France

Muriel Derrien
Danone Nutricia Research, Palaiseau, France

Mo´ nica V. Orellana
Institute for Systems Biology, Seattle, Washington, United States of America
Polar Science Center, Applied Physics Laboratory, University of Washington, Seattle, Washington, United States of America

Wyming L. Pang
Institute for Systems Biology, Seattle, Washington, United States of America
Genomatica, Inc., San Diego, California, United States of America

Pierre M. Durand
Department of Molecular Medicine, University of the Witwatersrand and National Health Laboratory Service, Parktown, South Africa
Department of Ecology and Evolutionary Biology, University of Arizona, Tucson, Arizona, United States of America

Kenia Whitehead
Institute for Systems Biology, Seattle, Washington, United States of America
Integral Consulting Inc., Seattle, Washington, United States of America

Nitin S. Baliga
Institute for Systems Biology, Seattle, Washington, United States of America
Department of Microbiology, University of Washington, Seattle, Washington, United States of America

Shun'ichi Ishii
J. Craig Venter Institute, San Diego, California, United States of America
Biomedical Research Institute, National Institute of Advanced Industrial Science and Technology (AIST), Tsukuba, Ibaraki, Japan
Japan Society for the Promotion of Science (JSPS), Chiyoda-ku, Tokyo, Japan

Shino Suzuki
J. Craig Venter Institute, San Diego, California, United States of America

Trina M. Norden-Krichmar
J. Craig Venter Institute, San Diego, California, United States of America

Kenneth H. Nealson
J. Craig Venter Institute, San Diego, California, United States of America
University of Southern California, Los Angeles, California, United States of America

Yuji Sekiguchi
Biomedical Research Institute, National Institute of Advanced Industrial Science and Technology (AIST), Tsukuba, Ibaraki, Japan

Yuri A. Gorby
University of Southern California, Los Angeles, California, United States of America

Orianna Bretschger
J. Craig Venter Institute, San Diego, California, United States of America

Preface

Immobilized microbial cells have been used extensively in various industrial and scientific endeavours. However, immobilized cells have not been used widely for environmental applications. The text *Analytical Applications of Immobilized Microbial Cells* examines many of the scientific and technical aspects involved in using immobilized microbial cells in environmental applications, with a particular focus on cells encapsulated in biopolymer gels. Some advantages and limitations of using immobilized cells in bioreactor studies are also discussed. In first chapter, we demonstrate a new process for carbazole biodegradation employing magnetically immobilized cells. An improved and simple method for the immobilization of *Sphingomonas* sp. strain XLDN2-5 in magnetic gellan gel beads was developed, and the stability and activity of the biocatalyst for the degradation of carbazole were also evaluated. Second chapter focuses on toxicity of DON3G using yeast and algae, and a comparison between type B trichothecenes was conducted to reveal their toxic character. The aim of third chapter is to compare the pendimethalin degradation by freely suspended and immobilized cells of *Bacillus lehensis* XJU on various matrices in batch and semi-continuous degradation, and to evaluate the effect of pH, temperature, and storage stability of pendimethalin degradation rate by polyurethane foam (PUF)-immobilized bacterial cells. In fourth chapter, yeast cell physiological activity was assessed on the basis of the in situ activity of two important enzymes, succinate dehydrogenase and pyruvate decarboxylase. In fifth chapter, we addressed the question whether the cytosol of eukaryotic host cells can provide a suitable environment for the formation of S-layer self assembly products, despite the presence of numerous chaperons and proteases. Sixth chapter shows that continuous culture with cell immobilization is a valid approach for selecting cells adapted to hydrogen peroxide. Elucidation of H_2O_2 adaptation mechanisms in HPR2 could be helpful to develop oxygen resistant bifidobacteria. The main objective of seventh chapter is to analyze upstream bioprocess and comparative studies of l-Lysine production by free cells of *Corynebacterium glutamicum* ATCC 13032 and MH 20-22 B in stirred tank bioreactor. In eighth chapter, we present a proof-of-concept immobilization approach that allows exoelectrogenic activity of cells on an electrode based on applying a layer of latex to hold bacteria on surfaces. The aim of ninth chapter is to investigate the use of immobilized fecal microbiota to develop different designs of continuous colonic fermentation models mimicking elderly gut fermentation. Immobilization of fecal microbiota obtained from three different donors was performed independently. Tenth chapter focuses on the impact of programmed cell death (PCD) on a population's growth as well as its role in the exchange of carbon between two naturally co-occurring halophilic organisms. In eleventh chapter, we address the use of microbial fuel cells (MFCs) for the degradation of carbon sources in primary clarifier effluents from a conventional wastewater treatment plant. Last chapter shows the efficacy of alkali ballast water treatment in reducing ballast water microbial diversity and demonstrated the application of new Ion Torrent sequencing techniques to microbial community studies.

Chapter 1

DEGRADATION OF CARBAZOLE BY MICROBIAL CELLS IMMOBILIZED IN MAGNETIC GELLAN GUM GEL BEADS

Xia Wang[1], Zhonghui Gai[2], Bo Yu[3], Jinhui Feng[1], Changyong Xu[1], Yong Yuan[1], Zhixin Lin[2], and Ping Xu[1,2,3]

[1]State Key Laboratory of Microbial Technology, Shandong University, Jinan 250100, People's Republic of China

[2]Key Laboratory of Microbial Metabolism, Ministry of Education, College of Life Science and Biotechnology, Shanghai Jiao Tong University, Shanghai 200240, People's Republic of China

[3]Institute of Microbiology, Chinese Academy of Sciences, Beijing 100080, People's Republic of China

ABSTRACT

Polycyclic aromatic heterocycles, such as carbazole, are environmental contaminants suspected of posing human health risks. In this study, we investigated the degradation of carbazole by immobilized *Sphingomonas* sp. strain XLDN2-5 cells. Four kinds of polymers were evaluated as immobilization supports for *Sphingomonas* sp. strain XLDN2-5. After comparison with agar, alginate, and κ-carrageenan, gellan gum was selected as the optimal immobilization support. Furthermore, Fe_3O_4 nanoparticles were prepared by a coprecipitation method, and the average particle size was about 20 nm with 49.65-electromagnetic-unit (emu) g^{-1} saturation magnetization. When the mixture of gellan gel and the Fe_3O_4 nanoparticles served as an immobilization support, the magnetically immobilized cells were prepared by an ionotropic method. The biodegradation experiments were carried out by employing free cells, nonmagnetically immobilized cells, and magnetically immobilized cells in aqueous phase. The results showed that the magnetically immobilized cells presented higher carbazole biodegradation activity than nonmagnetically immobilized cells and free cells. The highest biodegradation activity was obtained when the concentration of Fe_3O_4 nanoparticles was 9 mg ml^{-1} and the saturation magnetization of magnetically immobilized cells was 11.08 emu g^{-1}. Additionally, the recycling experiments demonstrated that the degradation activity of magnetically immobilized cells increased gradually during the eight recycles. These results support developing efficient biocatalysts using magnetically immobilized cells and provide a promising technique for improving biocatalysts used in the biodegradation of not only carbazole, but also other hazardous organic compounds.

The pollution of soil, river sediments, and ground water by hazardous organic compounds has been gaining increasing attention in the field of environmental remediation. Carbazole and its derivatives are common nitrogen compounds found in environments contaminated by coal tar, crude oil, and creosote (13). When these nitrogen compounds are combusted, nitrogen oxides, NO$_x$, are released into the atmosphere, which causes acid rain and air pollution (23). Moreover, carbazole and its derivatives have been found to be toxic (24, 34) and mutagenic (10, 27), and they readily undergo radical chemistry to generate the more poisonous hydroxynitrocarbazoles (1). Soil, river sediments, and ground water polluted by carbazole have become a great threat to the environment. Therefore, it is necessary to establish effective methods to clean up carbazole and its derivatives to protect the environment.

Many researchers have focused their studies on the isolation and identification of carbazole-degrading microorganisms, such as *Pseudomonas* (7, 25), *Sphingomonas* (32), *Ralstonia* (29), *Bacillus* (14), *Gordonia*(28), and *Mycobacterium* and *Xanthamonas* (8). Our laboratory has also isolated and constructed several bacteria that can degrade these heterocyclic compounds (6, 18, 19, 20, 43, 44). The isolates degrade carbazole by following similar pathways, in which carbazole is initially attacked at the angular position by dioxygenation, followed by spontaneous conversion of the dihydroxylated intermediate to 2'-aminobiphenyl-2,3-diol. The extradiol dioxygenase attacks the hydroxylated ring at the *meta* position to give 2-hydroxy-6-(2-aminophenyl)-6-oxo-2,4-hexadienoic acid. This *meta*-cleavage product is hydrolyzed to produce anthranilic acid, which is then mineralized (42). Nevertheless, current studies are mostly focused on the pathways of such chemical metabolisms, as well as the genes and enzymes involved, and rarely on the development of an immobilization method for bioremediation.

The use of immobilized microorganisms rather than free cells in biotransformation is advantageous to enhance the stability of the biocatalyst and to facilitate its recovery and reuse. These advantages have encouraged researchers to investigate the application of immobilized cells in the biodegradation of toxic compounds, such as phenol, pyridine, dibenzothiophene, and quinoline (5, 15, 17, 37, 38). However, mass transfer limitation involved in substrate diffusion to the reaction system is still the major drawback in the application of an entrapment technique.

Nanoparticles represent a new generation of environmental-remediation technologies that could provide cost-effective solutions to some of the most challenging environmental clean-up problems. There are two factors that contribute to the capabilities of nanoparticles as an extremely versatile remediation tool. First, the size (1 to 100 nm) that characterizes nanoparticles provides them with large specific surfaces and high specific energies. Second, their flexibility makes them versatile both in situ and ex situ (12, 45). Recent laboratory research has largely established nanoparticles as an effective method for removing a broad range of environmental contaminants, such as chlorinated ethenes, heavy metal ions, dibenzothiophene, and polycyclic aromatic hydrocarbons (PAHs) (16, 26, 30, 31, 35). Therefore, the exploitation of nanoscale technology in environmental applications appears very promising.

In this study, we demonstrate a new process for carbazole biodegradation employing magnetically immobilized cells. An improved and simple method for the immobilization of *Sphingomonas* sp. strain XLDN2-5 in magnetic gellan gel beads was developed, and the stability and activity of the biocatalyst for the degradation of carbazole were also evaluated.

MATERIALS AND METHODS

Microorganism and Cultivation

Sphingomonas sp. strain XLDN2-5, which can use carbazole as the sole source of carbon, nitrogen, and energy, was cultivated in mineral salts medium (MSM) as previously described (6). Cells were harvested in the exponential phase (the optical density was about 0.68 to 0.70 at 620 nm) by centrifugation at 12,000 $\times$ g for 10 min. The pellet was washed twice with distilled water and then resuspended in distilled water. The inactive controls were made using heat-killed cells (autoclaved at 115°C for 20 min).

Chemicals

Analytical grade carbazole was purchased from Sigma-Aldrich (St. Louis, MO). Gellan gum was prepared as previously described (39, 40). All other chemicals were of analytical grade and commercially available.

Preparation of Fe_3O_4 Nanoparticles

The Fe_3O_4 nanoparticles were prepared by the conventional coprecipitation method (21) with some modifications: 23.5 g $FeCl_3 \cdot 6H_2O$ and 8.6 g $FeCl_2 \cdot 4H_2O$

were dissolved in 600 ml distilled water at 30°C. Before reaction, N_2 gas flowed through the reaction medium to prevent possible oxidation. $NH_3 \cdot H_2O$ (8 M) was then slowly injected into the mixture of $FeCl_3$ and $FeCl_2$ with vigorous stirring until the pH reached 10. After precipitation, the Fe_3O_4 particles were repeatedly washed until the pH was constant and then lyophilized for 48 h under vacuum to form powder. Fe_3O_4 powder (1.5 g) was put into 10 ml distilled water to form the Fe_3O_4 particle suspension. After ultrasonic disruption (25 KHz; 10 min; BUG25-06; BRANSON) of the suspension, the Fe_3O_4 nanoparticles were well dispersed in distilled water to form a stable suspension, which we called the magnetic suspension. The average diameter of the Fe_3O_4 particles was about 20 nm (Fig. 1a), and their saturation magnetization was 49.65 electromagnetic units (emu) g^{-1} (Fig. 1b1). The Fourier transform infrared spectroscopy (FTIR) absorption band of the Fe—O bond was about 603 cm^{-1} (Fig. 1c).

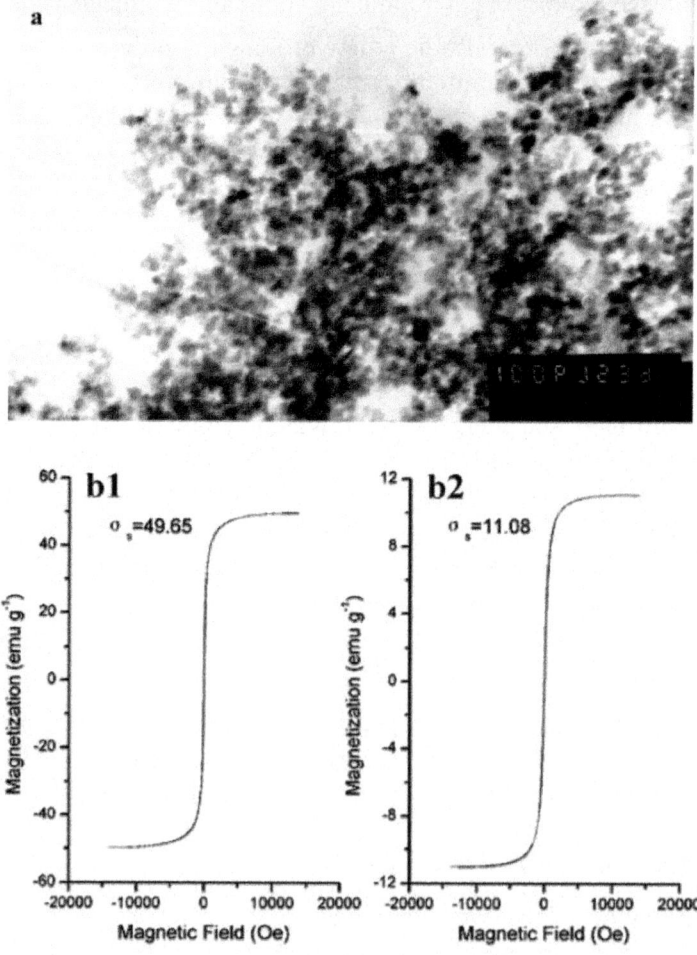

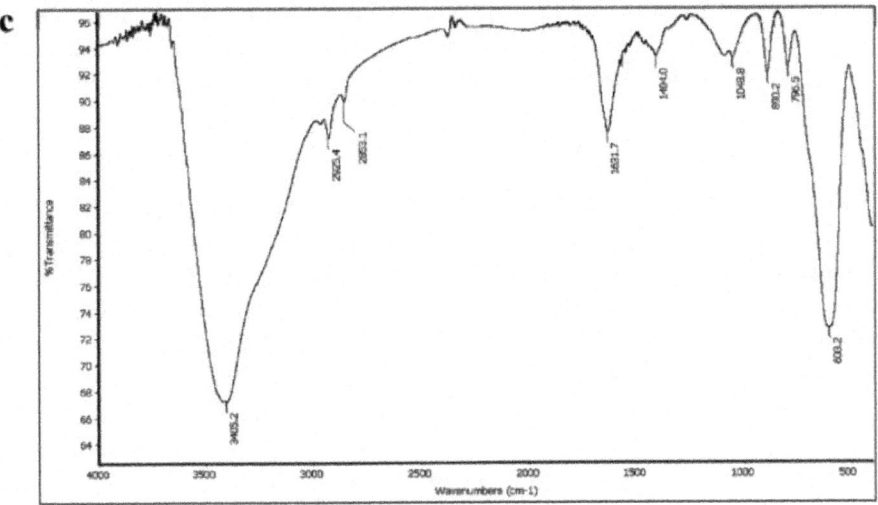

Figure. 1: The nature of Fe_3O_4 nanoparticles. (a) Transmission electron microscopy photograph of Fe_3O_4 nanoparticles (magnification, ×100,000). (b1) Magnetic curve for Fe_3O_4 nanoparticles. (b2) Magnetic curve for magnetically immobilized cells (σ_s, saturation magnetization; Oe, Oersted). (c) FTIR spectra of Fe_3O_4 nanoparticles.

Preparation of Gel Beads and Nonmagnetically and Magnetically Immobilized Cells

The ionotropic method described by Woodward was used to form gel beads from gellan gum and alginate (41). The interphase technique described by López et al. was used to form gel beads from agar and κ-carrageenan (22).

The gellan gel (1% [wt/vol]) and cell suspension were mixed at a ratio of cell wet weight to dry gellan gum powder of 3 (wt/wt). Nonmagnetically immobilized cells were formed by extruding the mixture through a syringe into 0.2 M $CaCl_2$ and letting it solidify for 2 h. For preparing magnetically immobilized cells, an appropriate Fe_3O_4 particle suspension was added to the above-mentioned mixture of gellan gel and cell suspension, and the procedure was the same as that for nonmagnetically immobilized cells. Nonmagnetically immobilized inactive cells and magnetically immobilized inactive cells were prepared as described above.

Adsorption Experiments

Agar, alginate, κ-carrageenan, and gellan gum were used as adsorbents in adsorption experiments. All adsorption experiments were carried out in 100-ml flasks containing 10 ml MSM at 30°C on a reciprocal shaker at 180 rpm.

In each experiment, 3,340 μg of carbazole was added to MSM, and gel beads made with 2 ml gel served as the adsorbent. All the subsequent experiments contained the same amounts of carbazole and gel beads described above.

Biodegradation Experiments

In biodegradation experiments, the initial content of carbazole and the incubation conditions were the same as those of the adsorption experiments. Nonmagnetically immobilized cells and magnetically immobilized cells were added to MSM with carbazole as a biocatalyst. The controls were gellan gel beads, nonmagnetically immobilized inactive cells, and magnetically immobilized inactive cells incubated in MSM with carbazole. In the recycling experiments, after each biodegradation batch, the magnetically immobilized cells were collected by application of a magnetic field and then were washed once with MSM to remove the free cells. After the MSM was drained, 10 ml of fresh MSM containing carbazole was added to repeat the cycle. Additionally, the same amount of cells (the cell wet weight was 60 mg) was used in all batch biodegradation experiments, including the first cycle of reuse experiments. All experiments were performed in triplicate. The specific biodegradation rate was expressed as the amount of carbazole (in μg) consumed by 1 g (wet weight) of cells per hour.

Analytical Methods

After each batch of biodegradation, the biodegradation mixture was filtered through glass wool to separate the gel beads from the supernatant. Then, 20 ml and 4 ml ethanol were added to the supernatant and the gel beads, respectively, followed by centrifugation (12,000 rpm for 20 min) and filtration. The residual carbazole contents were determined using high-performance liquid chromatography performed with an Agilent 1100 series (Hewlett-Packard) instrument equipped with a reverse-phase C_{18} column (4.6 mm by 150 mm; Hewlett-Packard). The mobile phase was a mixture of methanol and deionized water (90:10 [vol/vol]) at a flow rate of 0.5 ml min^{-1}, and carbazole was monitored at 254 nm with a variable-wavelength detector.

The sizes and morphologies of the magnetic nanoparticles were determined by transmission electron microscopy (JEM-100cx II; JEOL, Japan). Each sample was prepared by evaporating a drop of properly diluted nanoparticle suspension on a carbon copper grid. The morphology of cells immobilized in gel beads was determined using a scanning electron microscope (SEM) (S-570; Hitachi, Japan). Magnetization curves for the magnetically immobilized cells were obtained with a vibrating sample magnetometer (MicroMag 2900/3900).

The spectra of FTIR were obtained on a NEXUS 380 (Nicolet). To determine the average size of the beads, a direct measurement was carried out. The diameters of 30 randomly chosen beads were measured with a Vernier caliper. The breakage of beads was determined as the quotient of the number of broken beads divided by the total number of beads.

RESULTS

Selection of Immobilization Supports

Four kinds of polymers were evaluated as immobilization supports for *Sphingomonas* sp. strain XLDN2-5. Adsorption experiments showed that the highest adsorption of carbazole was presented by magnetic gellan gel beads (24.10 mg g^{-1}), followed by gellan gel beads (12.15 mg g^{-1}), agar (4.44 mg g^{-1}), and κ-carrageenan (0.77 mg g^{-1}) at equilibrium. In contrast, all calcium alginate gel beads were broken. In order to investigate the most suitable immobilization support, the physical properties (size, formation, and breakage) of gel beads were also studied. As shown in Table 1, all gel beads were spherical and homogeneous. The gel beads of gellan gum, κ-carrageenan, and agar were more robust (breakages were zero) and presented higher breakage resistance than those of calcium alginate.

Table 1. Properties of gel beads prepared with different polymers

Polymer	Concn (% [wt/vol])	Bead-forming procedure	Form	Bead size (mm)^a	Breakage (%)
Gellan gum	1	Ionotropic	Spherical beads	2.57 ± 0.12	0
Agar	2	Interphase	Spherical beads	2.98 ± 0.16	0
Alginate	3	Ionotropic	Spherical beads	2.25 ± 0.09	100
κ-Carrageenan	2	Interphase	Spherical beads	2.93 ± 0.17	0

^{a}Values are the means ± standard deviations of three separate determinations.

We also investigated the biodegradation activities of nonmagnetically immobilized *Sphingomonas* sp. strain XLDN2-5 in gellan gum, κ-carrageenan, and agar. Figure 2a shows that 3,340 μg carbazole could be degraded in 20 h by free cells. The equivalent amount of carbazole could be degraded in 36 h by immobilized cells when gellan gum served as the immobilization support. In contrast, when beads of immobilized inactive cells and beads without cells were used for the degradation reaction, no decrease in the total content of carbazole was detected (Fig. 2b). The activities of cells immobilized by κ-carrageenan and agar were lower, and the residual contents of carbazole were 2,381 μg and 1,875 μg after 48 h of incubation, respectively (Fig. 2c and d). Therefore, gellan gum was chosen as the most suitable support in the subsequent experiments.

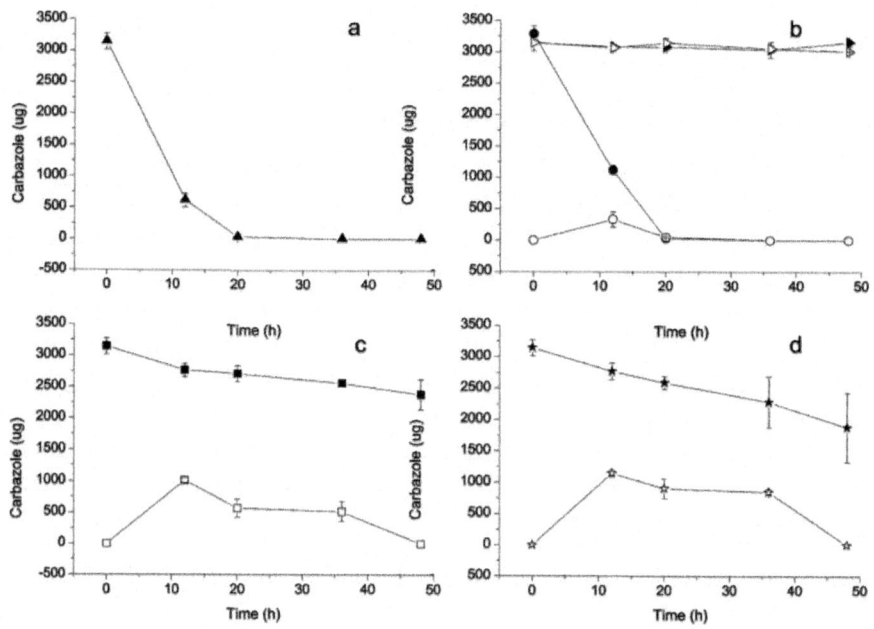

Figure. 2: Carbazole contents of the supernatant and gel beads during biodegradation by immobilized cells in different supports. (a) Biodegradation by free cells. (b) Biodegradation by immobilized cells in gellan gum. (c) Biodegradation by immobilized cells in κ-carrageenan. (d) Biodegradation by immobilized cells in agar. The solid symbols, including free cells (▲), cells immobilized by gellan gum (•), cells immobilized by κ-carrageenan (▪), and cells immobilized by agar (★), represent the supernatant; the open symbols, including gellan beads (○), κ-carrageenan beads (□), and agar beads (⋆), represent the gel beads. Gellan gel beads without cells (▷) and nonmagnetically immobilized inactive cells (▶) were controls. The same amounts of cells (the cell wet weight was 60 mg) were used in all experiments. The error bars represent standard deviations.

Biodegradation of Carbazole by Magnetically Immobilized Cells

The effects of different concentrations of Fe_3O_4 nanoparticles (3, 6, 9, and 12 mg ml^{-1}) on the activity of immobilized cells were studied. The biodegradation of carbazole was conducted in MSM by free cells, nonmagnetically immobilized cells, and magnetically immobilized cells. Figure 3 shows that the highest biodegradation activity for carbazole was presented by the magnetically immobilized cells, and 3,340 µg carbazole could be degraded completely in 16 h to 18 h. While the equivalent amount of carbazole could be degraded completely in 20 h by free cells, the residual carbazole content was 1,035 µg at 20 h with nonmagnetically immobilized cells. In contrast, no decrease of the total carbazole content was observed when magnetically immobilized inactive cells and nonmagnetically immobilized inactive cells served as biocatalysts.

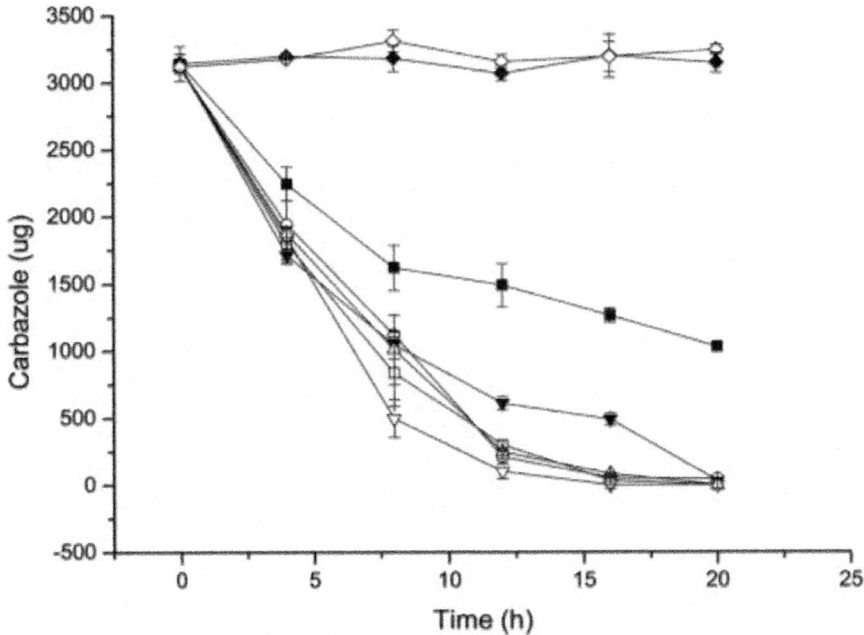

Figure. 3: Biodegradation of carbazole by magnetically immobilized cells at different concentrations of Fe_3O_4 nanoparticles. Free cells, ▾; nonmagnetically immobilized cells, ▪; magnetically immobilized cells at an Fe_3O_4 nanoparticle concentration of 3 mg ml^{-1}, ○; magnetically immobilized cells at an Fe_3O_4 nanoparticle concentration of 6 mg ml^{-1}, □; magnetically immobilized cells at an Fe_3O_4 nanoparticle concentration of 9 mg ml^{-1}, ▿; magnetically immobilized cells at an Fe_3O_4 nanoparticle concentration of 12 mg ml^{-1}, ▵; control nonmagnetically immobilized inactive cells, ◇; control magnetically immobilized inactive cells, ◆. The error bars represent standard deviations.

Reuse of Magnetically Immobilized Cells and Nonmagnetically Immobilized Cells for Carbazole Biodegradation

The activities of magnetically immobilized cells (at the optimal Fe_3O_4 nanoparticle content of 9 mg ml^{-1}) and nonmagnetically immobilized cells were tested repeatedly. As shown in Fig. 4a, from the first to the fifth cycle,

3,340 µg carbazole was completely consumed by magnetically immobilized cells in 16 h; from the sixth to the eighth cycle, the same amount of carbazole was completely consumed in only 12 h. In contrast, from the first to the fifth cycle, the residual carbazole contents were about 123 µg to 232 µg at 20 h with nonmagnetically immobilized cells, and from the sixth to the eighth cycle, the residual carbazole contents were about 210 µg to 324 µg (Fig. 4b).

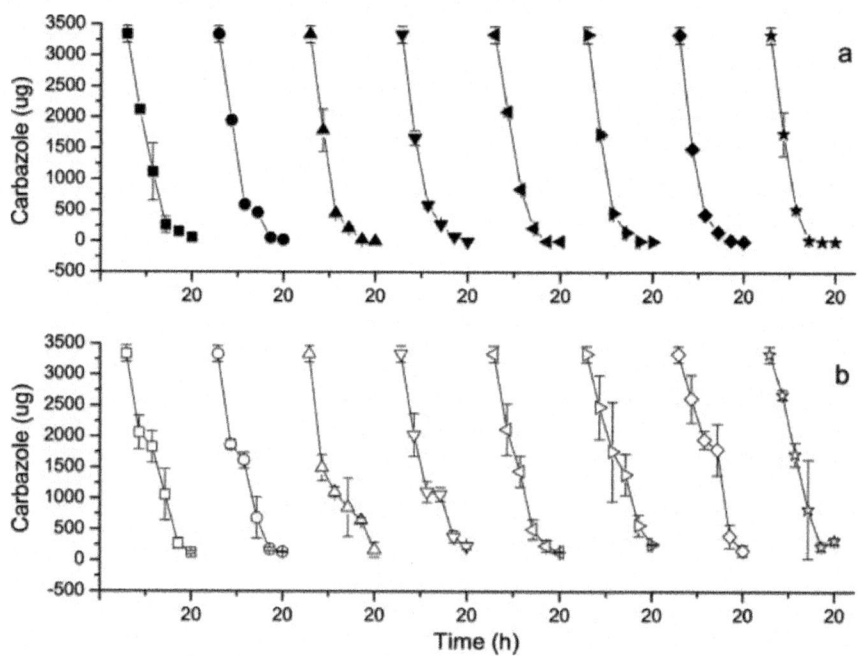

Figure. 4: Reuse of magnetically immobilized cells (a) and nonmagnetically immobilized cells (b) for carbazole biodegradation. Symbols for magnetically immobilized cells: first cycle, ■; second cycle, •; third cycle, ▲; fourth cycle, ▼; fifth cycle, ◄; sixth cycle, ►; seventh cycle, ◆; eighth cycle, ★. Symbols for nonmagnetically immobilized cells: first cycle, □; second cycle, ○; third cycle, △; fourth cycle, ▽; fifth cycle, ◁; sixth cycle, ▷; seventh cycle, ◇; eighth cycle, ☆. The error bars represent standard deviations.

SEM Images of *Sphingomonas* sp. Strain XLDN2-5 Immobilized in Gellan Gel Beads and Magnetic Gellan Gel Beads

SEM images of *Sphingomonas* sp. strain XLDN2-5 are shown in Fig.5. The *Sphingomonas* sp. strain XLDN2-5 cells can be clearly observed on the surfaces of the gellan gel beads (Fig. 5a1), while inside the gellan gel beads, the sheets of gellan gum matrix were tightly bound together and bacterial cells were not clearly observed (Fig. 5a2). The SEM images of *Sphingomonas* sp. strain XLDN2-5 immobilized on the surfaces of magnetic gellan gel beads are almost the same as those of the surfaces of gellan gel beads (Fig. 5b1). However, the sheets of gellan gum matrix were loosely bound together inside magnetic gellan gel beads, and many pores existed between the sheets of gellan gum matrix (Fig. 5b2). Figures 5c1 and 5c2 are SEM images of *Sphingomonas* sp. strain XLDN2-5 immobilized in magnetic gellan gel beads after eight cycles of the biodegradation experiments. On the surfaces and inside of magnetic gellan gel beads, the amounts of cells evidently increased. Moreover, the sheets of gellan gum matrix were also loosely bound together inside magnetic gellan gel beads, and many pores also existed between the sheets of gellan gum matrix.

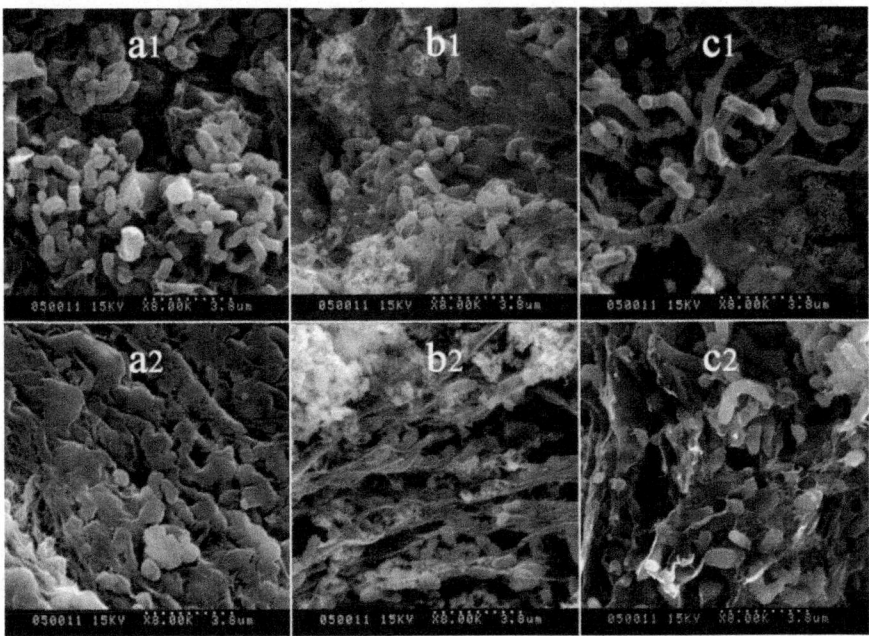

Figure. 5: SEM images of *Sphingomonas* sp. strain XLDN2-5 immobilized in gellan gel beads and magnetic gellan gel beads. (a1 and a2) SEM images of *Sphingomonas* sp. strain XLDN2-5 immobilized in gellan gel beads: surfaces of gellan gel beads (a1); inside of gellan gel beads (a2). (b1 and b2) SEM images of *Sphingomonas* sp. strain XLDN2-5 immobilized in magnetic gellan gel beads: surfaces of magnetic gel-lan gel beads (b1); inside of magnetic gellan gel beads (b2). (c1 and c2) SEM images of *Sphingomonas* sp. strain XLDN2-5 immobilized in magnetic gellan gel beads after eight cycles of degradation experiments: surfaces of magnetic gellan gel beads (c1); inside of magnetic gellan gel beads (c2).

DISCUSSION

Microbial degradation of hydrophobic compounds, such as chlorophenols and PAHs, in soil and sediments is thought to be limited by their mass transfer to the aqueous phase (2, 9, 36). The bioavailability of such hydrophobic compounds is controlled by a number of physicochemical processes, such as adsorption and desorption, diffusion, and dissolution (2). Adsorption of these hydrophobic compounds to extracellular polymeric substances has been reported previously. For example, PAHs, benzene, toluene, and xylene have been adsorbed to exopolysaccharides to promote their bioavailability or to remove them from water (11, 33). In addition, Dohse et al. reported that microbial polymers, acting as phenanthrene adsorbents and carriers, may partly facilitate the transport of phenanthrene in sand columns (4). Carbazole and its derivatives are hydrophobic aromatic compounds, and there are problems of mass transfer in their biodegradation processes in the aqueous phase. Considering this, we investigated the adsorption of carbazole by several polymers, including gellan gum, κ-carrageenan, agar, and alginate. The results showed that gellan gum presented the highest adsorption activity for carbazole at equilibrium. However, for immobilization supports, high adsorption of carbazole alone is not enough to ensure a high biocatalyst activity. Therefore, the biodegradation activities of *Sphingomonas* sp. strain XLDN2-5 in immobilization supports were also investigated. The results indicated that the biodegradation activities of *Sphingomonas* sp. strain XLDN2-5 immobilized by κ-carrageenan and agar were very low (Fig. 2c and d). The reason may be that an oily phase was used during the gel bead preparation of κ-carrageenan and agar by the interphase technique, and the oil adsorbed by gel beads may also result in impeding of the degradation rates for carbazole. Moreover, the gel bead-forming procedure was complex and ineffective, which made them less suitable for industrial biodegradation processes. In contrast, the *Sphingomonas* sp. strain XLDN2-5 cells immobilized by gellan gum presented a high specific biodegradation rate (1,546 μg g cell wet weight^{-1} h^{-1}) Fig.2b). The carbazole contents of the supernatant and gellan gel beads could not be detected at the same time, which may be due to the immediate consumption of carbazole adsorbed by gellan gel beads in the biodegradation process. Additionally, no decrease of carbazole content was observed when nonmagnetically immobilized inactive cells and gellan gel beads without cells served as biocatalysts, which confirmed that the removal of carbazole was due to biodegradation by the *Sphingomonas* sp. strain XLDN2-5 cells. It is clear that gellan gum was the optimal immobilization support due to high carbazole adsorption, superiority in maintaining the high biodegradation activity of the biocatalyst, and a simple gel bead-forming procedure. Moreover, gellan gum produced by *Sphingomonas paucimobilis* ATCC 31461 is considered natural, nontoxic, and compatible with the environment (39, 40). These results were also consistent with previous reports that the semicolloid gellan gum may enhance the aqueous solubility of fluoranthene, which would in turn lead to increased mineralization rates (11). The specific biodegradation rate of cells immobilized by gellan gum was still lower than that of free cells (1,546 µg g cell wet weight^{-1} h^{-1} to 2,783 µg g cell wet weight^{-1} h^{-1}) (Fig. 2a and b). This may be due to a mass transfer limitation of immobilized cells, which may have somewhat reduced the bioavailable concentration in the inner spaces of the beads in contrast to that in the bulk liquid.

Nanoscale particles have large specific surfaces and high surface reactivity, which gives them the potential to address some of the challenges of environmental remediation. Recently, there have been reports that the remediation of Cr(VI) and Pb(II) was carried out by nanoscale zero-valent iron (26) and that engineered polymeric nanoparticles were used in remediation of soil contaminated with PAHs (35). Figure 1a and b1 show that the average particle diameter of Fe$_3$O$_4$ nanoparticles was about 20 nm, and their saturation magnetization was 49.65 emu g^{-1}, which provides the particles with superparamagnetic properties so that the Fe$_3$O$_4$ nanoparticles could be easily separated and recycled by an external magnetic field. In the biodegradation processes, the specific biodegradation rate of nonmagnetically immobilized cells was only 1,761 µg g cell wet weight^{-1} h^{-1} (Fig.3). The reason may be that bacterial cells were tightly trapped in the gellan gum matrix, which resulted in impeding of the mass transfer of substrate from the environment to the central reaction site (Fig. 5a2). In contrast, high specific biodegradation rates (from 3,092 µg g cell wet weight^{-1} h^{-1} to 3,479 µg g cell wet weight^{-1} h^{-1}) were obtained when magnetically immobilized cells served as the biocatalyst, which may be supported by the existence of nanoparticles, the loose binding of the sheets of gellan gum matrix, and the existence of many pores between the sheets of gellan gum matrix (Fig. 5b2). When nonmagnetically immobilized inactive cells and magnetically immobilized inactive cells were incubated as biocatalysts, no degradation of carbazole was observed, which also confirmed that the removal of carbazole was not due to adsorption but to biodegradation by *Sphingomonas* sp. strain XLDN2-5. Among the different concentrations of Fe$_3$O$_4$ nanoparticles, the biodegradation rate was slightly higher at an Fe$_3$O$_4$ nanoparticle concentration of 9 mg ml^{-1}, especially in the period from 4 h to 8 h, and the saturation magnetization of magnetically immobilized cells was 11.08 emu g^{-1} (Fig. 1b2), which made it possible to solve the problem of recovering immobilized cells with a magnetic field. Additionally, the saturation magnetization of magnetic gellan gel beads increased with the number of Fe$_3$O$_4$ nanoparticles added, while the resistance to breakage of magnetic gellan gel beads decreased. These results revealed that the biodegradation activity of the immobilized *Sphingomonas* sp. strain XLDN2-5 cells was significantly enhanced by adding Fe_3O_4 nanoparticles, which may be due to the reduction or elimination of mass transfer problems.

In an industrial bioremediation process, the recycling of the biocatalysts could be an important factor that determines the effectiveness of degradation over time. From the first to the eighth cycle, magnetically immobilized cells presented higher biodegradation activity (the specific biodegradation rates increased from 3,479 μg g cell wet weight^{-1} h^{-1} to 4,638 μg g cell wet weight^{-1} h^{-1}), as shown in Fig. 4a. Nevertheless, the specific biodegradation rates of nonmagnetically immobilized cells decreased from 2,680 μg g cell wet weight^{-1} h^{-1} to 2,513 μg g cell wet weight^{-1} h^{-1} during the recycling processes (Fig. 4b). The high biodegradation activity may be supported by the good growth of cells in the magnetic gellan gel beads, as shown in Fig. 5c1 and c2. The increase of biomass was most pronounced on the surfaces of the magnetic gellan gel beads, and the concentration of cells inside the magnetic gellan gel beads also increased. These results also confirmed that the removal of carbazole was due to biodegradation by *Sphingomonas* sp. strain XLDN2-5, because *Sphingomonas* sp. strain XLDN2-5 could grow with carbazole as the sole source of carbon, nitrogen, and energy. These results were also consistent with a previous report that the growth of cells in celite beads was considered to have enhanced the desulfurization rate in the subsequent batch (3).

In conclusion, magnetically immobilized cells were evaluated as a novel aspect of the industrialization of cell immobilization. Gellan gum, as an immobilization support, required simple gel bead-forming procedures and presented high carbazole adsorption, which led to an increased mineralization rate. Moreover, magnetic (Fe_3O_4) nanoparticles, as one component of the magnetic immobilization support, have a large specific surface and superparamagnetic properties, which not only reduced the mass transfer resistance of traditional immobilization processes, but also facilitated the recovery of immobilized cells in the reuse processes. Additionally, the recycling experiments demonstrated that the degradation activity of magnetically immobilized cells was still high after eight cycles. These results support the development of efficient biocatalysts using magnetically immobilized cells and provide a promising technique for improving the biocatalysts used in the biodegradation of not only carbazole, but also other hazardous organic compounds.

ACKNOWLEDGMENTS

This work was partially supported by grants from the National Natural Science Foundation of China (grant numbers 20377026 and 20577031). We also acknowledge financial support and chemical analysis by Shanghai Apple Flavor and Fragrance Co., Ltd., and Shanghai Jiao Tong University (China).

REFERENCES

1. Benedik, M. J., P. R. Gibbs, R. R. Riddle, and R. C. Willson. 1998. Microbial denitrogenation of fossil fuels. Trends Biotechnol. 16:390-395.

2. Bosma, T. N. P., P. J. M. Middeldorp, G. Schraa, and A. J. B. Zender. 1997. Mass transfer limitation of biotransformation: quantifying bioavailability. Environ. Sci. Technol. 31:248-252.

3. Chang, J. H., Y. K. Chang, H. W. Ryu, and H. N. Chang. 2000. Desulfurization of light gas oil in immobilized-cell systems of *Gordona* sp. CYKS1 and *Nocardia* sp. CYKS2. FEMS Microbiol. Lett.182:309-312.

4. Dohse, D. M., and L. W. Lion. 1994. Effect of microbial polymers on the sorption and transport of phenanthrene in a low-carbon sand. Environ. Sci. Technol. 28:541-548.

5. Dwyer, D. F., M. L. Krumme, S. A. Boyd, and J. M. Tiedje. 1986. Kinetics of phenol biodegradation by an immobilized methanogenic consortium. Appl. Environ. Microbiol. 52:345-351.

6. Gai, Z. H., B. Yu, L. Li, Y. Wang, C. Q. Ma, J. H. Feng, Z. X. Deng, and P. Xu. 2007. Cometabolic degradation of dibenzofuran and dibenzothiophene by a newly isolated carbazole-degrading *Sphingomonas*sp. strain. Appl. Environ. Microbiol. 73:2832-2838.

7. Gieg, L. M., A. Otter, and P. M. Fedorak. 1996. Carbazole degradation by *Pseudomonas* sp. LD2: metabolic characteristics and the identification of some metabolites. Environ. Sci. Technol. 30:575-585.

8. Grosser, R. J., D. Warshawsky, and J. R. Vestal. 1991. Indigenous and enhanced mineralization of pyrene, benzo[*a*]pyrene, and carbazole in soils. Appl. Environ. Microbiol. 57:3462-3469.

9. Harms, H., and T. N. P. Bosma. 1997. Mass transfer limitation of microbial growth and pollutant degradation. J. Ind. Microbiol. Biotechnol. 18:97-105.

10. Jha, A. M., and M. K. Bharti. 2002. Mutagenic profiles of carbazole in the male germ cells of Swiss albino mice. Mutat. Res. 500:97-101.

11. Johnsen, A. R., and U. Karlson. 2004. Evaluation of bacterial strategies to promote the bioavailability of polycyclic aromatic hydrocarbons. Appl. Microbiol. Biotechnol. 63:452-459.

12. Kamat, P. V., and D. Meisel. 2003. Nanoscience opportunities in environmental remediation. C. R. Chimie 6:999-1007.

13. Kilbane, J. J., II, A. Daram, J. Abbasian, and K. J. Kayser. 2002. Isolation and characterization of*Sphingomonas* sp. GTIN11 capable of carbazole metabolism in petroleum. Biochem. Biophys. Res. Commun. 297:242-248.

14. Kobayashi, T., R. Kurane, N. Kenji, Y. Nakamura, K. Kirimura, and S. Usami. 1995. Isolation of bacteria degrading carbazole under microaerobic conditions, i.e., nitrogen gas substituted conditions. Biosci. Biotechnol. Biochem. 59:932-933.

15. Lee, S. T., S. K. Rhee, and G. M. Lee. 1994. Biodegradation of pyridine by freely suspended and immobilized *Pimelobacter* sp. Appl. Microbiol. Biotechnol. 41:652-657.

16. Lein, H., and W. Zhang. 2001. Nanoscale iron particles for complete reduction of chlorinated ethenes.Colloid Surface A 191:97-105.

17. Li, F. L., P. Xu, J. H. Feng, L. Meng, Y. Zheng, L. L. Luo, and C. Q. Ma. 2005. Microbial desulfurization of gasoline in a *Mycobacterium goodii* X7B immobilized-cell system. Appl. Environ. Microbiol. 71:276-281.

18. Li, L., P. Xu, and C. Q. Ma. 2002. Isolation and degradation characteristics of carbazole-degradation bacteria. J. Chem. Ind. Eng. 53:127-129.

19. Li, L., P. Xu, and H. D. Blankespoor. 2004. Degradation of carbazole in the presence of non-aqueous phase liquids by *Pseudomonas* sp. Biotechnol. Lett. 26:581-584.

20. Li, L., Q. G. Li, F. L. Li, Q. Shi, B. Yu, F. R. Liu, and P. Xu. 2006. Degradation of carbazole and its derivatives by a *Pseudomonas* sp. Appl. Microbiol. Biotechnol. 73:941-948.

21. Liu, Z. L., Y. J. Liu, K. L. Yao, Z. H. Ding, J. Tao, and X. Wang. 2002. Synthesis and magnetic properties of Fe_3O_4 nanoparticles. J. Mater. Synth. Process. 10:83-87.

22. López, A., N. Lázaro, and A. M. Mosbach. 1997. The interphase technique: a simple method of cell immobilization in gel-beads. J. Microbiol. Methods 30:231-234.

23. Nakagawa, H., K. Kirimura, T. Nitta, K. Kino, R. Kurane, and S. Usami. 2002. Recycle use of*Sphingomonas* sp. CDH-7 cells for continuous degradation of carbazole in the presence of $MgCl_2$. Curr. Microbiol. 44:251-256.

24. O'Brien, T., J. Schneider, D. Warshawsky, and K. Mitchell. 2002. In vitro toxicity of 7H-dibenzo[c,g]carbazole in human liver cell lines. Toxicol. In Vitro 16:235-243.

25. Ouchiyama, N., Y. Zhang, T. Omori, and T. Kodama. 1993. Biodegradation of carbazole by*Pseudomonas* spp. CA06 and CA10. Biosci. Biotechnol. Biochem. 57:455-460.

26. Ponder, S. M., J. G. Darab, and T. E. Mallouk. 2000. Remediation of Cr(VI) and Pb(II) aqueous solutions using supported nanoscale zerovalent iron. Environ. Sci. Technol. 34:2564-2569.

27. Reddy, M. Y., and K. Randerath. 1990. A comparison of DNA adduct formation in white blood cells and internal organs of mice exposed to benzo[a]pyrene, dibenzo[c,g]carbazole, safrole and cigarette smoke condensate. Mutat. Res. 241:37-48.

28. Santos, S. C. C., D. S. Alviano, C. S. Alviano, M. Pádula, A. C. Leitão, O. B. Martins, C. M. S. Ribeiro, M. Y. M. Sassaki, C. P. S. Matta, J. Bevilaqua, G. V. Sebastián, and L. Seldin. 2006. Characterization of *Gordonia* sp. strain F.5.25.8 capable of dibenzothiophene desulfurization and carbazole utilization. Appl. Microbiol. Biotechnol. 71:355-362.

29. Schneider, J., R. J. Grosser, K. Jayasimhulu, W. Xue, B. Kinkle, and D. Warshawsky. 2000. Biodegradation of carbazole by *Ralstonia* sp. RJGII.123 isolated from a hydrocarbon contaminated soil. Can. J. Microbiol. 46:269-277.

30. Shan, G. B., H. Y. Zhang, W. Q. Cai, J. M. Xing, and H. Z. Liu. 2005. Improvement of biodesulfurization rate by assembling nano-sorbents on the surface of microbial cells. Biophys. J. 89:L58-L60.

31. Shan, G. B., J. M. Xing, H. Y. Zhang, and H. Z. Liu. 2005. Biodesulfurization of dibenzothiophene by microbial cells coated with magnetic nanoparticles. Appl. Environ. Microbiol. 71:4497-4502.

32. Shepherd, J. M., and G. Lloyd-Jones. 1998. Novel carbazole degradation genes of *Sphingomonas*CB3: sequence analysis, transcription and molecular ecology. Biochem. Biophys. Res. Commun. 247:129-135.

33. Späth, R., and S. Wuertz. 1998. Sorption properties of biofilms. Water Sci. Technol. 37:207-210.

34. Sverdrup, L. E., J. Jensen, A. E. Kelley, P. H. Krogh, and J. Stenersen. 2002. Effect of eight polycyclic aromatic compounds on the survival and reproduction of *Enchytraeus crypticus* (Oligochaeta, Clitellata). Environ. Toxicol. Chem. 21:109-114.

35. Tungittiplakorn, W., L. W. Lion, C. Cohen, and J. Y. Kim. 2004. Engineered polymeric nanoparticles for soil remediation. Environ. Sci. Technol. 38:1605-1610.

36. Volkering, F., A. M. Breure, A. Sterkenburg, and J. G. van Andel. 1992. Microbial degradation of polycyclic aromatic hydrocarbons: effect of substrate availability on bacterial growth kinetics. Appl. Microbiol. Biotechnol. 36:548-552.

37. Wang, J. L., L. P. Han, H. C. Shi, and Y. Qian. 2001. Biodegradation of quinoline by gel immobilized *Burkholderia* sp. Chemosphere 44:1041-1046.

38. Wang, J. L., X. C. Quan, L. P. Han, Y. Qian, and H. Werner. 2002. Microbial degradation of quinoline by immobilized cells of *Burkholderia pickettii*. Water Res. 36:288-296.

39. Wang, X., P. Xu, Y. Yuan, C. L. Liu, D. Z. Zhang, Z. T. Yang, C. Y. Yang, and C. Q. Ma. 2006. Modeling for gellan gum production by *Sphingomonas paucimobilis* ATCC 31461 in a simplified medium.Appl. Environ. Microbiol. 72:3367-3374.

40. Wang, X., Y. Yuan, K. N. Wang, D. Z. Zhang, Z. T. Yang, and P. Xu. 2007. Deproteinization of gellan gum produced by *Sphingomonas paucimobilis* ATCC 31461. J. Biotechnol. 128:403-407.

41. Woodward, J. 1988. Methods of immobilization of microbial cells. J. Microb. Methods 8:91-102.

42. Xu, P., B. Yu, F. L. Li, X. F. Cai, and C. Q. Ma. 2006. Microbial degradation of sulfur, nitrogen, and oxygen heterocycles. Trends Microbiol. 14:398-405.

43. Yu, B., C. Q. Ma, W. J. Zhou, S. S. Zhu, Y. Wang, J. Y. Qu, F. L. Li, and P. Xu. 2006. Simultaneous biodetoxification of S, N, and O pollutants by engineering of a carbazole-degrading gene cassette in a recombinant biocatalyst. Appl. Environ. Microbiol. 72:7373-7376.

44. Yu, B., P. Xu, S. S. Zhu, X. F. Cai, Y. Wang. L. Li, F. L. Li, X. Y. Liu, and C. Q. Ma. 2006. Selective biodegradation of S and N heterocycles by a recombinant *Rhodococcus erythropolis* strain containing carbazole dioxygenase. Appl. Environ. Microbiol. 72:2235-2238.

45. Zhang, W. X. 2003. Nanoscale iron particles for environmental remediation: an overview. J. Nanopart. Res. 5:323-332.

Chapter 2

LOW TOXICITY OF DEOXYNIVALENOL-3-GLU-COSIDE IN MICROBIAL CELLS

Tadahiro Suzuki and Yumiko Iwahashi

Applied Microbiology Division, National Food Research Institute, 2-1-12 Kannondai, Tsukuba, Ibaraki 305-8642, Japan

ABSTRACT

Host plants excrete a glucosylation enzyme onto the plant surface that changes mycotoxins derived from fungal secondary metabolites to glucosylated products. Deoxynivalenol-3-glucoside (DON3G) is synthesized by grain uridine diphosphate-glucosyltransferase, and is found worldwide, although information on its toxicity is lacking. Here, we conducted growth tests and DNA microarray analysis to elucidate the characteristics of DON3G. The *Saccharomyces cerevisiae PDR5* mutant strain exposed to DON3G demonstrated similar growth to the dimethyl sulfoxide control, and DNA microarray analysis revealed limited differences. Only 10 genes were extracted, and the expression profile of stress response genes was similar to that of DON, in contrast to metabolism genes like *SER3*, which encodes 3-phosphoglycerate dehydrogenase. Growth tests with *Chlamydomonas reinhardtii* also showed a similar growth rate to the control sample. These results suggest that DON3G has extremely low toxicity to these cells, and the glucosylation of mycotoxins is a useful protective mechanism not only for host plants, but also for other species.

INTRODUCTION

Disease-causing fungi like *Fusarium* species produce type B trichothecenes, and these mycotoxins contaminate grains worldwide. Because these mycotoxins or fungi cause substantial economic losses, pesticides are used to decrease fungal infection at cultivation. Plants possess resistance to fungi and mycotoxins by suppressing fungal growth and mycotoxin toxicity. To this end, some plants synthesize glycosyltransferases that metabolize secondary metabolites (plant secondary product glycosyltransferases; PSPGs) [1]. PSPGs contribute to the metabolism of pesticides or products that are secreted by microorganisms.

However, the activity is strictly regulated by the specific structure. Thus, glucosyltransferase, which is a type of glycosyltransferase, has been reported to be a glucosylation enzyme of mycotoxins [2,3]. Recently, the existence of various glucosylated mycotoxins has been revealed [4]. However, at present, studies on these mycotoxins are developing; therefore, only a limited number of toxicity and detection studies have been reported. Some contamination studies on deoxynivalenol-3-glucoside (DON3G; Figure 1) have been published [5,6]. DON3G was detected at a level of up to 30 mol % in wheat, and up to 50 mol % in soybean, compared with DON [7]. This study also reported that DON3G was detected at higher levels than acetylated-DONs. Meanwhile, the same amount of DON3G compared with DON was detected in low-alcohol-content beer [8]. In that study, DON3G was also reported to be at a higher level than acetylated-DONs. Despite the contamination risk, glucosylated mycotoxin cannot be detected easily and is also not regulated; therefore it is often referred to as a masked mycotoxin.

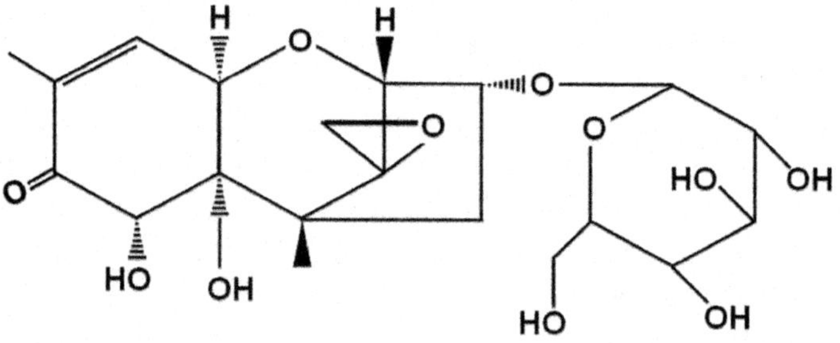

Figure 1. Deoxynivalenol-3-glucoside.

The toxicity of DON3G varies from DON in several ways. For example, *Arabidopsis thaliana* decreases DON toxicity by expressing barley uridine diphosphate (UDP)-glucosyltransferase. In a DON exposure study with UDP-glucosyltransferase-inserted seeds, the level of DON3G significantly increased compared with the control sample in the plant extract [3]. Barley UDP-glucosyltransferase-inserted yeast cells also acquired DON resistance, and the level of DON3G increased in the culture media over time [2]. This means that regardless of the organism species, UDP-glucosyltransferase, which metabolizes DON to DON3G, influences mycotoxin resistance. Therefore, by secreting that enzyme onto the plant cell surface, feed and food products might be able to protect themselves from some, but not all, mycotoxins.

The toxicity and risk profile of DON3G is unknown. Berthiller *et al.* [9] have reported that lactic acid bacteria, such as *Enterococcus durans*, *Enterococcus mundtii* and *Lactobacillus plantarum*, regenerate DON from DON3G by hydrolysis. In a rat feeding study with DON3G, the recovery rates of DON3G from the urine and feces were extremely low, and more than half of the mycotoxin recovered from feces was metabolized to DON [10,11]. Also, it has been reported that the microflora in the feces metabolize DON3G to DON in several hours, and in 1 day they metabolize it to de-epoxy-DON. However, the nutrient transition time from enterocytes to mammalian cells is not always the same, and the composition of the microflora is different depending on circumstances and/or organisms. Thus, intake of DON3G has a risk of DON exposure; hence, similarly to DON, DON3G is thought to be a potential threat. However, the toxicity of DON3G has not gathered much attention. Plant and yeast studies have not elucidated sufficiently the toxic character of DON3G, though its phenotypes were well observed. Here, we evaluated the toxicity of DON3G using yeast and algae, and a comparison between type B trichothecenes was conducted to reveal their toxic character. Yeast growth and DNA microarray analysis demonstrated that DON3G has lower toxicity than DON and acetyl-DON, and the algae study suggests that DON3G has low toxicity to various species.

RESULTS

Yeast Growth with DON3G

The yeast *PDR5* gene encodes the Pdr5 protein, which is a multidrug resistance ATP-binding cassette (ABC) transporter that localizes on the plasma membrane. Using a deletion mutant, it has been demonstrated that Pdr5 contributes to resistance against trichothecene mycotoxins [12,13]. In these studies, we used a *PDR5* mutant, $\Delta pdr5$, to evaluate DON3G. Here, first, a yeast growth test was conducted. When 30–160 µM mycotoxins were used, DON at >80 µM caused growth inhibition (Figure 2a). Conversely, DON3G did not change the growth rate at any concentration, although the DON3G samples did not completely correspond with the control sample, which was treated with dimethyl sulfoxide (DMSO) (Figure 2c). Because the normal culture medium did not show an adequate difference, we next used growth media containing a low concentration of sodium dodecyl sulfate (SDS) to increase membrane permeability [14]. The growth curves with 0.01% SDS demonstrated that the DON exposure clearly caused growth inhibition compared with the normal media (Figure 2b). In contrast, even with SDS, the DON3G samples completely corresponded with each growth rate, although the control growth was marginally inhibited (Figure 2d). Inhibition rates of DON exposure samples exhibited concentration dependent growth inhibition. Conversely, DON3G did not cause any growth change, or was associated with slightly increased growth (Table 1). Regardless of the existence of SDS, DON caused significant growth inhibition ($p < 0.05$), whereas there was no growth inhibition with DON3G. The growth test results indicated that DON3G has extremely low toxicity against yeast cells.

DON3G Influences a Small Number of Genes

In the previous study, we conducted DNA microarray analysis to compare the toxicities of type B trichothecenes at 25 mg/kg, because the molecular weight of DON, nivalenol (NIV) and their acetylated products are relatively similar to each other. However, the molecular weight of DON3G is nearly 150% compared with the other trichothecenes used in the previous study. Therefore, we applied 80 μM DON3G, which is a concentration that caused evident growth delay under DON exposure, and used the entry to yeast cells for clustering analysis. DNA microarray with t-test analysis ($p < 0.05$, >1.5-fold) between DON3G exposure and the control revealed only 10 genes. *SER3*, encoding 3-phosphoglycerate dehydrogenase, which functions in the serine/glycine biosynthesis pathway, showed decreased expression. Similarly, the gene expression of *MCH2* (monocarboxylate permease-like protein), *EEB1* (acyl-coenzyme A), YMR230W-A, and an unknown protein also decreased after exposure to DON3G. The stress response genes, *ALD6*, *STP4*, *SRX1* and *ECM13*, were induced by DON3G. *HXT2* and *RGS2*, protein coding genes that are sensitive to changes in glucose level, also were induced by DON3G (Table 2).

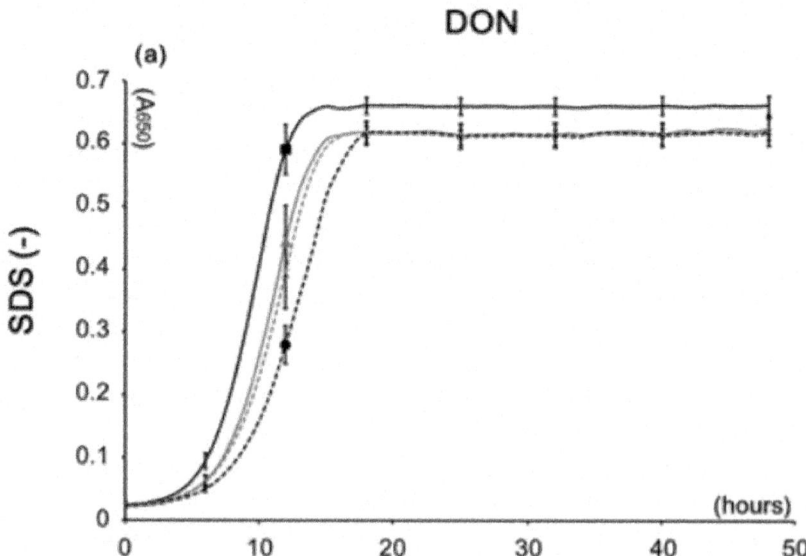

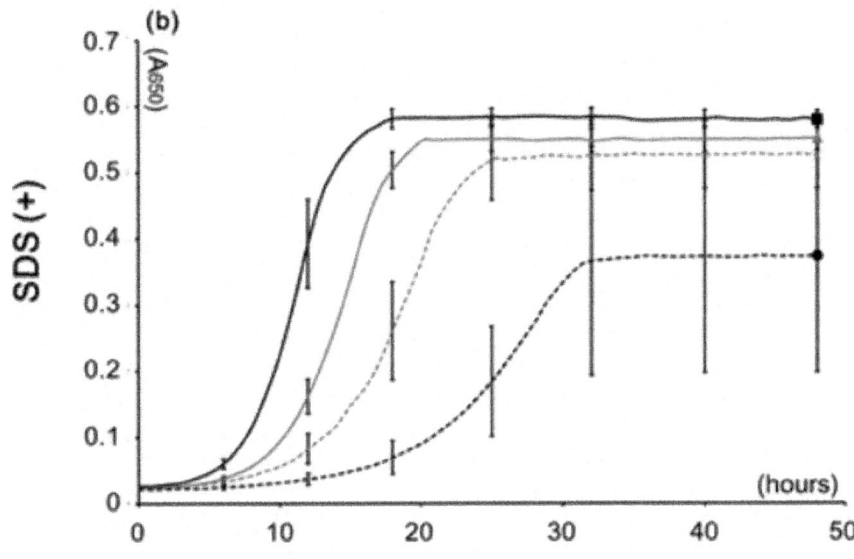

DON3G

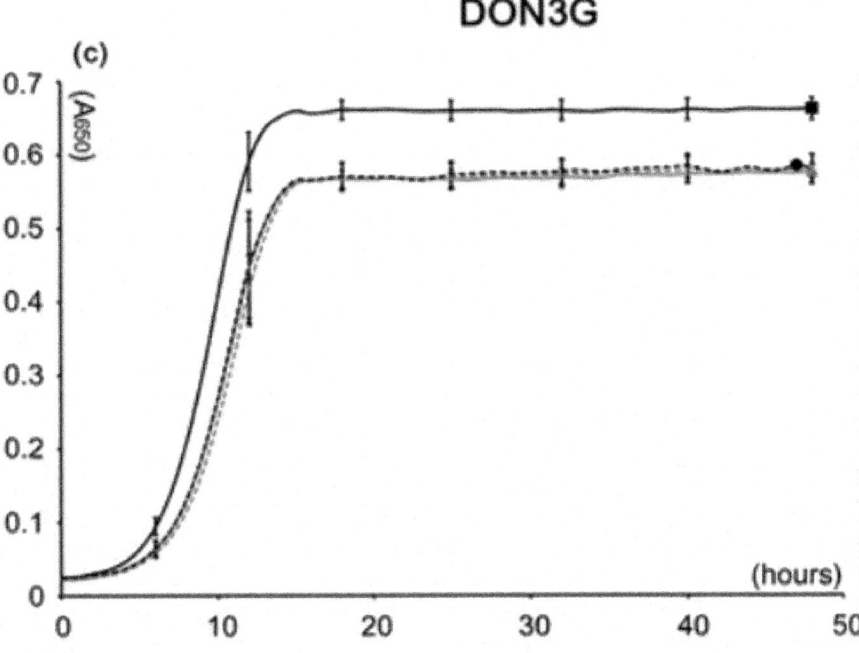

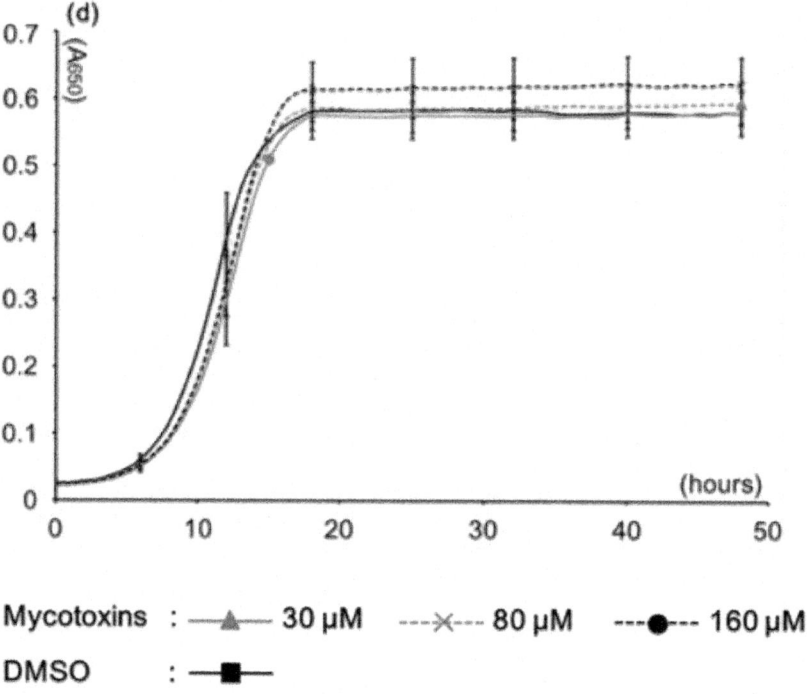

Figure 2. Effect of Deoxynivalenol (DON) and Deoxynivalenol-3-glucoside (DON3G) on yeast cell growth. (**a**) Growth curve under DON exposure in yeast extract peptone dextrose (YPD) medium. (**b**) Growth curve under DON exposure in YPD medium containing 0.01% SDS. (**c**) Growth curve under DON3G exposure in YPD medium. (**d**) Growth curve under DON3G exposure in YPD medium containing 0.01% sodium dodecyl sulfate (SDS). A650, absorbance at 650 nm; SDS (+): 0.01% SDS in YPD medium; Bars = SE; $n = 3$.

Table 1. Inhibitory rate (%) and inhibitory concentration of each condition

Medium condition	Mycotoxin	Concentrations (µM)			IC20 * (µM)
		30	80	160	
SDS (−)	DON	13.3 ± 0.64 **	14.4 ± 0.15	18.9 ± 0.08	>160
	DON3G	19.7 ± 0.98	20.2 ± 0.84	19.3 ± 2.09	- ***
SDS (+)	DON	21.1 ± 1.58	37.4 ± 8.72	71.7 ± 13.54	30.4
	DON3G	9.7 ± 3.84	7.7 ± 1.03	2.7 ± 2.45	-

* IC20: 20% inhibitory concentration of cell growth. The end time of log phase at DMSO was applied for each control of the calculation.

** ± numbers mean standard error $n = 3$.

*** IC was not calculated because of unusable parameters.

Table 2. Significant gene expression changes revealed by DNA microarray

Systematic No.	Gene Symbol	Fold Change	Gene Title	Description
YER081W	SER3	0.31	3-phosphoglycerate dehydrogenase	Serine and glycine biosynthesis
YMR230W-A	-	0.35	Putative protein of unknown function	-
YKL221W	MCH2	0.41	Protein with similarity to mammalian monocarboxylate permeases	Monocarboxylate transport
YPL095C	EEB1	0.47	Acyl-coenzymeA: ethanol O-acyltransferase	Fatty acid ethyl ester biosynthesis
YPL061W	ALD6	2.06	Cytosolic aldehyde dehydrogenase	Oxidative stress response
YDL048C	STP4	2.25	Protein containing a Kruppel-type zinc-finger domain	DNA replication stress
YKL086W	SRX1	2.57	Sulfiredoxin	Oxidative stress resistance DNA replication stress
YOR107W	RGS2	3.15	Negative regulator of glucose-induced cAMP signaling	Glucose response
YBL043W	ECM13	3.39	Non-essential protein of unknown function	UVA irradiation response
YMR011W	HXT2	5.66	High-affinity glucose transporter	Induced by low levels of glucose Repressed by high levels of glucose

Real-Time Polymerase Chain Reaction Analysis

DNA microarray produces false-positive data that correspond to the significance level. In our study, a commercial gene chip in which approximately 5700 target gene probes were mounted was applied, and the significance level was set to 5%. Under this condition, statistical analysis permits the presence of about 280 false-positive data. Nonetheless, only 10 gene expression changes were extracted in this study. To avoid false data, we confirmed the expression profiles by semi-quantitative real-time polymerase chain reaction (PCR) analysis. As shown in Figure 3, the genes that showed decreased expression after DON3G exposure did not present significant changes compared with the control samples. In contrast, these genes were induced by exposure to DON, indicating an opposite trend, except for *MCH2*, which showed the same trend in the real-time PCR assay. *SER3* expression showed a significant difference between the DON condition and the DON3G condition. As for the DON3G-induced genes, DON induced a relatively smaller expression increase than DON3G compared with the control, and the differences were significant (unpaired *t*-test, $p < 0.05$).

Clustering Analysis

In the type B trichothecene studies, we focused on DON, NIV and their acetylated products as the main contaminants rather than DON3G. These compounds were the focus of our previous study in which we obtained microarray data using *PDR5* yeast cells. Therefore, we compared our previous microarray data with the current DON3G data. Because the sampling conditions were not identical, each control sample was integrated, and the clustering analysis was processed. Hierarchical clustering indicated that the trend of gene expression changes after DON3G exposure was similar to that after 3-acetyl-deoxynivalenol (3AcDON) or NIV exposure (Figure 4).

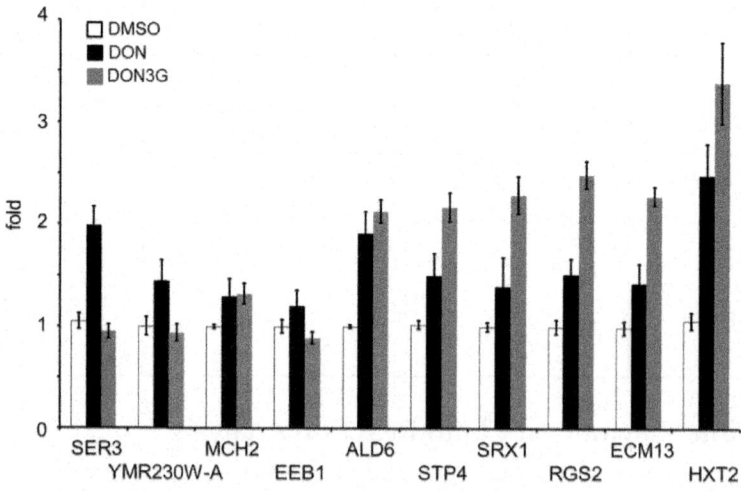

Figure 3. Semi-quantitative analysis of DNA microarray data. Total RNA that was used in the DNA microarray analysis and DON-treated RNA samples were prepared for synthesizing cDNA templates. Bars = SE; $n = 3$.

Figure 4. Clustering analysis of DNA microarray expression data. Except for DON3G, the gene expression data of DON, 3AcDON, 15AcDON, NIV and FusX [15] were applied by integrating the control condition samples of each condition.

Growth Test with Algal Cells

C. reinhardtii is a single cell green alga that is an autotroph and a heterotroph. Hence, it is used in photosynthesis and photoresponse studies as a model cell. It has also been used as a phytotoxicity evaluation model for trichothecenes, and NIV has demonstrated moderately high toxicity in these studies [16,17]. Sensitivity to NIV has not been reported in yeast cell studies, although a number of mammalian cell studies have indicated NIV sensitivity. In our yeast cell study, DON3G also did not exhibit significant toxicity (Figure 2). However, *C. reinhardtii* was anticipated to show DON3G sensitivity, because it has differential sensitivity to some mycotoxins. Thus, we next conducted a DON3G-exposure test on *C. reinhardtii*, and examined its growth rate (Figure 5a).

The lighting intensity can be adjusted to a photosynthetic photon flux density (PPFD) of 400 to 700 nm, and at the same time the wavelength for the lighting source needs to be selected. The growth of *C. reinhardtii* does not show a significant difference between the different wavelength conditions, although the red light (660 nm) contributes to the most efficient growth [17]. In this study, therefore, we applied red lighting. DON, acetylated DON (3-acetyl-DON; 3AcDON, 15-acetyl-DON; 15Ac-DON), NIV and FusX exposures were examined, as well as DON3G. The cell growth with 80 μM DON3G showed a similar curve to that of the DMSO control sample. All the other compounds, except 3AcDON, demonstrated reduced growth, whereas DON exposure caused complete growth inhibition (Figure 5a). When cell growth was examined at the end of the log phase, DON and FusX exposure caused significant growth inhibition while NIV and 15AcDON exposure showed a moderate level of inhibition (unpaired *t*-test, $p < 0.01$, Figure 5b). Conversely, DON3G and 3AcDON did not cause cell growth inhibition.

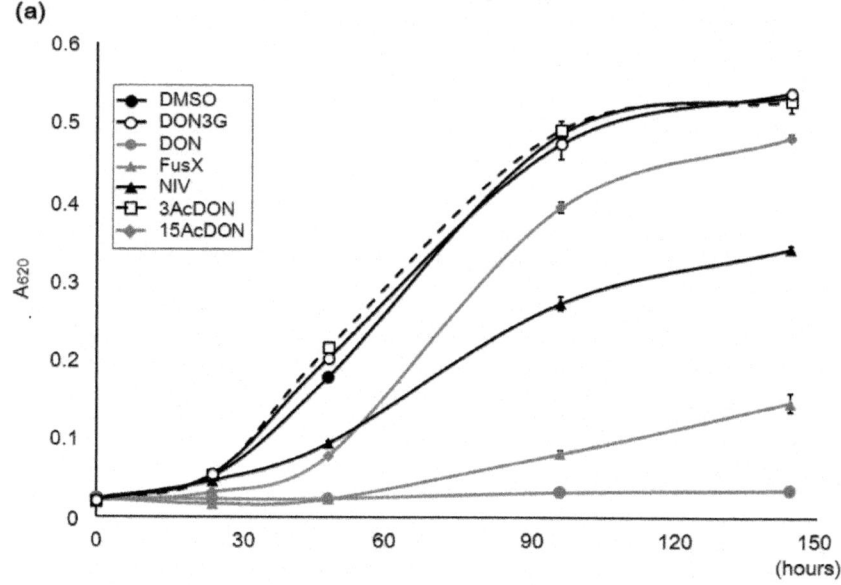

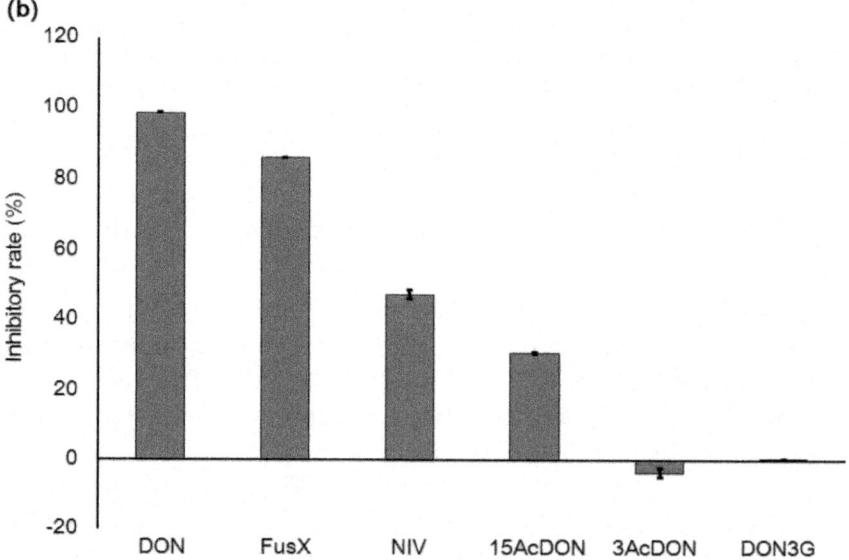

Figure 5. Algal cell growth test with 80 μM of mycotoxins under red light (120 μmol·m⁻²·s⁻¹). The details of the photosynthetic photon flux density (PPFD; μmol·m⁻²·s⁻¹) are described in the Materials and Methods. NIV, nivalenol; 3Ac-DON, 3-acetyl-DON; 15AcDON, 15-acetyl-DON; FusX, fusarenon-X (4-acetylnivalenol). (a) Growth curves of each test exposure. (b) Inhibition rates of algal growth at the end of the log phase (144 h). Bars = SE; $n = 3$.

DISCUSSION

Mycotoxins are secondary metabolites produced by pathogenic fungi. They are often glucosylated by an enzymatic reaction with β-glucosyl-transferase, which is excreted by the host plant onto the plant surface. When glycosylated, mycotoxins exhibit considerably less toxicity; therefore, mycotoxin glucosylation is clearly a host plant resistance response. If a mycotoxin causes cellular damage, it is natural that the host plant would evolve a mechanism for decreasing the toxicity. For example, it has been identified in plant studies that glucosylated DON is less toxic than DON [2,3]. However, it is unclear whether the attenuation in toxicity by glucosylation is a plant cell-specific character, because a sufficient amount of refined glucosylation product is difficult to obtain for toxicity studies. DON3G has been synthesized by an organic chemical technique [5,18], and by a molecular biological technique, which is inserted into microbes or plants [19]. However, its availability for study is still lower than other mycotoxins; hence, this research is useful for understanding the features of DON3G.

In our previous study, we examined gene expression changes after exposure to DON, NIV and their acetylated products [15]. Therefore, in this study, 80 μM, which was easy to compare with the other data, was used for growth tests and DNA microarray analysis. Additionally, 160 μM DON3G was applied to elucidate the influence of concentration change on growth. However, DON3G did not cause significant growth differences, as opposed to those caused by DON, especially with the addition of 0.01% SDS to the medium. It is thought that *S. cerevisiae* Pdr5 is the transporter for trichothecenes like DON and T-2 toxin. In contrast, 3AcDON, which is the less toxic product to various cell lines, and NIV, which shows high toxicity to mammalian cells, are not affected by Pdr5 deletion. It is likely that yeast cells have a number of transporters for stress resistance. Under a multiple transporter scenario, it is possible that deletion of Pdr5 would not affect DON3G.

The real-time PCR data of *SER3* did not indicate a significant difference between the control and DON3G treatment, whereas the DNA microarray data did, demonstrating a slight decrease; thus, it seems that *SER3* expression may not have changed in response to DON3G exposure as expected from the microarray result. On the other hand, *SER3* was significantly induced by DON exposure; hence, the gene expression patterns after DON and DON3G exposure were different. YMR230W-A and *EEB1* had a similar expression trends; however, the differences between the treatments were lower than that of *SER3*. The other tested genes showed similar expression patterns between DON and DON3G, although the expression levels were different. Taken together, the gene expression pattern of *SER3* may be useful for discriminating the toxicity differences between DON and DON3G. 3-Phospho-glycerate dehydrogenase, encoded by *SER3*, plays a role in the metabolism of 3-phospho-glycerate derived from the glycolytic system. As the gene expression profile of the glycolytic system changes as a result of mycotoxin exposure [20], the change in expression of *SER3* may be attributable to that influence. However, the induction level of *HXT2*, which encodes a glucose transporter induced by low glucose conditions, was higher by DON3G than by DON. Generally, it is thought that the DON-induced gene expression change is larger than that of DON3G; hence, the gene expression pattern of *HXT2* is also valuable. Most of the detected genes do not encode proteins that localize in mitochondria. Inhibition of mitochondrial function causes cell death, which is easy to observe at lethal toxic agent exposures. In fact, highly toxic type B trichothecenes like DON decrease the expression of mitochondria ribosomal genes [15]. Meanwhile, less toxic mycotoxins like 3AcDON or NIV do not cause such radical expression patterns. Taken together, the data show that DON3G does not affect critical metabolic functions. Robust function of mitochondria might be a key point of the more highly induced genes after DON3G exposure compared with that of DON. However, more detailed information is necessary to elucidate the results of reverse-transcription PCR. A DON trap model by the yeast cell wall layer has been reported [21]. Moreover, DON3G has a larger molecular weight and a different structure than DON. Therefore, it is possible that the trap efficiency is different. However, we cannot exclude the possibility that DON3G affects intercellular components, because DNA stress response genes were induced, and cellular metabolism genes like *SER3* were repressed. Nevertheless, the stress response was significantly limited, suggesting only a marginal strain. Thus, the role of these genes in this study remains unresolved. Clarifying their role will elucidate the cellular response mechanism to the toxicity of DON3G. When compared with other type B trichothecene mycotoxins, DON3G belonged to the low toxicity cluster, which includes 3AcDON and NIV, suggesting that DON3G has a different character from the highly toxic mycotoxins DON, 15AcDON and FusX. Taken together, these results indicate that the toxicity of DON3G to yeast cells is low.

The toxicity of DON3G to *C. reinhardtii* was also not significant. *C. reinhardtii* is more sensitive to type B trichothecene mycotoxins than yeast; especially NIV, which caused significant growth inhibition in the toxicity study. This character is similar to that of mammalian cell lines. Moreover, *C. reinhardtii* is a model cell for photo reactions or photosynthesis [22]. Additionally, *C. reinhardtii* is sensitive to continuously high intensity of light, fast water flow, and somewhat high concentrations of DMSO [23]. Regardless, the toxicity to *C. reinhardtii* corresponded to that of yeast. It has been reported that glucosylation of mycotoxins contributes to the attenuation of toxicity [3].

Therefore, considering the results of this study, it appears that the attenuation effect is not exclusive to host plants.

Finally, there are some concerns. Barley UDP-glucosyltransferase-introduced *A. thaliana* increases the resistance to DON [3]. On the other hand, when measuring lethality (LD50) using *A. thaliana* leaf, the toxicity of both NIV and FusX was low, while DON and 3AcDON showed a similar level of toxicity [24]. Many studies using mammalian or yeast cells indicated a difference between DON and 3AcDON, and yeast cells were sensitive to FusX but not to NIV. Conversely, NIV sensitivity has been reported in a study using mammalian cells. It is apparent that sensitivity to mycotoxins depends on the species. Therefore, toxicity evaluation of glucosylated mycotoxins to different species and cell lines is even more necessary.

There are several recent reports regarding glucosylated and other masked mycotoxins [25,26,27]. It is thought that the stress response of the host plant functions to attenuate the toxicity of various mycotoxins. Mycotoxin-derived cellular lesions are important for fungal invasion of host plants; hence, glucosylation may well be one component of the plant's antifungal resistance arsenal. Because of their abundance and potential risk, glucosylated mycotoxins will be the subject of increased study in the near future.

MATERIALS AND METHODS

Chemicals

Commercial products of the mycotoxins, deoxynivalenol (DON; Santa Cruz, Dallas, TX, USA), NIV (Wako, Osaka, Japan), 3AcDON (Sigma, St. Louis, MO, USA), 15AcDON (Santa Cruz), FusX (Sigma) and DON3G (Wako) were dissolved in DMSO (Wako) to prepare sample solutions.

Biomaterials

A glycerol stock of yeast *Saccharomyces cerevisiae* Δ*pdr5*—plasma membrane ABC transporter Pdr5 deletion mutant strain of *S. cerevisiae* BY4743 (MATa/α his3Δ1/his3Δ1 leu2Δ0/leu2Δ0 LYS2/lys2Δ0 met15Δ0/MET15 ura3Δ0/ura3Δ0; Open Biosystems, Huntsville, AL, USA)—was thawed, and cells were transferred by an inoculation needle into 5 mL of YPD medium (1% yeast extract, 2% peptone, and 2% dextrose) in glass tubes. Triplet samples were pre-incubated with 150 rpm rotation at 25°C for 2 days. *Chlamydomonas reinhardtii* wild-type strain 137C was provided by Prof. M. Tsuzuki (Department of Applied Life Science, Tokyo University of Pharmacy and Life Science, Tokyo, Japan). Algal cells were picked from a slant culture on Tris-acetate-phosphate (TAP) medium agar [28] and were inoculated into 100 mL of TAP medium. Culture media were pre-incubated with 100 rpm rotation at 25°C, and constant lighting for at least 3 days.

Yeast Growth

Pre-incubated *S. cerevisiae PDR5* mutant was diluted in YPD medium. At approximately OD_{650} = 0.01, the cell culture containing DON, DON3G or DMSO was dispensed into a 96-well plate. Each volume of mycotoxin was adjusted by adding DMSO. SDS was also added into culture media, and test plates were incubated at 25 °C. Optical density, which indicates growth rate, was measured by a plate reader (Filter Max F5, Molecular Devices, Sunnyvale, CA, USA). For measuring the inhibition rate of cell growth, each sum of area under the growth curve was calculated until the time that corresponded to the end of the log phase of growth of the control sample (36 h), and it was divided by that of DMSO data, which was set as a control. Inhibition rates were tested by unpaired *t*-test analysis ($p < 0.05$) compared with DMSO control, respectively. Twenty percent of inhibition concentration (IC20) was calculated by using asymptotic line that was derived from time dependent inhibition rates.

DNA Microarray Analysis

Pre-incubated cultures were diluted to OD_{650} of about 0.85–1.0. Cultures were centrifuged at 1580× *g* for 5 min, and cell pellets were prepared for the RNA study. Total RNA was prepared using a commercial kit (FastRNA Pro Red kit, MP Biomedicals, Irvine, CA, USA) following the supplier's instructions. Contaminated genomic DNA was removed by RNeasy mini kit (Qiagen, Venlo, The Netherlands). The quality of the total RNA was evaluated with a nucleic acid analyzer (Experion, Bio-Rad, Hercules, CA, USA). RNA samples were used for synthesizing labeled RNA with 3'IVT Express kit, and DNA microarrays (GeneChip Yeast Genome 2.0 array) were processed according to the manufacturer's instructions (Affymetrix, Santa Clara, CA, USA). The DNA microarray data were transferred into GeneSpring analysis software (ver. 12, Agilent Technologies, Santa Clara, CA, USA). Cluster analyses were performed for each condition. After using the MAS5 algorithm to obtain summarized probeset-level expression data, the average expression of triplicates was normalized to the control condition. An unpaired *t*-test was used for statistical analysis, and significant differences in gene expression were selected using a *p* value < 0.05. To avoid detection of false positives, a multiple testing correction (Benjamini-Hochberg FDR) was applied to obtain corrected *p* values. Each character of a selected gene was confirmed according to the Saccharomyces Genome Database (SGD; http://www.yeastgenome.

org/). The microarray data set has been assigned the accession number, GSE63663, in the Gene Expression Omnibus Database (GEO; http://www. ncbi.nlm.nih.gov/geo/).

Real-Time PCR Analysis

One microgram of genomic DNA-free total RNA, which was used for the microarray analysis, was reverse transcribed (PrimeScript first cDNA synthesis kit, Takara Bio, Shiga, Japan). Similarly treated DON-exposed samples were also prepared. First strand cDNA was diluted five times with Tris-EDTA, pH 8.0. Each 1 μL of cDNA solution was used as a DNA template. Primer sets were designed on primer3 (Table 3). The primer set of *ACT1*, which encodes the actin protein, was used for internal control, and the *PDR5* primer set was used for negative control. The *ACT1* primer set, which has been previously reported [29], was prepared. The same volume of mycotoxin-treated sample templates was also dispensed on the plate. A total of 19 μL of reaction mix [0.4 μL of 10 μM primer each, 8.2 μL of distilled water, 10 μL of 2× Master mix (KAPA SYBR FAST qPCR kit, Kapa Biosystems, Woburn, MA, USA)] was added into a PCR plate. Sample plates were processed at 95 °C for 2 min, followed by 40 cycles of 95 °C for 3 s and 60 °C for 20 s in a thermal cycler (MX3000P, Agilent Technologies, Santa Clara, CA, USA). The amplified *ACT1* product was used as an internal control, and each triplicate was averaged. Unpaired *t*-test analyses were conducted between control and DON3G samples ($p < 0.05$). One-way analysis of variance analyses were also conducted between control, DON, and DON3G samples ($p < 0.05$) to consider gene expressions of DON.

C. reinhardtii Growth

Light-emitting diode (LED) conditions were manually constructed on an LED platform (SPL-100-CC; Revox, Kanagawa, Japan) with red (660 nm) diodes, and the photon flux density was modulated by a pulse-width modulation dimmer controller. The spectrum of LED irradiation was measured by an illuminance spectrophotometer (CL-500A; Konica Minolta, Tokyo, Japan) as irradiances (W/m^2). The total irradiances of spectra from 400 to 700 nm were counted and the PPFD of each spectrum condition was calculated with the following formula: PPFD (μmol·m^{-2}·s^{-1}) = [irradiance (W/m^2) × spectrum (m) × 10^{-9}]/[Planck's constant (6.626×10^{-34}; J.s) × speed of light (2.998×10^{8}; m/s) × Avogadro constant (6.022×10^{23}; mol^{-1})] × 10^6. Mycotoxins were added into TAP medium at a final concentration of 80 μM. The same volume of DMSO was added as a control. For measuring the inhibition rate of cell growth, each sum of area under the growth curve was calculated until the time that corresponded to the end of the log phase of growth of the control sample (144 h), and it was divided by that of DMSO data, which was set as a control. Inhibition rates were tested by unpaired t-test analysis ($p < 0.05$) compared with DMSO control, respectively.

Table 3. Primer sets for real-time PCR.

Gene symbol	Direction	Sequence	Product size (bp)
SER3	forward	TCACCAAAATGTACCAGGTGT	227
	reverse	TGACAGTATGAGAATCGATTGCA	
YMR230W-A	forward	GGATGTGTTACGATGCAGACA	170
	reverse	ATATGGCGCGTTCTTGAAGG	
MCH2	forward	CGTGGGTTTTGCGTACTTTG	194
	reverse	GTGACCTTTAGACTCTCCTAGGT	
EEB1	forward	TCCAGTTACAGGTGAAAACGT	164
	reverse	ACTCATCAAAGCTGCCCAAG	
ALD6	forward	AGATGTTGAAGGCCGGTACC	185
	reverse	TGACGGAAAGAAATGCAGGT	
STP4	forward	TTTGCATTTCGAGTACCCGC	183
	reverse	TGTGTGTATGTATGAGTCGGTG	
SRX1	forward	CTGGGCGTGCGAGTCAAG	176
	reverse	ATGTCGAGACTGCTGCCC	
RGS2	forward	AGGAATTCTCAACTCGGGGA	173
	reverse	TCCACAGATGATGAAGAGGCT	
ECM13	forward	CGAGCAGACGGATGAACTTG	214
	reverse	TACGGAACCATCGTCGACAT	
HXT2	forward	GGGTATGTCTTCATGGGCTGT	180
	reverse	TATAATCTCTTATTCCTCGGAAACTC	
PDR5	forward	ACAGTGAGAGATGGAGAAATTATGG	170
	reverse	GTCCATCTTGGTAAGTTTCTTTTCTT	
ACT1	forward	ATTGCCGAAAGAATGCAAAAGG	220
	reverse	CGCACAAAAGCAGAGATTAGAAACA	

ACKNOWLEDGMENTS

We are grateful to Mikio Tsuzuki and Shoko Fujiwara (Department of Applied Life Science, Tokyo University of Pharmacy and Life Science, Tokyo, Japan) for providing the *C. reinhardtii* strain and technical information. We are also grateful to Katsuhide Fujita and Makiko Fukuda [National Institute of Advanced Industrial Science and Technology (AIST), Technology Research Association for Single Wall Carbon Nanotubes (TASC)] for supporting microarray analysis. This work was supported by Japan Society for the Promotion of Science (JSPS) KAKENHI Grant Number 25871096.

AUTHOR CONTRIBUTIONS

Tadahiro Suzuki conceived, designed and performed the experiments, and wrote the manuscript. Yumiko Iwahashi supervised the experiments and revised the manuscript.

REFERENCES

1.   Kunikane, S.; Nakayama, T. Recent advances in plant secondary product glycosyltransferase research. *Seikagaku* 2008,*80*, 1033–1038.

2.   Schweiger, W.; Boddu, J.; Shin, S.; Poppenberger, B.; Berthiller, F.; Lemmens, M.; Muehlbauer, G.J.; Adam, G. Validation of a candidate deoxynivalenol-inactivating UDP-glucosyltransferase from barley by heterologous expression in yeast. *Mol. Plant Microbe Interact.* 2010, *23*, 977–986.

3.   Shin, S.; Torres-Acosta, J.A.; Heinen, S.J.; McCormick, S.; Lemmens, M.; Paris, M.P.; Berthiller, F.; Adam, G.; Muehlbauer, G.J. Transgenic *Arabidopsis thaliana* expressing a barley UDP-glucosyltransferase exhibit resistance to the mycotoxin deoxynivalenol. *J. Exp. Bot* 2012, *63*, 4731–4740.

4.   Berthiller, F.; Crews, C.; Dall'Asta, C.; Saeger, S.D.; Haesaert, G.; Karlovsky, P.; Oswald, I.P.; Seefelder, W.; Speijers, G.; Stroka, J. Masked mycotoxins: A review. *Mol. Nutr. Food Res.* 2013, *57*, 165–186.

5.   Berthiller, F.; Dall'Asta, C.; Schuhmacher, R.; Lemmens, M.; Adam, G.; Krska, R. Masked mycotoxins: Determination of a deoxynivalenol glucoside in artificially and naturally contaminated wheat by liquid chromatography-tandem mass spectrometry. *J. Agric. Food Chem.* 2005, *53*, 3421–3425.

6.   Ovando-Martínez, M.; Ozsisli, B.; Anderson, J.; Whitney, K.; Ohm, J.B.; Simsek, S. Analysis of deoxynivalenol and deoxynivalenol-3-glucoside in hard red spring wheat inoculated with *Fusarium graminearum*. *Toxins* 2013, *5*, 2522–2532.

7.   Berthiller, F.; Dall'asta, C.; Corradini, R.; Marchelli, R.; Sulyok, M.; Krska, R.; Adam, G.; Schuhmacher, R. Occurrence of deoxynivalenol

and its 3-β-D-glucoside in wheat and maize. *Food Addit. Contam. Part A* 2009, *26*, 507–511.

8.   Kostelanska, M.; Hajslova, J.; Zachariasova, M.; Malachova, A.; Kalachova, K.; Poustka, J.; Fiala, J.; Scott, P.M.; Berthiller, F.; Krska, R. Occurrence of deoxynivalenol and its major conjugate, deoxynivalenol-3-glucoside, in beer and some brewing intermediates. *J. Agric. Food Chem.* 2009, *57*, 3187–3194.

9.   Berthiller, F.; Krska, R.; Domig, K.J.; Kneifel, W.; Juge, N.; Schuhmacher, R.; Adam, G. Hydrolytic fate of deoxynivalenol-3-glucoside during digestion. *Toxicol. Lett.* 2011, *206*, 264–267.

10.   Nagl, V.; Schwartz, H.; Krska, R.; Moll, W.D.; Knasmüller, S.; Ritzmann, M.; Adam, G.; Berthiller, F. Metabolism of the masked mycotoxin deoxynivalenol-3-glucoside in rats. *Toxicol. Lett.* 2012, *213*, 367–373.

11.   Gratz, S.W.; Duncan, G.; Richardson, A.J. The human fecal microbiota metabolizes deoxynivalenol and deoxynivalenol-3-glucoside and may be responsible for urinary deepoxy-deoxynivalenol. *Appl. Environ. Microbiol.*2013, *79*, 1821–1825.

12.   Suzuki, T.; Sirisattha, S.; Mori, K.; Iwahashi, Y. Mycotoxin toxicity in *Saccharomyces cerevisiae* differs depending on gene mutations. *Food Sci. Technol. Res.* 2009, *15*, 453–458.

13.   Ghaffari, M.R.; Mardi, M.; Ehya, F.; Farsad, L.K.; Hosseini, S.; Ghareyazie, B. Mapping and expression analysis of a*Fusarium* head blight resistance gene candidate pleiotropic drug resistance 5 (PDR5) in wheat. *Iran. J. Biotechnol.* 2010,*8*, 112–116.

14.   Sirisattha, S.; Momose, Y.; Kitagawa, E.; Iwahashi, H. Toxicity of anionic detergents determined by *Saccharomyces cerevisiae* microarray analysis. *Water Res.* 2004, *38*, 61–70.

15.   Suzuki, T.; Iwahashi, Y. Comprehensive gene expression analysis of type B trichothecenes. *J. Agric. Food Chem.* 2012,*60*, 9519–9527.

16.   Alexander, N.J.; McCormick, S.P.; Ziegenhorn, S.L. Phytotoxicity of selected trichothecenes using *Chlamydomonas reinhardtii* as a model systemt. *Nat. Toxins* 1999, *7*, 265–269.

17.   Suzuki, T.; Iwahashi, Y. Phytotoxicity evaluation of type B trichothecenes using a *Chlamydomonas reinhardtii* model system. *Toxins* 2014, *6*, 453–463.

18.   Savard, M. Deoxynivalenol fatty acid and glucoside conjugates. *J. Agric. Food Chem.* 1991, *39*, 570–574.

19. Poppenberger, B.; Berthiller, F.; Lucyshyn, D.; Sieberer, T.; Schuhmacher, R.; Krska, R.; Kuchler, K.; Glössl, J.; Luschnig, C.; Adam, G. Detoxification of the *Fusarium* mycotoxin deoxynivalenol by a UDP-glucosyltransferase from*Arabidopsis thaliana*. *J. Biol. Chem.* 2003, *278*, 47905–47914.

20. Suzuki, T.; Iwahashi, Y. Gene expression profile of MAP kinase *PTC1* mutant exposed to Aflatoxin B1: Dysfunctions of gene expression in glucose utilization and sphingolipid metabolism. *Chem-Bio Informat. J.* 2009, *9*, 94–107.

21. Yiannikouris, A.; André, G.; Poughon, L.; François, J.; Dussap, C.G.; Jeminet, G.; Bertin, G.; Jouany, J.P. Chemical and conformational study of the interactions involved in mycotoxin complexation with β-D-glucans. *Biomacromolecules*2006, *7*, 1147–1155.

22. Fukuzawa, H.; Kubo, T.; Yamano, T. Genome of a green alga, *Chlamydomonas reinhardtii*, lights up key functions of plant and animal cells. (In Japanese). *Tanpakushitsu Kakusan Koso* 2008, *53*, 1133–1143.

23. Kam, V.; Moseyko, N.; Nemson, J.; Feldman, L.J. Gravitaxis in *Chlamydomonas reinhardtii*: Characterization using video microscopy and computer analysis. *Int. J. Plant. Sci.* 1999, *160*, 1093–1098.

24. Desjardins, A.E.; McCormick, S.P.; Appell, M. Structure-activity relationships of trichothecene toxins in an *Arabidopsis thaliana* leaf assay. *J. Agric. Food Chem.* 2007, *55*, 6487–6492.

25. Berthiller, F.; Schuhmacher, R.; Adam, G.; Krska, R. Formation, determination and significance of masked and other conjugated mycotoxins. *Anal. Bioanal. Chem.* 2009, *395*, 1243–1252.

26. Nakagawa, H.; Ohmichi, K.; Sakamoto, S.; Sago, Y.; Kushiro, M.; Nagashima, H.; Yoshida, M.; Nakajima, T. Detection of a new *Fusarium* masked mycotoxin in wheat grain by high-resolution LC-Orbitrap MS. *Food Addit. Contam. Part A*2011, *28*, 1447–1456.

27. Maragos, C.M.; Kurtzman, C.; Busman, M.; Price, N.; McCormick, S. Development and evaluation of monoclonal antibodies for the glucoside of T-2 toxin (t2-glc). *Toxins* 2013, *5*, 1299–1313.

28. Harris, E.H. A comprehensive guide to biology and laboratory use. In *The Chlamydomonas Sourcebook*; Academic Press: San Diego, CA, USA, 1989; pp. 1503–1504.

29. Iwahashi, H.; Odani, M.; Ishidou, E.; Kitagawa, E. Adaptation of *Saccharomyces cerevisiae* to high hydrostatic pressure causing growth inhibition. *FEBS Lett.* 2005, *579*, 2847–2852.

Chapter 3

ENHANCED DEGRADATION OF PENDIMETHALIN BY IMMOBILIZED CELLS OF BACILLUS LEHENSIS XJU

Veena S. More[1] Preeti N. Tallur[2] Francois N. Niyonzima[3] Sunil S. More[4]

[1] Department of Biochemistry, MSRCASC, Mattikere, MSRIT Post, Bangalore 560054, India

[2] Department of Chemistry, Government Arts and Science College, Karwar 581301, India

[3] Department of Maths, Science and PE, College of Education, University of Rwanda, Kigali 5039, India

[4] Department of Biochemistry, Center for Post Graduate Studies, Jain University, Bangalore 560011, India

ABSTRACT

A bacterium capable of degrading pendimethalin was isolated from the contaminated soil samples and identified as *Bacillus lehensis* XJU based on 16S rRNA gene sequence analysis. 6-Aminopendimethalin and 3,4-dimethyl 2,6-dinitroaniline were identified as the metabolites of pendimethalin degradation by the bacterium. The biodegradation of pendimethalin by freely suspended and the immobilized cells of *B. lehensis* on various matrices namely agar, alginate, polyacrylamide, and polyurethane foam was also investigated. The batch degradation rate was nearly the same for both free and immobilized cells in agar and alginate, whereas polyacrylamide- and PUF-immobilized cells degraded 93 and 100 of 0.1 % pendimethalin after 96 and 72 h, respectively. At higher concentration, the degradation rate of freely suspended cells decreased; whereas the same immobilized cells on polyurethane foam completely degraded 0.2 % pendimethalin within 96 h. The repeated batch degradation with the polyurethane foam-immobilized cells was reused for 35 cycles without losing the 0.1 % pendimethalin degrading ability. In contrast, agar-, alginate- and polyacrylamide-immobilized cells could be reused for 15, 18, and 25 cycles, respectively. When the pendimethalin concentration was increased to 0.2 %, the immobilized cells could be reused but the pendimethalin degradation rate was decreased. Polyurethane foam-immobilized cells exhibited better tolerance to pH and temperature alterations than freely suspended cells and could be stored for more than 3 months without losing pendimethalin degrading ability. The immobilization of cells capable of degrading pendimethalin may serve as an ideal technique for the complete degradation of the herbicide in the environment.

INTRODUCTION

Microorganisms are one of the tools used to detoxify toxic compounds present in the environment. Free suspended or immobilized microbial cells can be used for this purpose. However, the immobilized microbial cells have many advantages over free suspended cells under different conditions. For instance, the immobilization of whole cells increases degradation rate owing to increased cell population density, cell wall permeability, and extracellular microbial enzymes stability are improved, cells can be easily removed from the reaction mixture, higher operational stability and storage stability, reuse of immobilized cells in continuous reactors, and allows the bioreactors to operate at flow rates different from the growth rate of the microorganisms (Bettmann and Rehm 1984; Hall and Rao 1989; Cassidy et al. 1996; Ha et al. 2009; Zheng et al. 2009). In addition, the immobilized cell systems act as a protective cover in the presence of toxic compounds and are more resistant to pH or temperature changes. However, free suspended cells have better mass transfer aspects compared to immobilized bacterial or fungal cells (Trevors et al. 1992; Zheng et al.2009).

In the last two decades, there have been intensive researches on the use of immobilized microbial cells as biocatalysts, using numerous reactors like fed batch, semi-continuous fed batch, and continuous packed bed reactor. Each reactor type possesses its disadvantages and advantages, and the choice of a particular type of a reactor may depend on the operational conditions, and inexpensive and non-toxic support inert material for microbial cell immobilization, etc., (Zheng et al. 2009). Bacterial cells immobilized on various matrices have been used extensively for biodegradation of various toxic nitroaromatics such as trinitrotoluene (TNT) (Rho et al. 2001; Ullah et al. 2010), nitrobenzene (Zheng et al. 2009; Qi et al. 2012), 2-nitrotoluene (Mulla et al.2013), and 3-nitrobenzoate (Mulla et al. 2012).

Pendimethalin   [*N*-(1-ethyl   propyl)   2,6-dinitro-3,4-xylidine],   a common water and soil contaminant, herbicide of dinitroaniline group, is used to control weeds in various crop plants. The use of pendimethalin may adversely affect endangered species of terrestrial and semi-aquatic plants and invertebrates (Kole et al. 1994). One of the best strategies to degrade the hazardous compounds (including pendimethalin) is to use microorganisms. There are few reports on the degradation of pendimethalin by free cells of *Fusarium oxysporum* and *Paecilomyces variotii* (Singh and Kulshrestha 1991), *Azotobacter chroococcum* (Kole et al.1994), *Bacillus circulans* (Megadi et al. 2010), and fungus *Lecanicillium saksenae* (Pinto et al. 2012). However, there is no report on the degradation of pendimethalin by immobilized bacterial or fungal cells. The aim of the present investigation was therefore to compare the pendimethalin degradation by freely suspended and immobilized cells of *Bacillus lehensis* XJU on various matrices in batch and semi-continuous degradation, and to evaluate the effect of pH, temperature, and storage stability of pendimethalin degradation rate by polyurethane foam (PUF)-immobilized bacterial cells.

MATERIALS AND METHODS

Chemicals and Reagents

Pendimethalin was a generous gift from Rallis Agrochemicals India Ltd. (Mumbai, India). The 6-aminopendimethalin and 3,4-dimethyl 2,6-dinitroaniline standards were purchased from Sigma Chemical Co. (St. Louis, USA). Polyurethane foam and nylon meshes were purchased from Merck Specialities Private Limited (Mumbai, India). Acrylamide, bisacrylamide, ammonium persulfate, sodium alginate (SA), nutrient agar, agar–agar, sodium chloride, and calcium chloride were purchased from HiMedia Laboratories (Mumbai, India). All other chemicals used were of analytical grade and available commercially.

Isolation and Identification of the Bacterium

The bacterium was isolated from the contaminated soil samples by enrichment cultures with pendimethalin as a sole carbon source. It was identified based on morphological and biochemical aspects, as well as based on 16S rRNA gene sequence analysis. It was maintained on mineral salt (Seubert 1960) agar slants containing 0.1 % (w/v) pendimethalin.

Isolation and Identification of the Metabolites of Pendimethalin Degradation by the Isolate

The mineral salt medium (Seubert 1960) containing 0.1 % (w/v) pendimethalin was used as the production medium. The inoculated flasks were incubated on an orbital incubator (Model S150, Stuart, India) regulated at 150 rev/min for 24 h at room temperature (30 ± 2 °C). The bacterial growth was turbidometrically followed at 660 nm. The culture filtrate was collected by centrifugation, and ethyl acetate was used to extract the end products of pendimethalin degradation from the culture supernatant. The methanol was utilized to dissolve the residue from extraction. Thin layer chromatography (TLC) on silica gel G plates was used to analyze the residue for the presence of metabolites, using various solvent mixtures. An aqueous solution of $FeCl_3-K_3Fe(CN)_6$ was used to visualize the metabolites after exposure to I_2 vapors. The metabolites were also quantified by reversed-phase HPLC with a 5 μ spherisorb ODS (C_{18}) column (250 × 4.6 mm). The mobile phase was a mixture of acetonitrile and 50 mM phosphate buffer of pH 7.0 in the (70: 30, v/v) ratio. The flow rate was 1 mL/min, and the peaks were detected at 254 nm. UV–visible spectrophotometer was used to record the metabolites absorbance spectra. The metabolites were also further subjected to mass spectral studies (JEOL DX303).

Media used for the Degradation Studies

Two different media were used for this study. The mineral salt medium (MM1) consisting of (g/l) K_2HPO_4(6.30), KH_2PO_4 (1.83), NH_4NO_3 (1.00), $MgSO_4$·$7H_2O$ (0.10), $CaCl_2·2H_2O$ (0.10), $FeSO_4·7H_2O$ (0.10), $Na_2MoO_4·2H_2O$ (0.005), and $MnSO_4·H_2O$ (0.10) was used for the pre-cultivation of the bacterium. The medium was filtered and the pH was adjusted to 7.0. The medium was then dispersed in 100 mL quantities in 500 mL Erlenmeyer flasks and sterilized by autoclaving for 15 min at 15 psi. After sterilization by membrane filtration, pendimethalin was supplemented to the cultivation medium before bacterial inoculation. For the degradation studies, the mineral salt medium (MM2) comprising (g/l) K_2HPO_4 (6.30), NH_4NO_3(1.00), $MgSO_4·7H_2O$ (0.20), $CaCl_2·2H_2O$ (0.20), and $FeCl_3$ (0.05) was used. The medium was filtered and the pH was adjusted to 7.0. The medium was then dispersed in 100 mL quantities in 500 mL Erlenmeyer flasks and sterilized by autoclaving for 15 min at 15 psi; 0.1/0.2 % (w/v) pendimethalin was then added after MM2 sterilization. The bacterial cultures were then incubated at 30 °C with shaking at 150 rpm. The plate count method was used to find out the bacterial cell concentration.

Immobilization of Bacterial Cells in Various Matrices

Bacillus lehensis XJU was grown in the MM1 containing pendimethalin. During the mid-logarithmic growth, the bacterial cells were collected by centrifugation at 5000×g for 10 min at 15 °C, and washed two times with phosphate buffer (pH 7.0) of 50 mM strength. Alginate, agar, polyacrylamide (PA), and polyurethane foam (PUF) were then used as inert materials to immobilize the bacterial cells.

A procedure proposed by Bettmann and Rehm (1984) was used for sodium alginate entrapment of bacterial cells. 4 % (w/v) sodium alginate was mixed with 100 mL of distilled and autoclaved for 15 min. For biodegradation investigation, 20 g of wet bacterial cells was dissolved in 50 mL of MM2. The resulted suspension of *B. lehensis* XJU cells was mixed with 200 mL of alginate solution and stirred well with a magnetic stirrer. A combination of sodium alginate-bacterial cells was then added drop wise to a cold 0.2 M $CaCl_2$ solution with a burette. The formed gel beds of about 2.5 mm diameter were hardened for 5 h by re-suspending into a freshly sterile solution of $CaCl_2$ solution, and then frozen overnight at −18 °C. The beds were finally washed many times with sterilized distilled water and used for further studies.

The method proposed by Jonathan (1988) was followed for agar entrapment of bacterial cells. 4.5 mL of 0.9 % NaCl (w/v) was mixed with 100 mg of agar and heated at 100 °C till complete dissolution. The mixture was then cooled down at 40 °C. A slurry of *B. lehensis* XJU cells was taken and dissolved in 0.9 % (w/v) NaCl solution. 4.5 mL of the agar solution was mixed with 0.5 mL of bacterial cell slurry, poured on the nylon net placed on the glass plate, and cooled down to 5 °C. The 0.1 M phosphate buffer of pH 7.0 was used to store nylon membrane until used.

Jonathan (1988) method was also used for bacterial cells immobilization in polyacrylamide. 10 mL of distilled water was mixed with 15 g of wet bacterial cells and chilled in ice. 2.85 g acrylamide, 0.15 g bisacrylamide, and 10 mg ammonium persulfate were dissolved in chilled 10 mL of 0.2 M potassium phosphate buffer (pH 7.0). The bacterial cell suspension was immediately mixed with the buffer solution, poured into Petri plates, and allowed to polymerize for 1 h. 100 mL of 0.2 M potassium phosphate buffer (pH 7.0) was used to suspend the resulted sieved gels and allowed to settle down. After decantation, the gel was ready for degradation investigations.

Hall and Rao (1989) method was followed for polyurethane foam (PUF) immobilization. The PUF was cut into 5 mm cubes, washed with distilled water two times, and then dried. 4 g sterile foam cubes were placed in 100 mL of bacterial cell suspensions (9×10^9 cfu/mL) contained in 500-mL Erlenmeyer flasks, mixed for 2 h with the help of magnetic stirrer, and shaken for 1 h at 150 rpm. The conical flasks were left undistributed for 2 more hours. After medium removal, a saline solution was used to wash the immobilized foam cubes for further studies.

Pendimethalin Degradation Conditions

The batch degradations were carried out for both freely and immobilized *B. lehensis* XJU cells in four matrices. 3 mL of heat-killed cells along with 3 mL of freely suspended exponentially growing cells was added to 500 mL Erlenmeyer flasks containing 97 mL mineral salts medium (MM2) with different amount of substrates (0.1 and 0.2 %, w/v). For immobilized *B. lehensis* XJU cells, 11 g wet beads/foam cubes of the four matrices were supplemented to a 500 mL Erlenmeyer flasks containing 100 mL of mineral salts medium (MM2) with 0.1/0.2 % pendimethalin. The cell population in various matrices was in the 1.0 to 1.3×10^{10} cfu/g beads/foam cubes range. The degradation process by free and immobilized cells was carried out at room temperature (30 °C) on a rotary shaker (150 rpm). The various samples from culture medium were extracted at regular intervals for analysis of the residual substrate pendimethalin by high performance liquid chromatography (HPLC).

To evaluate the longevity of degrading activity of immobilized bacterial cells in various matrices, repetitive batch degradations were carried out. After each cycle of incubation period (96 h/cycle), the spent medium was decanted and beads/foam cubes were washed with double distilled water and transferred into a fresh MM2 medium containing pendimethalin. The degradation process was carried out under identical conditions and spent medium was used for the residual pendimethalin analysis by HPLC.

Effect of pH and Temperature on the Pendimethalin Degradation by Freely Suspended and PUF-Immobilized Cells

The rate of degradation of pendimethalin by freely suspended cells and PUF-immobilized cells of *B. lehensis* XJU at different pH (4.0–10.0) and temperatures (20–45 °C) was measured after 96 h of incubation.

Storage Stability of Freely Suspended and PUF-Immobilized Bacterial Cells in Degrading Pendimethalin

The storage stability of both free suspended and PUF-immobilized cells was evaluated at 4 °C every 10 days for a period 90 days.

Analytical Methods

The pendimethalin in the spent medium was quantified by reversed-phase HPLC with a 5 μ spherisorb ODS (C_{18}) column (250 × 4.6 mm). The mobile phase was a mixture of acetonitrile and 50 mM phosphate buffer of pH 7.0 in the (70: 30, v/v) ratio. The flow rate was 1 mL/min and the peaks were detected at 254 nm.

Statistical Analysis

Three independent experiments were conducted to evaluate pendimethalin degradation by freely suspended and immobilized bacterial cells. The means were compared by one-way analysis of variance (ANOVA) and means for groups in homogeneous subsets were given by Duncan's multiple range test (DMRT) at the 5 % significance level. The SPSS statistical package (PASW Statistics 18) was utilized for all statistical evaluations.

RESULTS

Biodegradation of Pendimethalin by *Bacillus lehensis* XJU

The bacterium isolated from the contaminated soil samples by enrichment cultures with pendimethalin as a sole carbon source was identified as *B. lehensis* XJU based on 16S rRNA gene sequence analysis (GenBank Accession Number: AY793550). The analysis of metabolites of the culture filtrates of *B. lehensis* XJU grown on pendimethalin showed the presence of two compounds whose chromatographic and spectral values corresponded well with that of authentic 3,4-dimethyl 2,6-dinitroaniline and 6-aminopendimethalin (Table 1).

Table 1: Chromatographic and spectral properties of metabolites resulted from pendimethalin degradation by *Bacillus lehensis* XJU

Property	Isolated metabolite 1	Authentic 6-ami-nopendi-methalin	Isolated metabolite 2	Authentic 3,4-dimeth-yl 2,6-dini-troaniline
TLC: R_f values in different solvent systems				
A: Hexane–ethyl acetate (1:1, v/v)	0.87	0.87	0.73	0.74
B: Toluene-dioxan-acetic acid (90:20:4, v/v)	0.93	0.94	0.83	0.83
UV absorption λ_{max} (nm)	229.43	229.43	234.42	234.42
HPLC retention time (min)	7.63	7.63	1.39	1.40
MS M^+ at m/z	236, 220, 191	236, 220, 191	181, 121, 55	181, 121, 55

Degradation of Pendimethalin by Freely Suspended and Immobilized Cells of *Bacillus lehensis* XJU in Batch Cultures

Batch degradation on shaken cultures with *B. lehensis* XJU showed that increasing concentration of pendimethalin was better tolerated and quickly degraded by immobilized cells than by free organisms. At low concentration (0.1 % w/v), the degradation rate was nearly the same for both free and immobilized cells in agar, alginate, and polyacrylamide after 96 h, with complete degradation in bacterial cells immobilized in PUF after 72 h (Fig. 1). However, with increasing concentration (0.2 % w/v), the degradation rate of free cells decreased and 20 % pendimethalin was only degraded; whereas the same immobilized cells in agar, alginate, and polyacrylamide matrices degraded 50–77 %. A complete degradation of pendimethalin was also observed for cells immobilized in PUF but within 96 h (Fig. 2). At the concentration of 0.7 % w/v, pendimethalin was not degraded by free cells, but was degraded by PUF-immobilized cells (data not shown), which were more effective than cells immobilized in agar, alginate, and polyacrylamide.

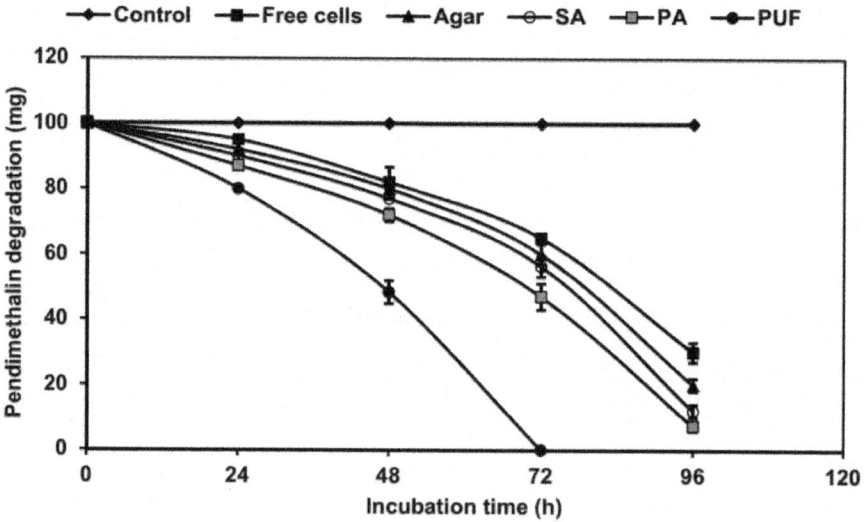

Figure. 1: Batch culture degradation of 0.1 % pendimethalin by cells of *Bacillus lehensis* XJU immobilized on polyurethane foam (PUF, *filled circle*), polyacrylamide (PA, *filled square*), sodium alginate (SA, *open circle*), agar (*filled triangle*), and by free suspended cells (*filled square*). The uninoculated culture served as control (*filled diamond*).

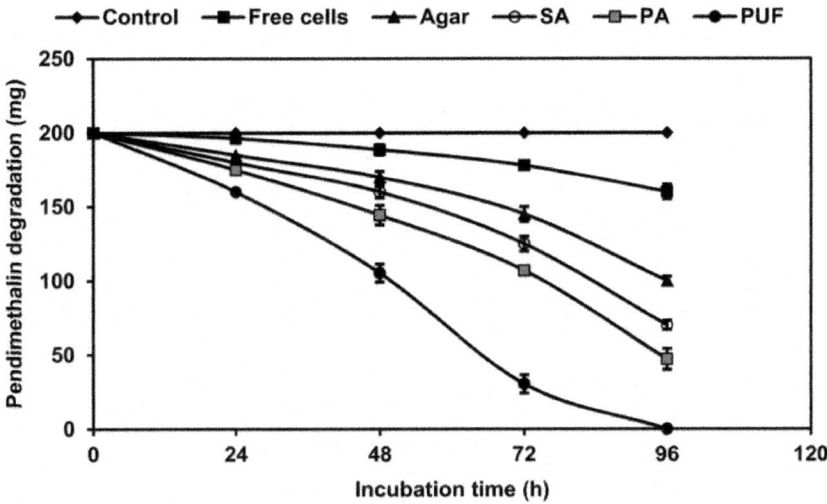

Figure. 2: Batch culture degradation of 0.2 % pendimethalin by cells of *Bacillus lehensis* XJU immobilized on polyurethane foam (PUF, *filled circle*), polyacrylamide (PA, *filled square*), sodium alginate (SA, *open circle*), agar (*filled triangle*), and by free suspended cells (*filled square*). The uninoculated culture served as control (*filled diamond*).

Semi-Continuous Degradation of Pendimethalin by Immobilized Cells of *B. lehensis* XJU

The degradation of pendimethalin by cells immobilized in polyurethane foam (PUF), alginate, agar, and polyacrylamide was carried out at different concentrations of pendimethalin (0.1 and 0.2 %) for 96 h. The polyurethane foam (PUF)-immobilized cells were reused for 35 cycles without losing the pendimethalin degrading capacity when the initial concentration of pendimethalin was 0.1 %. In contrast, agar-, alginate-, and polyacrylamide-immobilized cells were reused for 15, 18, and 25 cycles, respectively, (Fig. 3). When the initial concentration of pendimethalin was increased to 0.2 %, the immobilized cells could be reused but the rate of degradation of pendimethalin was decreased (Fig. 4).

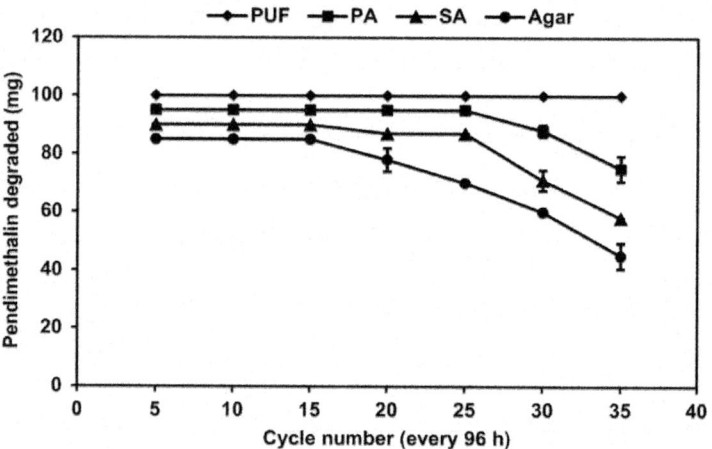

Figure. 3: Semi-continuous degradation of pendimethalin (0.1 %) by cells of *Bacillus lehensis* XJU immobilized on polyurethane foam (PUF, *filled diamond*), polyacryl-amide (PA, *filled square*), sodium alginate (SA, *filled triangle*), and agar (*filled circle*).

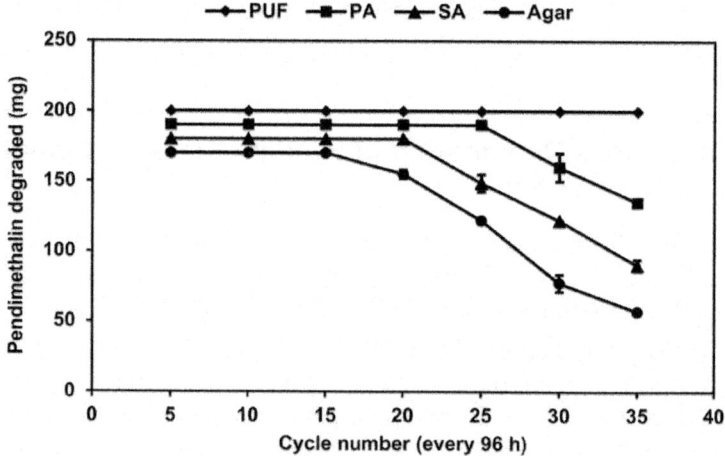

Figure. 4: Semi-continuous degradation of pendimethalin (0.2 %) by cells of *B. lehensis* XJU immobilized on polyurethane foam (PUF, *filled diamond*), polyacrylamide (PA, *filled square*), sodium alginate (SA, *filled triangle*), and agar (*filled circle*).

Effect of pH on Pendimethalin Degradation Rate by PUF-Immobilized Bacterial Cells

The effect of pH on degradation rates of pendimethalin by immobilized cells on PUF and free suspended cells was investigated in the pH 4.0–10.0 range. The initial pH alteration in the 6.0 and 8.0 range had no effect on the pendimethalin degradation by PUF-immobilized cells. A slight degradation rate was observed at pH 4.0, 9.0, and 10.0. However, freely suspended cells were active at pH 7.0, and other pH below or above 7.0 had adverse effects on both degradation rate and organism growth (Fig. 5).

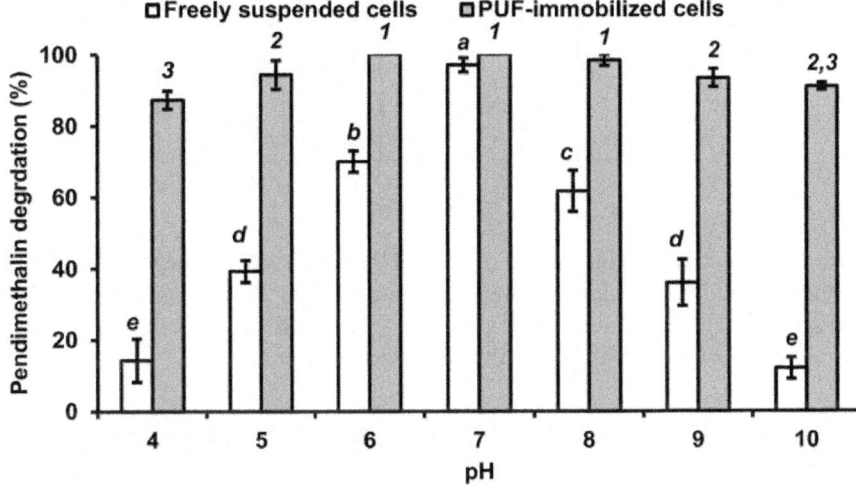

Figure. 5: Effect of pH on the degradation of 0.1 % pendimethalin by freely suspended (*open square*) and PUF-immobilized (*filled square*) *Bacillus lehensis* XJU cells. The pendimethalin degradation values with different numbers or alphabets on the *error bars* significantly differ from each other at $P_{0.05}$

Effect of Temperature on Pendimethalin Degradation Rate by PUF-Immobilized Bacterial Cells

The effect of temperature on the degradation of pendimethalin by PUF-immobilized cells was analyzed. The pendimethalin degradation was seen in the 20–45°C range with optimum at 30°C, although statistically at par with 25, 35, and 40°C for immobilized bacterial cells. However, the temperatures below and above 30°C were not suitable for degradation of pendimethalin by freely suspended cells (Fig. 6).

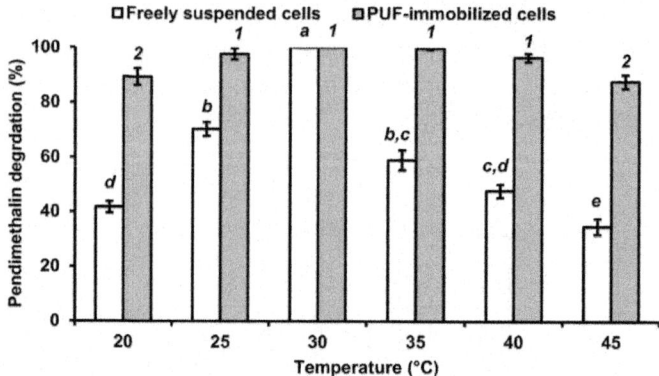

Figure. 6: Influence of temperature on the 0.1 % pendimethalin degradation by free-ly suspended cells (*open square*) and PUF-immobilized *Bacillus lehensis* XJU cells (*filled square*). The pendimethalin degradation values with different numbers or alphabets on the *error bars* significantly differ from each other at $P_{0.05}$

Storage Stability of the PUF-Immobilized Cells Degrading Pendimethalin

PUF-immobilized cells were stored at 4°C for 90 days and no decline in degradation ability observed for 50 days, and a marginal decrease of less than 10 % was observed after 90 days. However, a significant gradual decrease in pendimethalin degradation ability was seen with freely suspended bacterial cells and no degradation observed after 3 months (Fig. 7).

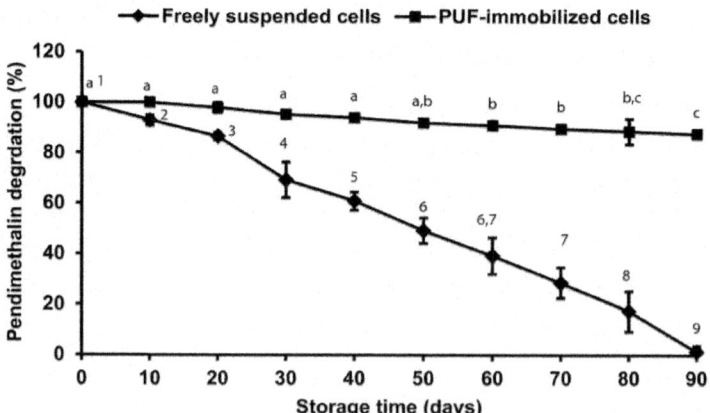

Figure. 7: Storage stability of freely suspended (*filled diamond*) and PUF-immobilized (*filled square*) *Bacillus lehensis* XJU cells grown on 0.1 % pendimethalin. The pendimethalin degradation values with different numbers or alphabets on the *error bars* significantly differ from each other at $P_{0.05}$

DISCUSSION

The degradation of pendimethalin by *B. lehensis* XJU based on TLC, reverse phase HPLC, UV visible, and mass spectral studies has resulted in the formation of 3,4-dimethyl 2,6-dinitroaniline and 6-aminopendimethalin. Similarly, the biodegradation of the herbicide pendimethalin by *B. circulans*, resulted in the formation of two metabolites, viz., 6-aminopendimethalin by pendimethalin reduction, and 3,4-dimethyl 2,6-dinitroaniline by pendimethalin oxidative dealkylation (Megadi et al. 2010). Biodegradation of pendimethalin by freely suspended cells of *B. lehensis* XJU and by cells immobilized in agar, SA, PUF, and polyacrylamide was compared with respect to their degradation rate and tolerance against increasing concentrations of pendimethalin. In batch cultures, the freely suspended cells degraded pendimethalin comparatively well at lower concentrations (0.1 %) and degradation rate decreased at higher concentrations. The cells immobilized in SA, PUF, agar, and polyacrylamide matrices were able to survive and degrade pendimethalin at higher concentrations. The cells immobilized in PUF degraded pendimethalin up to the concentration of 0.7 % (w/v) in batch cultures. The present results revealed that the toxicity of pendimethalin at high concentration level could inhibit the metabolism resulting in lower removal efficiency by the free cells. In addition, it indicates that in an immobilized cell culture, the carrier material act as a protective cover against the toxicity of pendimethalin. The enhanced pendimethalin degradation rate can also be ascribed to higher cell population density and higher activity of the cells immobilized in or on these carrier inert materials. Similarly, the inert materials served as a protective cover towards 2-nitrotoluene (Mulla et al. 2013) and 3-nitrobenzoate (Mulla et al. 2012). Likewise, the higher degradation rates of various toxic nitroaromatics were attributed to higher cell density in the matrices (Cassidy et al. 1996; Qi et al. 2012). Increased concentration of pendimethalin was better tolerated and degraded by PUF-immobilized cells than *B. lehensis* XJU cells immobilized in agar, SA, and polyacrylamide, as well as free cells. The cells of Micrococcus *luteus* Z3 immobilized in PUF had the higher nitrobenzene degradation capacity compared to free suspended cells (Qi et al. 2012). The moderate degradation rate observed with bacterial cells immobilized in agar, alginate, and polyacrylamide can be attributed to the leakage that may result from the mechanical instability and monomer and/or radical toxicity (Hall and Rao 1989; Trevors et al. 1992).

The semi-continuous degradation data showed that polyacrylamide and PUF-immobilized cells retained the pendimethalin degrading ability for a longer period since they could be reused for 26 and 35 cycles, respectively. But when the concentration of pendimethalin increased, polyurethane foam (PUF)-immobilized cells degraded pendimethalin faster than polyacrylamide-immobilized cells, which suggest that the PUF immobilization is a better technique for the degradation of toxic herbicides in the environment. However, the bacterial cells immobilized in SA and agar exhibited lower degradation ability of pendimethalin with increased cycle numbers. This was ascribed to the mechanical instability and gradual cell leakage from the inert porous beads (Trevors et al. 1992; Ha et al. 2009). Similarly, polyurethane foam was also an excellent support for the degradation of numerous nitroaromatics by bacterial cells (Zheng et al. 2009; Mulla et al.2012, 2013). The increased degradation rate was ascribed to adsorbing capacity, stability, mechanical strength, and high porosity for the PUF-immobilized cells (Romaškevic et al. 2006).

The immobilized cells on PUF exhibited an excellent resistance to pH and temperature alterations, and had the better storage stability of more than 3 months than freely suspended cells in degrading pendimethalin. Similarly, the free and PUF-immobilized *Micrococcus* cells were active in the pH 7.0–8.0 at 30–35 °C and 5.0–10.0 at 25–40 °C, respectively, in degrading 2-nitrotoluene (Mulla et al. 2013). Likewise, a narrow range of pH (6.5–7.5) and temperature (30–35°C), and a broad range of pH (5.0 to 10.0) and temperature (20–40 °C) were observed for freely suspended and PUF-immobilized cells, respectively, in degrading 3-nitrobenzoate (Mulla et al. 2012). The immobilized cells could be stored more than 3 months without losing too much pendimethalin degradation capacity. Similarly, the PUF-immobilized cells of *Bacillus flexus* strain XJU-4 and *Micrococcus* sp. strain SMN-1 were stored for 60 and 70 days at 4°C without losing the capacity to degrade 3-nitrobenzoate (Mulla et al. 2012) and 2-nitrotoluene (Mulla et al. 2013), respectively. An effective biodegradation of elevated concentration of dinitroaniline herbicides (such as pendimethalin) and various nitroaromatic compounds can be achieved using immobilized microbial technology.

CONCLUSION

The present investigation has revealed the biodegradation of pendimethalin by the bacterial isolate *B. lehensis* XJU. It also showed the pendimethalin biodegradation by freely suspended and immobilized *B. lehensis* XJU on different matrices. The microorganisms capable of degrading toxic compounds can therefore be immobilized by entrapment in an inexpensive supports and the immobilized cells retain their ability over a considerable period of time, especially for PUF. Thus, the immobilized microbial cell systems may find applications in the treatment of various contaminated environment sites. However, prior to large-scale application of such systems, further studies are needed for determining the optimal operating conditions.

ACKNOWLEDGMENTS

The authors acknowledge the Jain University for financial support.

REFERENCES

1.   Bettmann H, Rehm HJ. Degradation of phenol by polymer entrapped microorganisms. Appl Microbiol Biotechnol. 1984;20:285–290. doi: 10.1007/BF00270587.

2.   Cassidy MB, Lee H, Trevors JT. Environmental applications of immobilized microbial cells: a review. J Ind Microbiol. 1996;16:79–101. doi: 10.1007/BF01570068.

3.   Ha J, Engler CR, Wild JR. Biodegradation of coumaphos, chlorferon, and diethylthiophosphate using bacteria immobilized in Ca-alginate gel beads. Bioresour Technol. 2009;100:1138–1142. doi: 10.1016/j.biortech.2008.08.022.

4.   Hall DO, Rao KK. Immobilized photosynthetic membranes and cells for the production of fuels and chemicals. Chim Oggi. 1989;1:41–47.

5.   Jonathan W. Methods of immobilization of microbial cells. J Microbiol Meth. 1988;8:91–102. doi: 10.1016/0167-7012(88)90041-3.

6.   Kole RK, Saha J, Pal S, Chaudhuri S, Chowdhuri A. Bacterial degradation of the herbicide pendimethalin and activity evaluation of its metabolites. Bull Environ Contam Toxicol. 1994;52:779–786. doi: 10.1007/BF00195503.

7.   Megadi VB, Tallur PN, Hoskeri RS, Mulla SI, Ninnekar HZ. Biodegradation of pendimethalin by *Bacillus circulans*. Indian J Biotech. 2010;9:173–177.

8.   Mulla SI, Talwar MP, Hoskeri RS, Ninnekar HZ. Enhanced degradation of 3-nitrobenzoate by immobilized cells of *Bacillus flexus* strain XJU-4. Biotechnol Bioprocess Eng. 2012;17:1294–1299. doi: 10.1007/s12257-012-0211-2.

9.   Mulla SI, Talwar MP, Bagewadi ZK, Hoskeri RS, Ninnekar HZ. Enhanced degradation of 2-nitrotoluene by immobilized cells of *Micrococcus* sp. strain SMN-1. Chemosphere. 2013;90:1920–1924. doi: 10.1016/j.chemosphere.2012.10.030.

10. Pinto AP, Serrano C, Pires T, Mestrinho E, Dias L, Teixeira DM, Caldeira AT. Degradation of terbuthylazine, difenoconazole and pendimethalin pesticides by selected fungi cultures. Sci Total Environ. 2012;435–436:402–410. doi: 10.1016/j.scitotenv.2012.07.027.

11. Qi Y, Zheng CL, Zhang YT. Microbial degradation of nitrobenzene by immobilized cells of *Micrococcus luteus*. Adv Mat Res. 2012;599:52–59. doi: 10.4028/www.scientific.net/AMR.599.52.

12. Rho D, Hodgson J, Thiboutot S, Ampleman G, Hawari J. Transformation of 2,4,6-trinitrotoluene (TNT) by immobilized *Phanerochaete chrysosporium* under fed-batch and continuous TNT feeding conditions. Biotechnol Bioeng. 2001;73:271–281. doi: 10.1002/bit.1060.

13. Romaškevič T, Budrienė S, Pielichowski K, Pielichowski J. Application of polyurethane-based materials for immobilization of enzymes and cells: a review. Chemija. 2006;17:74–89.

14. Seubert W. Degradation of isoprenoid compounds by microorganisms I. Isolation and characterization of isoprenoid-degrading bacterium *Pseudomonas citronellolis*, n. sp. J Bacteriol. 1960;79:426–434.[PMC free article]

15. Singh SB, Kulshrestha G. Microbial degradation of pendimethalin. J Environ Sci Health B. 1991;26:309–321. doi: 10.1080/03601239109372737.

16. Trevors JT, Van Elsas JD, Lee H, Van Overbeek LS. Use of alginate and other carriers for encapsulation of microbial cells for use in soil. Microb Releases. 1992;1:61–69.

17. Ullah H, Shah AA, Hasan F, Hameed A. Biodegradation of trinitrotoluene by immobilized *Bacillus* sp. YRE1. Pak J Bot. 2010;42:3357–3367.

18. Zheng C, Zhou J, Wang J, Qu B, Wang J, Lu H, Zhao H. Aerobic degradation of nitrobenzene by immobilization of *Rhodotorula mucilaginos*a in polyurethane foam. J Hazard Mater. 2009;168:298–303. doi: 10.1016/j.jhazmat.2009.02.029.

Chapter 4

PHYSIOLOGICAL TESTS FOR YEAST BREWERY CELLS IMMOBILIZED ON MODIFIED CHAMOTTE CARRIER

Joanna Berlowska, Dorota Kregiel, and Wojciech Ambroziak

Institute of Fermentation Technology and Microbiology, Technical University of Lodz, ul. Wolczanska 171/173, 90-924 Lodz, Poland

ABSTRACT

In this study yeast cell physiological activity was assessed on the basis of the in situ activity of two important enzymes, succinate dehydrogenase and pyruvate decarboxylase. FUN1 dye bioconversion and cellular ATP content were also taken as important indicators of yeast cell activity. The study was conducted on six brewing yeast strains, which were either free cells or immobilized on a chamotte carrier. The experimental data obtained indicate clearly that, in most cases, the immobilized cells showed lower enzyme activity than free cells from analogous cultures. Pyruvate decarboxylase activity in immobilized cells was higher than in planktonic cell populations only in the case of the *Saccharomyces pastorianus* 680 strain. However, in a comparative assessment of the fermentation process, conducted with the use of free and immobilized cells, much more favorable dynamics and carbon dioxide productivity were observed in immobilized cells, especially in the case of brewing lager yeast strains. This may explain the higher total cell density per volume unit of the fermented medium and the improved resistance of immobilized cells to environmental changes.

INTRODUCTION

Current research on the application of immobilized yeast cells in brewing technology has three main focuses: the production of alcohol-free beer, and conducting either main fermentation or green beer maturation in continuous systems. Only a few of the technologies proposed in the literature have resulted in pilot-scale attempts or industrial implementation (Verbelen et al. 2006; Willaert 2000; Brányik et al. 2012). One of the main difficulties is maintaining the desired physiological state in immobilized microbial cells. The outcome of a brewery fermentation depends on wort composition, ambient technological conditions as well as on variations in pitching yeast activity. Therefore, for the end product quality, monitoring of yeast physiological state is very essential.

The term "physiological activity" could describe various important parameters: fermentation potential, stress tolerance, aging, growth or reproduction abilities. The physiology of immobilized cells is affected by the microenvironment and the supply of nutrients and metabolic products (internal and external mass transfer) (Junter et al. 2002; Pajić-Lijakovic et al. 2007; Gonga et al. 2010). Cellular stress at the stage of immobilization may also have a significant impact on the physiological state, and according to Smart (2001) this physiological history may determine a cell's efficiency during the technological process. The general aim is to maintain the greatest viability and metabolic activity of the cells, allowing to the process to be carried out with high efficiency for the longest time possible. Continuous system technologies in beer production require immobilized yeast cells to be kept for several months in bioreactors. In the case of yeast plankton populations, over time the linear dimensions of cells increase, there is a longer generation time and their metabolic activity decreases.

It has been shown that free and immobilized yeast cells differ in chemical composition and ploidy (Verbelen et al. 2006). Immobilized cells, in comparison with free cells, have a higher content of glycogen, trehalose, structural polysaccharides (glucans and mannan), fatty acids and DNA. Immobilization also causes changes in the proteome of a cell, in the level of gene expression, and has a significant impact on the quantitative composition and organization of the cytoplasmic membrane and cell wall structures (Brányik et al. 2008; Parascandola et al. 1997). Many studies have reported an increase in metabolic activity (increased rate of sugar uptake and productivity of selected metabolites) in immobilized cells (Junter et al. 2002; Norton and D'Amore 1994; Angelova et al. 2000; Talebnia and Taherzadeh 2007; Plessas et al. 2007; Li et al. 2007; Behera et al. 2011). Adsorption of *Saccharomyces carlsbergensis* on porous glass and *S. cerevisiae* on ceramic support resulted in increased production of ethanol and reduced production of CO_2 (Kourkoutas et al. 2004). In yeast entrapped in alginate matrices, a slight decrease was noticed in intracellular pH due to increased enzymatic activity. This promotes the permeability of membranes, which in turn leads to an increase in proton transport and ATP use, stimulating glycolysis processes (Galazzo and Bailey 1990). Higher efficiency in the pentose phosphate pathway and of glycolytic flux may also be explained by the increased activity of alcohol dehydrogenase and by more efficient regeneration of the NADH and NADPH cofactors Brányik et al. (2008).

Many changes have been made in the modern beer-brewing process since the first recorded beer production by mankind. However, despite all these changes, one constant factor is the requirement for good quality brewing yeast (Lodolo et al. 2008). Knowing the physiological state of immobilized yeast cells is important not only from a theoretical perspective. It is important to verify the efficacy of cell-carrier systems, as well as to monitor the continuous process. The precise evaluation of yeast physiology is rather difficult and sometimes problematic—the type of information gathered depends on the kind of analytical method applied. Therefore, monitoring of yeast physiology should be multi-parametric.

Observation, analytics and diagnostics of biofilms formed on abiotic surfaces are usually complicated, and often expensive. Visualization of the spaces colonized by the microorganisms and the architecture of the three-dimensional structure formed is made possible by such techniques as magnetic resonance imaging, optical coherence tomography, confocal laser scanning microscopy and fluorescence microscopy (Chandra et al. 2001; Nott et al. 2005; Xi et al. 2006). The coupling of fluorescent in situ hybridization (FISH) and microautoradiography allows for the consumption by the tested microorganisms of different substrates to be determined precisely (Lee et al. 1999; Kindaichi et al. 2004). In the description of the interactions between immobilized microorganisms and their metabolic characteristics, the use of microelectrodes and a combination of FISH, mass spectroscopy and isotopic labeling techniques can be of significant help (Jang et al. 2003; Majors et al. 2005). Many techniques used in the evaluation of the microbial physiology of immobilized cells require their detachment (Uppuluri et al. 2006). Multiple cell rinsing and centrifugation can have a significant impact on the value of the parameter under evaluation (Brányik et al. 2005). In the current study, we propose determining succinate dehydrogenase (SDH) and pyruvate decarboxylase (PDC) activity in situ in immobilized cells with increased membrane permeability. The aim of noninvasive analysis (in situ) is to avoid disturbing the normal functioning of the cells and so lowering their physiological activity. An in situ enzyme activity assay based on chemical changes in membrane permeability allowing migration of low molecular weight compounds (substrates, products, cofactors), while the enzymes and other macromolecules are kept in constant concentrations (Freire et al. 1998). The cytoplasmic membrane forms a barrier with low permeability and enzyme activity is determined in whole cells (Cordeiro and Freire 1995; Kippert 1995; Bindu et al. 1998; Kondo et al. 2008; Crotti et al. 2001; Gough et al. 2001; Chelico and Khachatourians2003; Berlowska et al. 2006, 2009; Miranda and Ferreira 2008).The advantage is that enzyme activity is determined for a specific physiological state of the cell. Using this type of enzyme assay, cellular regulatory effects can be observed and enzyme activities determined for cells immobilized on solid supports.

We focused our research on the immobilization of yeast brewery strains on chamotte ceramic carriers. Our study is the continuation of previous research on yeast adhesion to native and modified chamotte tablets (Kregiel et al. 2012, Berlowska et al. 2013), which led us to study chamotte modification as a way to enhance yeast cell adhesion efficiency. Enhanced in this way yeast immobilization and proper selected conditions of this process give opportunity to conduct physiological tests for adhered cells.

The aim of the present study was to determine the effect of the immobilization process on brewing yeast cell physiological activity. Multi-parameter physiological activity evaluation was conducted for free and immobilized yeast cells, which allowed the nature, characteristics, fermentation abilities and technological suitability of the tested strains to be described. The basis of physiological activity assessment was the activity assay in situ of two enzymes, SDH and PDC, important yeast metabolic pathways (the Krebs cycle and glycolysis respectively). PDC (EC 4.1.1.1) is an important enzyme for yeast fermentation. Measuring PDC activity allows the fermentation activity of individual yeast strains to be monitored. On the other hand, SDH (EC 1.3.5.1) is essential for the aerobic utilization of carbon sources and plays a crucial role in the supply of energy for the physiological activity of yeast cells (Kregiel et al. 2013). Therefore, SDH and PDC activity assays may be the important methods to evaluate yeast activity and control different biotechnological processes. ATP content and fermentative activity were also monitored.

This paper is the first to describe the physiological activity of brewery yeasts immobilized on inexpensive, porous chamotte covered by active organo-silanes.

MATERIALS AND METHODS

Carriers

During the experiment, solid carriers were used made from inexpensive chamotte material [mainly Al_2O_3(36 %) with SiO_2 (58 %) and Fe_2O_3 (2.6 %)]. The ceramic chamotte tablets were prepared from water and chamotte fire clay (50–1,000 μm) (Boleslawiec Refractory Plant BZMO Ltd., Poland) in the ratio 1:2. Chamotte carriers were made in our laboratory by firing chamotte fire clay at a temperature of 1,100 °C. The height and diameter of the chamotte tablets were 5 and 15 mm, respectively.

The carriers were modified in the Centre of Molecular and Macromolecular Studies of the Polish Academy of Science. The chamotte tablets were immersed in 10 % H_3PO_4, left for 2 h and then washed and dried at room temperature. The dry tablets were placed in 10 % 3-(N,N-dimethyl-N-2-hydroxy-ethyl) ammonium propyldimethoxysilane or (3-glycidoxy propyl) trimethoxysilane chloride isopropanol solution. The flask was repeatedly connected and disconnected from the vacuum pump to remove the air from and draw the silane solution into the pores. The prepared media were pre-dried to evaporate the isopropanol and then incubated at 80 °C for 12 h. For the (3-glycidoxy propyl) trimethoxysilane modification, dry carriers were laid on a Buchner funnel so as to form a single layer and their location was changed periodically (by turning them upside down). Under a funnel, 200 mL of 25 % ammonia-water was heated and stirred for 2 h. Free ammonia caused the collapse of the epoxide ring on the surface and creation of 3-(3-amino-2-hydroxy-1-propoxy) propyldimethoxysilane groups.

Yeast Strains

In our research, to assess physiological state of immobilized brewery yeasts, the six various top and bottom-fermenting *Saccharomyces* stains were used (Table 1.) These strains were selected according to their adhesion properties in the previous studies (Kregiel et al. 2012; Berlowska et al. 2013).

Table 1: Applied biological material and immobilization methods

Strain	Type	Collection	Immobilization medium
S. pastorianus W 34/70	Lager	Hafebank Weihenstephan (DE)	Ringer's solution
S. pastorianus 680		National Collection of Yeast Cultures (GB)	Wort broth
S. pastorianus B4		Collection LOCK105 (PL)	Wort broth
S. cerevisae TT	Ale		Ringer's solution
S. cerevisae 1017		National Collection of Yeast Cultures (GB)	Wort broth
S. cerevisae 1183			Ringer's solution

The yeast strains were stored on wort agar slants (Merck) at room temperature. Directly prior to the experiment, they were activated by placing them on fresh agar slants and incubated at 30 °C for 48 h.

Culture Conditions

The yeasts were propagated in liquid wort broth (Merck). The cultivation was carried out in a 500 mL round bottom flask filled with 50 mL of medium with 1 % v/v yeast suspension added. It was performed using the shaking flask method (220 rpm) at a temperature of 30 °C.

For the purpose of the experiment, starved cultures were also required. These were prepared from stationary phase cultures which were rinsed twice with Ringer's solution (Merck) and resuspended in the same solution.

Fermentations

To assess fermentation performance on a small laboratory scale, for consistent and reproducible procedures the simple, well defined glucose-based medium was used. Static fermentations were conducted in 50 mL medium ([$(NH_4)_2 \cdot SO_4$ 3 g/L; KH_2PO_4 1 g/L; $MgSO_4 \cdot 7H_2O$ 0.5 g/L; yeast extract (Difco) 0.5 g/L; $CaCO_3$ 3 g/L] with 12 % glucose) sealed with a fermentation lock containing paraffin oil. The fermentations, carried out both for adhered and free yeast cells, were conducted over 7 days at the appropriate temperature (top fermenting yeast 20 °C, bottom fermenting yeast 10 °C).

Yeast Cell Immobilization

Only the high efficiency of immobilization let us to evaluate physiological state of adhered yeast cells. Therefore, the parameters of adhesion process have been optimized for each strain on the base of previous studies results. Both character of chemical surface modifications and immobilization medium were taken into consideration (Kregiel et al. 2012; Berlowska et al. 2013). The yeast cells were harvested when they reached the appropriate phase of growth or physiological state, at which point they were standardized. In 50-mL sterile Erlenmeyer flasks, 5 mL of cells and medium suspensions with a density of 5×10^7 cells/mL were prepared. For dilutions, sterile basic cultivation (wort broth) medium or Ringer solution was used (Table 1). Next, sterile carriers were introduced into each of the previously prepared suspensions and incubated at 30 °C with agitation (75 rpm) for 24 h.

Determination of the Number of Immobilized Yeast Cells

After the adhesion process, five pieces of chamotte tablet were selected from each experimental sample. The tablets were suspended in 5 mL of 5 % H_2SO_4. Then, tubes filled with tablets suspended in appropriate solutions were boiled for 2 min. and vortexed for 15 min, after which the carriers were removed. The remaining solution was analyzed for the number of microorganisms. The density of the yeast suspensions was determined using the fluorimetric method based on DAPI staining (Kregiel et al. 2012).

Succinate Dehydrogenase (SDH) Activity

The in situ assay measured SDH activity in whole cells. After pre-incubation with digitonin, a permeabilization agent blue tetrazolium salt (BT), in the presence of phenazine methosulfate and sodium azide, was reduced intracellularly to colored formazan crystals. The amount of the formed formazan was determined spectrophotometrically after DMSO extraction (Kręgiel et al. 2008). Free cells were measured in standardized suspensions (3×10^8 cells/mL). In the case of the immobilized yeasts, 6 chamotte carriers with adhered cells were used (no centrifugation for cells separation was required). Knowing the number of immobilized yeast cells, the SDH activity values were recalculated appropriately.

ATP Content

Intracellular ATP content was determined in relative light units (RLU) on the basis of the luciferin/luciferase method using a Hy-Lite2 luminometer (Merck) (41). The measurement of free cells was conducted in standardized suspensions (1×10^4 cells/mL). Tablets with immobilized cells were washed with sterile distilled water, and 1 mL of Somatic Cell ATP Releasing Reagent (SIGMA-ALDRICH) was added to each carrier. After 5 min, the solution with released ATP was diluted to an equivalent concentration of 1×10^4 cells/mL (having determined the number of immobilized yeast cells) and analyzed. The readings in RLU were calculated on the basis of the standard curve and expressed in fg/dm^3 of ATP.

Pyruvate Decarboxylase (PDC) Activity

PDC activity was measured in situ in whole cells with digitonin permeabilized membranes. Sodium pyruvate (0.05 M) solution was used as a reaction substrate. The acetaldehyde formed was detected using the GC technique with a Headspace Autosampler (Berlowska et al. 2009). Measurements of the free cells were conducted in standardized suspensions (2×10^8 cells/mL) and in the case of immobilized cells using four carriers (no centrifugation for cell separation was required). Once the number of immobilized yeast cells had been determined, the PDC activity values obtained were recalculated appropriately.

Fermentation Activity

The fermentation activity of yeast populations was evaluated by a quantitative determination of the carbon dioxide production in grams per 100 mL of fermentative medium.

FUN1 Staining

Tablets with immobilized cells were rinsed with distilled water to wash out the medium that remained on the surface. To stain the cells with FUN1 they were first soaked in a solution of 2 % glucose in 10 mM Na-HEPES. Then, 200 μL of 0.1 μg/mL FUN1 was poured onto the surface of each tablet. After 5 min incubation at 30 °C, the tablets were left to dry and then examined under a fluorescence microscope OLYMPUS BX 41 equipped with the appropriate filter (excitation wavelength 470–490 nm).

Statistical Method

Each experiment was performed in triplicate and each datum was the arithmetic mean (a.m.) of three measurements. The standard deviations (SD) were calculated and the results given as am ± SD.

RESULTS AND DISCUSSION

Yeast Immobilization

For the purposes of the present study, six brewing yeast strains were immobilized on chamotte carriers with chemically modified surfaces (Fig. 1a). The number of cells per carrier was assayed fluorometrically using DAPI. This cationic dye specifically binds to DNA in places rich in adenine–thymine pairs. It is also accumulated in small grooves of the DNA double helix (Barker and Smart 1996). According to the authors' own research, the amount of emitted light, measured spectrofluorimetrically, is proportional to the number of stained, heat denatured, yeast cells. The effectiveness of the adhesion processes ranged from 2.6 to 4.0×10^7 cells per cm^2. The spatial distribution of immobilized microorganisms was imaged using a scanning electron microscope HITACHI S-3000N (Fig. 1b).

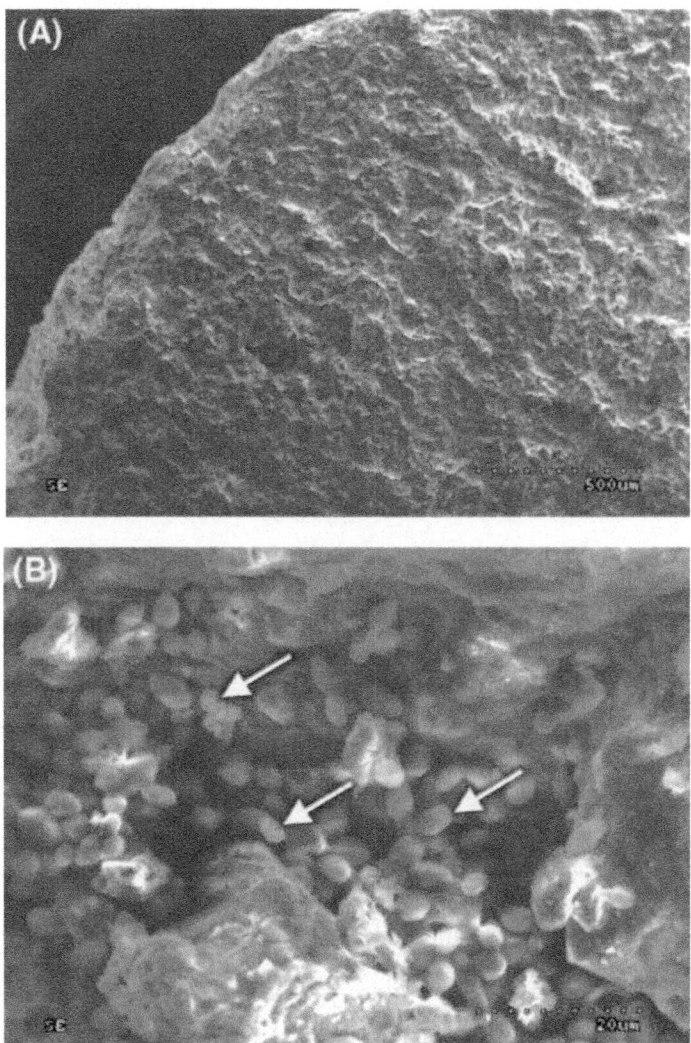

Figure. 1: Chamotte surface: **a** native, **b** with immobilized cells.

SDH Activity

Evaluating the activities of dehydrogenases is used more and more commonly as a method of determining the physiological state of microbial cells. Water-soluble, colorless tetrazolium salts are reduced to color formazans by dehydrogenases coupled with the electron transport system. SDH (EC 1.3.99.1), integrally connected to the inner mitochondrial membrane, catalyzes the dehydrogenation of succinate to fumarate. Cells that reduce tetrazolium dyes are treated as alive, and while those that do not reduce these salts cannot unambiguously be considered dead, they have lost their respiratory function (Breeuwer and Abee 2000; De Nooijer et al. 2005).

The highest relative SDH activity of immobilized cells was estimated for *S. pastorianus* W 34/70, and *S. cerevisiae* 1017. Higher enzyme activity after cell adhesion in comparison to the cells from young free aerobic cultures was observed only in the case of *S. pastorianus* W 34/70. For the remaining five microorganisms, a significant decrease in SDH activity was associated with cell adsorption on solid surfaces. These proportions were changed during the 7-day incubation period in the medium in which the process of immobilization was carried out. Aging of immobilized microorganisms suspended in wort broth resulted in higher levels of enzyme activity than in the case of free cells (Table 2).

Table 2: SDH activity of free and immobilized yeast cells

Strain	SDH activity (amount of μmol of formazan/3×10^8 cells)			
	Cells from stationary phase	Cells after adhesion	Free cells after 7-day incubation	Immobilized cells after 7-day incubation
(A) Adhesion in wort broth				
S. cerevisiae 1017	0.81 (±0.09)	0.15 (±0.02)	0.03 (±0.01)	0.08 (±0.01)
S. pastoria-nus 680	0.82 (±0.08)	0.02 (±0.00)	0.03 (±0.01)	0.08 (±0.01)
S. pastorianus B4	0.04 (±0.01)	0.02 (±0.00)	0.03 (±0.01)	0.05 (±0.01)
(B) Adhesion in Ringer's solution				
S. cerevisiae TT	0.13 (±0.02)	0.04 (±0.01)	0.04 (±0.01)	0.03 (±0.01)
S. cerevisiae 1183	0.33 (±0.05)	0.07 (±0.02)	0.06 (±0.01)	0.03 (±0.01)
S. pastorianus W 34/70	0.24 (±0.02)	0.28 (±0.03)	0.04 (±0.01)	0.03 (±0.01)

Similar phenomena were described by Tsukatani et al. (2003) in studies of yeast vitality conducted with WST-1 tetrazolium salt. The amount of formazan reached its maximum value in the stationary phase of growth and fell to a quarter on the sixth day of the culture.

For the immobilized microorganisms suspended in Ringer's solution, lower or similar activity was observed than in the case of free cells (Table 2). These differences can be explained by the stronger nutrient deficit. For the same reasons, within this group of strains, a decrease in SDH was observed after 7 days with regard to the values noted after adhesion. The activity determined after a 7-day aerobic incubation of *S. pastorianus*680 and B4 was higher than it was at the beginning of the process. This fact may be associated with the specific growth and aging patterns of brewing lager yeast strains.

ATP Content

It is well known that with the abandonment of adenosine triphosphate synthesis, the existing ATP rapidly degrades. This feature makes ATP a good marker of cell physiological activity (Imai and Ohno 1995; Sato et al. 2008; Guillou et al. 2003; Osorio et al. 2004). For the purpose of this study, a common method based on the luciferin/luciferase system was also adopted. This widely used procedure of ATP determination in solutions or cell suspensions was modified to measure the level of this nucleotide in yeast cells immobilized on solid supports through the use of a 'somatic cell ATP-releasing reagent'.

For all the tested strains, similar relationships were observed between the ATP content of the free and immobilized cells. However, a substantial decrease in ATP concentration throughout the adhesion process, and after longer incubation under aerobic conditions, was noted.

It is worth of noting that ATP values determined from the free cells in the stationary phase were even several times greater than for those from the immobilized cells (Table 3).

Table 3: ATP concentration in free and immobilized yeast cells

Strain	ATP (fg/cell)			
	Cells from stationary phase	Cells after adhesion	Free cells after 7-day incubation	Immobilized cells after 7-day incubation
(A) Adhesion in wort broth				
S. cerevisiae 1017	168.89 (±2.75)	13.44 (±1.14)	21.81 (±2.22)	0.00 (±0.00)
S. pastorianus 680	24.18 (±0.21)	7.86 (±0.68)	0.00 (±0.00)	0.00 (±0.00)
S. pastorianus B4	49.22 (±3.98)	8.94 (±0.97)	2.75 (±0.33)	0.00 (±0.00)
(B) Adhesion in Ringer's solution				
S. cerevisiae TT	58.27 (±5.21)	9.19 (±1.09)	1.57 (±0.22)	0.00 (±0.00)
S. cerevisiae 1183	51.97 (±4.43)	6.34 (±0.77)	3.98 (±0.51)	0.00 (±0.00)
S. pastorianus W 34/70	32.91 (±2.89)	3.99 (±0.45)	11.36 (±1.15)	0.00 (±0.00)

At the cellular level, ATP depletion is the earliest cell-damaging factor. In vivo, severe depletion of ATP leads to dysfunction, destabilization, and aggregation of many cellular proteins, including enzymes (Kabakov et al. 2002). Sustained lack of ATP is obviously lethal for the cell. On the contrary, a transient (reversible) drop in cellular ATP confers tolerance to the next energy-depriving exposure. Therefore, we can assume that very low ATP level in yeast cells after immobilization may be a result of cell adaptation mechanisms to different environmental conditions.

PDC Activity and Fermentation Activity

Pyruvate decarboxylase (EC 4.1.1.1) is one of the key enzymes in anaerobic yeast metabolism, so determining its activity allows one to describe the physiological state of cells and their fermentation abilities. It is a homotetrameric enzyme (EC 4.1.1.1) that catalyses the decarboxylation of pyruvic acid into acetaldehyde and carbon dioxide in the cytoplasm (Pronk et al. 1996). PDC activity was determined in cells cultured under aerobic and oxygen-limited conditions. Specific competition occurring between PDC and pyruvate dehydrogenase complexes explains the simultaneous coexistence of two metabolic pathways (glycolysis and the Krebs cycle) and aerobic alcoholic fermentation (Van Hoek et al. 1998; Flikweert et al.1999). The highest PDC activity of immobilized cells incubated under aerobic conditions was observed in the *S. pastorianus* B4 strain. Slightly lower values were recorded for *S. cerevisiae* 1017 and *S. pastorianus* W 34/70. Except for *S. pastorianus* B4, the activity of the newly immobilized microorganisms was reduced considerably. In four out of the six examined strains (excluding *S. cerevisiae* 1017 and 1183), there was a further decrease in PDC activity after a 7-day incubation period. Nevertheless, comparing these values to the activity of 7-day populations of free cells, higher or comparable activity was observed in the cases of all immobilized cell strains adhered in wort broth and of the *S. pastorianus* W 34/70 strain (Table 4).

Table 4: PDC activity of free and immobilized yeast cells—aerobic cultivation

Strain	PDC activity (amount of μmol of acetaldehyde/1 × 10⁸ cells)			
	Cells from stationary phase	Cells after adhesion	Free cells after 7-day incubation	Immobilized cells after 7-day incubation
(A) Adhesion in wort broth				
S. cerevisiae 1017	4.50 (±0.39)	0.15 (±0.02)	0.53 (±0.04)	0.59 (±0.07)
S. pastorianus 680	1.60 (±0.09)	0.29 (±0.03)	0.04 (±0.01)	0.06 (±0.01)

S. pastorianus B4	4.06 (±0.51)	1.54 (±0.01)	0.02 (±0.00)	1.18 (±0.12)
(B) Adhesion in Ringer's solution				
S. cerevisiae TT	2.63 (±0.25)	0.04 (±0.01)	2.78 (±0.03)	0.04 (±0.01)
S. cerevisiae 1183	2.58 (±0.35)	0.01 (±0.00)	0.87 (±0.09)	0.23 (±0.03)
S. pastorianus W 34/70	4.72 (±0.55)	0.38 (±0.04)	0.00 (±0.00)	0.29 (±0.02)

The conditions of 7-day aerobic shaking cultures, associated with oxidative stress and cell aging, caused a decrease in the enzyme activity and ATP content to almost trace values. The smallest reduction in metabolic activity was observed for the top fermenting yeast. The foam that can be observed forming on the surface of the fermented broth with greater oxygenation can therefore be associated with a higher resistance to oxidative stress factors.

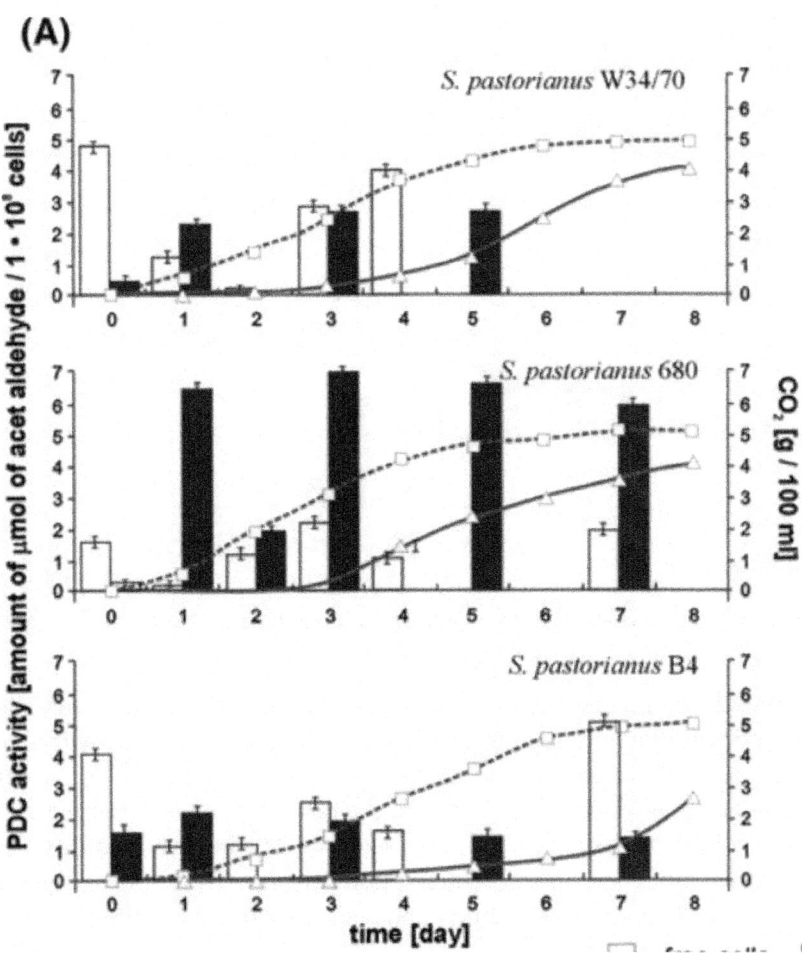

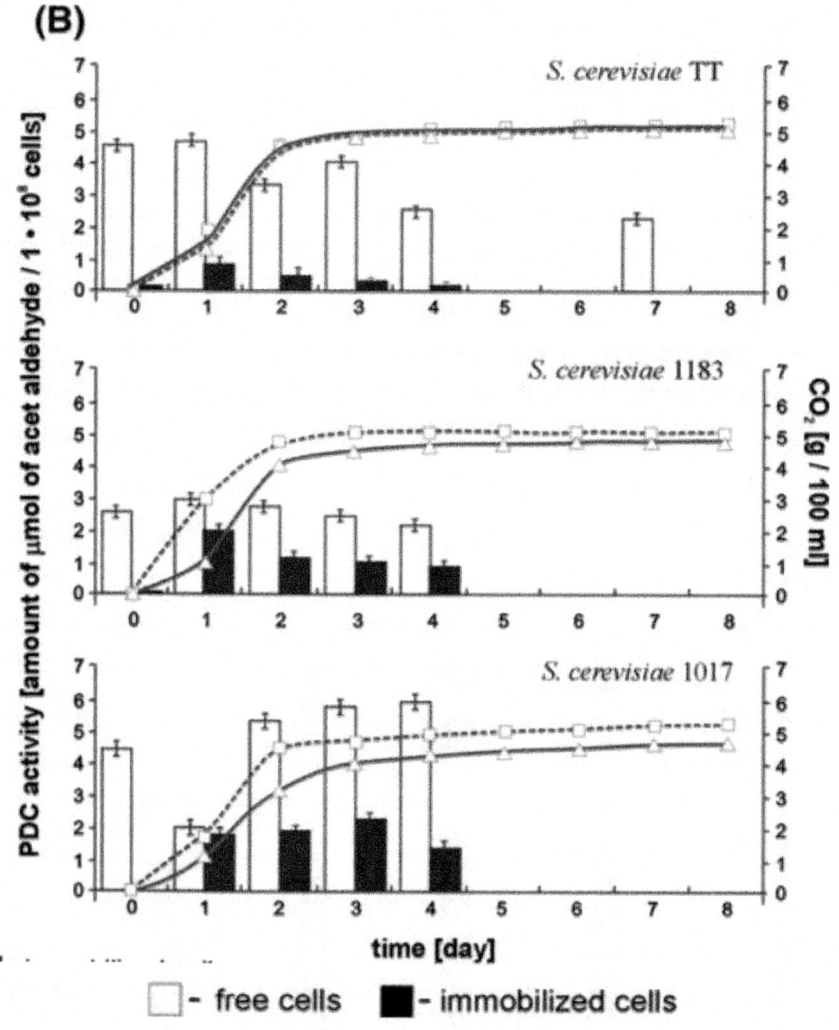

Figure. 2: PDC activity of free and immobilized yeast cells—fermentation process. **a** Lager strains, **b** ale strains.

Directly after adhesion, the carriers with immobilized cells were transferred to the fermentation medium. In parallel experiments conducted with free cells, changes in the composition of the medium and oxygen limitation caused a reduction in PDC activity (Fig. 2). On the other hand, we observed increased enzyme activity in the case of immobilized cells. The immobilization techniques also had no influence on the fermentation process. These phenomena can be explained by the increased stability and resistance to environmental changes of the immobilized yeast strains.

As in the previous tests, the PDC activity of immobilized microorganisms was lower in the case of most strains compared to the activity of planktonic populations. The exception was the immobilized cells of the *S. pastorianus* 680 strain, whose PDC activity was several times higher than in a free state. There is no known reason for this, but it could be due to an individual characteristic of the strain.

During the fermentation processes, the production of carbon dioxide was also measured. Changes in its values were similar for all *S. cerevisiae* strains (Fig. 2). Different fermentation dynamics were observed in the case of brewing lager yeast strains (Fig. 2a). These results suggest that high concentrations of immobilized cells per volume unit and their resistance to stress factors, including low temperature, could be responsible for a significant reduction in the adaptive phase and faster attenuation. Fermentation trials with immobilized cells showed that the immobilized yeasts adapted to the specific conditions. Despite the relatively low PDC activity of immobilized cells, the final fermentation effect, measured as the amount of CO_2 produced, was achieved in a shorter time. Numerous examples of reduced fermentation times achieved both in continuous technologies using yeast cells adsorbed on inorganic carriers and in batch processes carried out with cells immobilized on natural supports have also been cited by Kourkoutas et al. (2004).

FUN1 Staining

The metabolic activity (defined as the ability to fluorochrome bioconversion) of yeast strains was made visible using FUN1 stain. An enzyme activity assay and ATP determination of immobilized cells were conducted using more than one carrier. So the average value characterized the whole population. A large number of yeast cells with low enzyme activity were capable of bioconverting FUN1. Therefore, fluorescence staining revealed the diversity of metabolic activity in immobilized yeast cells (Fig. 3). This fact suggests an absence of physiological homogeneity in these populations.

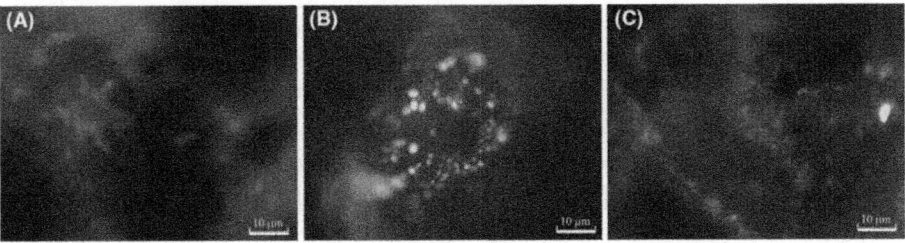

Figure. 3: FUN 1 staining. **a** Stationary phase, **b** after adhesion, **c** 7-day fermentation

Physiology of Immobilized Yeast Cells

Immobilization is ranked as one of the factors protecting yeast cells from adverse environmental effects (Junter et al. 2002; Norton et al. 1995; Kanda et al. 1998; Jirku 1999; Qun et al. 2002). In fermentations carried out with immobilized yeast cells, chemical changes in cellular composition were associated with an increased resistance to stress—especially to high concentrations of ethanol, and to the high gravity of the fermented wort (Verbelen et al. 2006; Brányik et al. 2008; Kourkoutas et al. 2004). Yeast cells immobilized in different polymer matrices were characterized by increased stability during freezing and freeze-drying. Osmotic stress also caused the production of polyols, and consequently an increase in resistance to toxic substances (Kourkoutas et al. 2004).

In our research, osmotic stress (after a change of wort broth or Ringer's solution for the fermentation medium) and limited oxygen conditions did not cause a deterioration of metabolic activity in the immobilized yeast cells tested.

The differences in the activity of key enzymes assayed on free and immobilized cells in the seventh day of aerobic culture provide evidence of different aging processes. Comparing the age that brewer's yeast reaches in traditional technologies with several months of continuous processes, the changes in the physiology of immobilized cells seem to be significant. A decrease in the viability and physiological activity of immobilized industrial yeast cells was also reported during the technological process (Calinescu et al. 2012). Both the aging of immobilized microorganisms and its impact on the quality of the resulting product are still unsolved issues (Brányik et al. 2005).

Metabolic changes caused by the immobilization process have been reported by many authors. An example is the increase in the amount of diacetyl in young beer obtained using immobilized yeast cells. This may be due to the inhibition of biosynthesis of certain amino acids, and to increased expression of acetohydroxyacid synthase, which is responsible for the conversion of pyruvic acid into its α-acetolactate diketones precursor (Brányik et al. 2008). Such changes are often associated with barriers to the mass transport of substrate and product, occurring in cases of encapsulation or entrapment. However, immobilization methods based on adhesion to solid surfaces have been described as less invasive in relation to the metabolic activity of immobilized microorganisms (Verbelen et al. 2006; Willaert 2000; Brányik et al. 2005,2008). For *S. cerevisiae* cells immobilized on DEAE-cellulose, used in the production of non-alcoholic beer, higher activity (in relation to free cells) of two glycolytic enzymes (hexokinase and PDC), was estimated (Van Iersel et al. 2000).

Many research studies have indicated an increase in the metabolic activity of immobilized cells that is manifested both by an increased rate of substrate use and by higher productivity of selected metabolites (Norton and D'Amore 1994; Angelova et al. 2000; Talebnia and Taherzadeh 2007; Nikoli et al. 2010; Bouallagui et al. 2013). In view of the volume of literature and data describing higher glucose flux (Junter et al. 2002; Plessas et al. 2007; Brányik et al. 2005) we would expect higher PDC activity in microorganisms immobilized than in suspended cells. In our study, we found higher PDC activity values only for *S. pastorianus* 680 during the fermentation with immobilized cells in comparison with free cells treated using the same process. However, in most cases, the experiments show that the immobilized microorganisms tested are characterized by a decrease in enzyme activity and ATP content. On the reduced metabolic activity of immobilized brewing yeast cells see also Masschelein et al. (1994).

The data obtained is very variable because it describes different type of yeasts. However, the our research complements not fully understood processes. The obtained results show that the activity of the immobilized cells may depend on the type, age or behavior of tested yeasts. Therefore, our study seems to be a valuable material for further studies on the area of physiology of immobilized yeasts.

CONCLUSIONS

The reduced metabolic activity of immobilized yeast cells does not preclude benefits to their technological applications. A suitably large density of yeast cells per unit volume, possible only in immobilized cell systems, and a very short attenuation time (due to their resistance to environmental changes) produce a more efficient fermentation process. Several theories have been proposed to explain the enhanced fermentation capacity as a result of immobilization. Adhesion has a major influence on the plasma membrane properties of the yeast, which can cause modifications of some solute transport systems. The enhanced fermentation properties of immobilized cell systems could also be explained by the CO_2 nucleation effect of the matrix (Verbelen et al. 2006). Moreover, a reduction in the ethanol concentration in the immediate microenvironment of the organism due to the specific adsorption of ethanol by the support may act to minimize end product inhibition. The greater volumetric productivity can be also a result of higher cell density in immobilized systems (Ivanova et al. 2011). However, our interesting results require a clear explanation in future studies.

ACKNOWLEDGMENTS

This work was supported by European Commission, Project NMP3-CT-504937 'PERCERAMICS' and the polish Ministry of Scientific Research and Information Technology, Project No. N312 018 32/1283. We would like to give our heartfelt thanks to Urszula Mizerska and Witold Fortuniak (the Centre of Molecular and Macromolecular Studies, the Polish Academy of Science) for the preparation of the modified carriers.

REFERENCES

1. Angelova MB, Pashova SB, Slokoska LS. Comparison of antioxidant enzyme biosynthesis by free and immobilized *Aspergillus niger* cells. Enzyme Microb Technol. 2000;26:365–375. doi: 10.1016/S0141-0229(00)00138-1.

2. Barker MG, Smart KA. Morphological changes associated with the cellular ageing of a brewing yeast strain. J Am Soc Brew Chem. 1996;54:121–126.

3. Behera S, Mohanty RCh, Ray RCh. Ethanol production from mahula (*Madhuca latifolia* L.) flowers with immobilized cells of *Saccharomyces cerevisiae* in *Luffa cylindrica* L. sponge discs. Appl Energy.2011;88:212–215. doi: 10.1016/j.apenergy.2010.07.035.

4. Berlowska J, Kregiel D, Klimek L, Orzeszyna B, Ambroziak W. Novel yeast cell dehydrogenase activity assay in situ. Pol J Microbiol. 2006;55:127–131. doi: 10.1099/jmm.0.46218-0.

5. Berlowska J, Kregiel D, Ambroziak W. Pyruvate decarboxylase activity assay in situ of different industrial yeast strains. Food Technol Biotechnol. 2009;47:96–100.

6. Berlowska J, Kregiel D, Ambroziak W. Enhancing adhesion of yeast brewery strains to chamotte carriers through aminosilane surface modification. World J Microbiol Biotechnol. 2013;29:1307–1316. doi: 10.1007/s11274-013-1294-4.

7. Bindu S, Somashekar D, Joseph R. A comparative study on permeabilization treatments for in situ determination of phytase of *Rhodotorula gracilis*. Lett Appl Microbiol. 1998;27:336–340. doi: 10.1046/j.1472-765X.1998.00451.x.

8. Bouallagui H, Touhami Y, Hanafi N, Ghariani A, Hamdi M. Performances comparison between three technologies for continuous ethanol production from molasses. Biomass Bioenergy. 2013;48:25–32. doi: 10.1016/j.biombioe.2012.10.018.

9. Brányik T, Vicente AA, Dostalek P, Teixteira JA. Continuous beer fermentation using immobilized yeast cell bioreactor systems. Biotechnol Prog. 2005;21:653–663. doi: 10.1021/bp050012u.

10. Brányik T, Vicente AA, Dostalek P, Teixeira JA. A review of flavour formation in continuous beer fermentations. J Inst Brew. 2008;114:3–13. doi: 10.1002/j.2050-0416.2008.tb00299.x.

11. Brányik T, Silva DP, Baszczynski M, Lehnert R, Almeida e Silva JB. A review of methods of low alcohol and alcohol-free beer production. J Food Eng. 2012;108:493–506. doi: 10.1016/j.jfoodeng.2011.09.020.

12. Breeuwer P, Abee T. Assessment of viability of microorganisms employing fluorescence techniques.Food Microbiol. 2000;55:193–200. doi: 10.1016/S0168-1605(00)00163-X.

13. Calinescu I, Corina T, Petre C, Adrian T, Elvira A, Ionescu A, Dima R (2012) Saccharomyces cerevisiae immobilization in polyacrylamide hydrogel obtained at low temperature. Rom Biotechnol Lett 17:7628–7638

14. Chandra J, Kuhn DM, Mukherjee PK, Hoyer LL, McCormick T, Ghannoum MA. Biofilm formation by the fungal pathogen *Candida albicans*: development, architecture, and drug resistance. J Bacteriol.2001;183:5385–5394.   doi: 10.1128/JB.183.18.5385-5394.2001.

15. Chelico   L,   Khachatourians   GG.   Permeabilization of *Beauveria bassiana* blastospores for in situ enzymatic assays. Mycologia. 2003;95:976–981. doi: 10.2307/3762025.

16. Cordeiro C, Freire AP. Digitonin permeabilization of *Saccharomyces cerevisiae* cell of in situ enzyme assay. Anal Biochem. 1995;229:145–148. doi: 10.1006/abio.1995.1394.

17. Crotti LB, Dragon T, Cabib E. Yeast cell permeabilization by osmotic shock allows determination of enzymatic activities in situ. Anal Biochem. 2001;292:8–16. doi: 10.1006/abio.2001.5051.

18. De Nooijer LJ, Duijnstee IAP, Van der Zwaan GJ. Novel application of MTT reduction: a viability assay for temperate shallow-water benthic foraminifera. J Foraminifer Res. 2005;36:195–200. doi: 10.2113/gsjfr.36.3.195.

19. Flikweert MT, Kuyper M, van Maris AJA, Kotter P, van Dijken JP, Pronk JT. Steady-state and transient-state analysis of growth and metabolite production in a *Saccharomyces cerevisiae* strain with reduced pyruvate-decarboxylase activity. Biotechnol Bioeng. 1999;66:42–50. doi: 10.1002/(SICI)1097-0290(1999)66:1<42::AID-BIT4>3.0.CO;2-L.

20. Freire AP, Martins AM, Cordeiro C. An experiment illustrating metabolic regulation in situ using digitonin permeabilized yeast cells. Biochem Educ. 1998;26:161–163. doi: 10.1016/S0307-4412(97)00124-6.

21. Galazzo J, Bailey JE. Growing *Saccharomyces cerevisiae* in calcium-alginate beads induces cell alterations, which accelerate glucose conversion to ethanol. Biotechnol Bioeng. 1990;36:417–426. doi: 10.1002/bit.260360413.

22. Gonga G-H, Houa Y, Zhaoa Q, Yua M-A, Liaob F, Jiangc L, Yanga X-L. A new approach for the immobilization of permeabilized brewer's yeast cells in a modified composite polyvinyl alcohol lens-shaped capsule containing montmorillonite and dimethyldioctadecylammonium bromide for use as a biocatalyst. Process Biochem. 2010;45:1445–1449. doi: 10.1016/j.procbio.2010.05.021.

23. Gough S, Deshpande M, Scher M, Rosazza JPN. Permeabilization of *Pichia pastoris* for glycolate oxidase activity. Biotechnol Lett. 2001;23:1535–1537. doi: 10.1023/A:1011601827503.

24. Guillou S, Besnard V, El Murr N, Federighi M. Viability of *Saccharomyces cerevisiae* cells exposed to low-amperage electrolysis as assessed by staining procedure and ATP content. Int J Food Microbiol.2003;88:85–89. doi: 10.1016/S0168-1605(03)00112-0.

25. Imai T, Ohno T. The relationship between viability and intracellular pH in the yeast *Saccharomyces cerevisiae*. Appl Environ Microbiol. 1995;61:3604–3608.

26. Ivanova V, Petrova P, Hristov J. Application in the ethanol fermentation of immobilized yeast cells in matrix of alginate/magnetic nanoparticles, on chitosan-magnetite microparticles and cellulose-coated magnetic nanoparticles. Int Rev Chem Eng. 2011;3:289–299.

27. Jang A, Bishop PL, Okabe S, Lee SG, Kim IS. Effect of dissolved oxygen concentration on the biofilm and in situ analysis by fluorescence in situ hybridization (FISH) and microelectrodes. Water Sci Technonol. 2003;47:49–57.

28. Jirku V. Whole-cell immobilization as a means of enhancing ethanol tolerance. J Ind Microbiol Biotechnol. 1999;22:147–151. doi: 10.1038/sj.jim.2900620.

29. Junter GA, Coquet L, Vilain S, Jouenne T. Immobilized-cell physiology: current data and the potentialities of proteomics. Enzyme Microb Technol. 2002;3:201–212. doi: 10.1016/S0141-0229(02)00073-X.

30. Kabakov AE, Budagova KR, Latchman DS, Kampinga HH. Stressful preconditioning and HSP70 overexpression attenuate proteotoxicity of cellular ATP depletion. Am J Physiol Cell Physiol.2002;283:C521–C534. doi: 10.1152/ajpcell.00503.2001.

31. Kanda T, Miyata N, Fukui T, Kawamoto T, Tanaka A. Doubly entrapped baker's yeast survives during the long-term stereoselective reduction of ethyl 3-oxobutanoate in an organic solvent. Appl Microbiol Biotechnol. 1998;49:377–381. doi: 10.1007/s002530051185.

32. Kindaichi T, Ito T, Okabe S. Ecophysiological interaction between nitrifying bacteria and heterotrophic bacteria in autotrophic nitrifying biofilms as determined by microautoradiography-fluorescence in situ hybridization. Appl Environ Microbiol. 2004;70:1641–1650. doi: 10.1128/AEM.70.3.1641-1650.2004.

33. Kippert F. A rapid permeabilization procedure for accurate quantitative determination of β-galactosidase activity in yeast cells. FEMS Microbiol Lett. 1995;128:201–206.

34. Kondo A, Liu Y, Futura M, Fujita Y, Matsumoto T, Fukuta H. Preparation of high activity whole cell biocatalyst by permeabilization of recombinant flocculent yeast with alcohol. Enzyme Microb Technol.2008;27:806–811. doi: 10.1016/S0141-0229(00)00304-5.

35. Kourkoutas Y, Bekatorou A, Banat IM, Marchant R, Koutinas AA. Immobilization technologies and support materials suitable in alcohol beverages production: a review. Food Microbiol. 2004;21:377–397. doi: 10.1016/j.fm.2003.10.005.

36. Kregiel D, Berlowska J, Ambroziak W. Adhesion of yeast cells to different porous supports, stability of cell-carrier systems and formation of volatile by-products. World J Microbiol Biotechnol.2012;28(12):3399–3408. doi: 10.1007/s11274-012-1151-x.

37. Kregiel D, Berlowska J, Ambroziak W. Growth and metabolic activity of conventional and non-conventional yeasts immobilized in foamed alginate beads. Enzyme Microb Technol. 2013

38. Kręgiel D, Berłowska J, Ambroziak W. Succinate dehydrogenase activity assay in situ with blue tetrazolium salt in crabtree-positive *Saccharomyces cerevisiae* strain. Food Technol Biotech.2008;46:376–381.

39. Lee N, Nielsen PH, Andreasen KH, Juretschko S, Nielsen JL, Schleifer KH, Wagner M. Combination of fluorescent in situ hybridization and micro autoradiography-a new tool for structure-function analyses in microbial ecology. Appl Environ Microbiol. 1999;65:1289–1297.

40. Li G-Y, Huang K-L, Jiang Y-R, Ding P. Production of (R)-mandelic acid by immobilized cells of*Saccharomyces cerevisiae* on chitosan carrier. Process Biochem. 2007;42:1465–1469. doi: 10.1016/j.procbio.2007.06.015.

41. Lodolo EJ, Kock JLF, Axcell BC, Brooks M. The yeast *Saccharomyces cerevisiae*-the main character in beer brewing. FEMS Yeast Res. 2008;8:1018–1036. doi: 10.1111/j.1567-1364.2008.00433.x.

42. Majors PD, Mclean JS, Pinchuk GE, Fredrickson JK, Gorby YA, Minard KR, Wind RA. NMR methods for in situ biofilm metabolism studies. J Microbiol Methods. 2005;62:337–344. doi: 10.1016/j.mimet.2005.04.017.

43. Masschelein ChA, Ryder DS, Simon J-P. Immobilized cell technology in beer production. Crit Rev Biotechnol. 1994;14:155–177. doi: 10.3109/07388559409086966.

44. Miranda HV, Ferreira AEN. Measuring intracellular enzyme concentrations. Biochem Mol Biol Edu.2008;36:135–138. doi: 10.1002/bmb.20166.

45. Nikoli S, Mojovic L, Pejin D, Rakin M, Vukasinovic M. Production of bioethanol from corn meal hydrolyzates by free and immobilized cells of *Saccharomyces cerevisiae* var. *ellipsoideus*. Biomass Bioenergy. 2010;34:1449–1456. doi: 10.1016/j.biombioe.2010.04.008.

46. Norton S, D'Amore T. Physiological effects of yeast cell immobilization: applications for brewing.Enzyme Microb Technol. 1994;16:365–375. doi: 10.1016/0141-0229(94)90150-3.

47. Norton S, Watson K, D'Amore T. Ethanol tolerance of immobilized brewers' yeast cells. Appl Microbiol Biotechnol. 1995;43:18–24. doi: 10.1007/BF00170616.

48. Nott KP, Heese FP, Paterson-Beedle M, Macaskie LE, Hall LD. Visualization of the function of a biofilm reactor by magnetic resonance imaging.Can J Chem Eng. 2005;83:68–72. doi: 10.1002/cjce.5450830112.

49. Osorio H, Moradas-Ferreira P, Günther Sillero MA, Sillero A. In *Saccharomyces cerevisiae*, the effect of H_2O_2 on ATP, but not on glyceraldehyde-3-phosphate dehydrogenase, depends on the glucose concentration. Arch Microbiol. 2004;181:231–236. doi: 10.1007/s00203-004-0648-6.

50. Pajić-Lijakovic I, Bugarski D, Plavsic M, Bugarski B. Influence of microenvironmental conditions on hybridoma cell growth inside the alginate-poly-l-lysine microcapsule. Process Biochem. 2007;42:167–174. doi: 10.1016/j.procbio.2006.07.023.

51. Parascandola P, de Alteriis E, Sentandreu R, Zueco J. Immobilization and ethanol stress induce the same molecular response at the level of cell wall in growing yeast. FEMS Microbiol Lett.1997;150:121–126. doi: 10.1016/S0378-1097(97)00107-9.

52. Plessas S, Bekatorou A, Koutinas AA, Soupioni M, Banat IM, Marchant R. Use of *Saccharomyces cerevisiae* cells immobilized on orange peel as biocatalyst for alcoholic fermentation. Bioresour Technol. 2007;98:860–865. doi: 10.1016/j.biortech.2006.03.014.

53. Pronk JT, Steensmays HY, Van Dijken JP. Pyruvate metabolism in *Saccharomyces cerevisiae*. Yeast.1996;12:1607–1633. doi: 10.1002/(SICI)1097-0061(199612)12:16<1607::AID-YEA70>3.0.CO;2-4.

54. Qun J, Shanjing Y, Lehe M. Tolerance of immobilized baker's yeast in organic solvents. Enzyme Microb Technol. 2002;30:721–725. doi: 10.1016/S0141-0229(02)00048-0.

55. Sato K, Yoshida Y, Hirahara T, Ohba T. On-line measurement of intracellular ATP of *Saccharomyces cerevisiea* and pyruvate during sake mashing. J Biosci Bioeng. 2008;90:294–301.

56. Smart KA. The management of brewing yeast stress. Proc Congr Eur Brew Conv. 2001;21:306–315.

57. Talebnia F, Taherzadeh MJ. Physiological and morphological study of encapsulated *Saccharomyces cerevisiae*. Enzyme Microb Technol. 2007;41:683–688. doi: 10.1016/j.enzmictec.2007.05.020.

58. Tsukatani T, Oba T, Ukeda H, Matsumoto K. Spectrophotometric assay of yeast vitality using 2,3,5,6-tetramethyl-1,4-benzoquinone and tetrazolium salts. Anal Sci. 2003;19:659–664. doi: 10.2116/analsci.19.659.

59. Uppuluri P, Sarmah B, Chaffin WL. *Candida albicans* SNO1 and SNZ1 expressed in stationary-phase planktonic yeast cells and base of biofilm. Microbiology. 2006;152:2031–2038. doi: 10.1099/mic.0.28745-0.

60. Van Hoek PV, Flikweert T, Steensma H, Van Dijken JP, Pronk JT. Effects of pyruvate decarboxylase overproduction on flux distribution at the pyruvate branch point in *Saccharomyces cerevisiae*. Appl Environ Microbiol. 1998;64:2133–2140.

61. Van Iersel MFM, Brouwer-Post E, Rombouts FM, Abee T. Influence of yeast immobilization on fermentation and aldehyde reduction during the production of alcohol-free beer. Enzyme Microb Technol. 2000;26:602–607. doi: 10.1016/S0141-0229(00)00140-X.

62. Verbelen PJ, De Schutter DP, Delvaux F, Verstrepen KJ, Delvaux FR. Immobilized yeast cell systems for continuous fermentation applications. Biotechnol Lett. 2006;28:1515–1525. doi: 10.1007/s10529-006-9132-5.

63. Willaert R. Beer production using immobilized cell technology. Minerva Biotechnol. 2000;12:319–330.

64. Xi Ch, Marks D, Schlachter S, Luo W, Boppart SA. High-resolution three-dimensional imaging of biofilm development using optical coherence tomography. J Biomed Opt. 2006;11:034001–034005. doi: 10.1117/1.2209962.

Chapter 5

EXPRESSION AND CYTOSOLIC ASSEMBLY OF THE S-LAYER FUSION PROTEIN MSBSC-EGFP IN EUKARYOTIC CELLS

Andreas Blecha[1], Kristof Zarschler[1], Klaas A Sjollema[2], Marten Veenhuis[2] and Gerhard Rödel[1]

[1]Institut für Genetik, Technische Universität Dresden, D-01062 Dresden, Germany

[2]Eukaryotic Microbiology, Groningen Biomolecular Sciences and Biotechnology Institute (GBB), University of Groningen, PO Box 14, NL-9750 AA Haren, The Netherlands.

ABSTRACT

Background

Native as well as recombinant bacterial cell surface layer (S-layer) protein of *Geobacillus (G.) stearothermophilus* ATCC 12980 assembles to supramolecular structures with an oblique symmetry. Upon expression in *E. coli*, S-layer self assembly products are formed in the cytosol. We tested the expression and assembly of a fusion protein, consisting of the mature part (aa 31–1099) of the S-layer protein and EGFP (enhanced green fluorescent protein), in eukaryotic host cells, the yeast*Saccharomyces cerevisiae* and human HeLa cells.

Results

Upon expression in *E. coli* the recombinant mSbsC-EGFP fusion protein was recovered from the insoluble fraction. After denaturation by Guanidine (Gua)-HCl treatment and subsequent dialysis the fusion protein assembled in solution and yielded green fluorescent cylindric structures with regular symmetry comparable to that of the authentic SbsC. For expression in the eukaryotic host *Saccharomyces (S.) cerevisiae* mSbsC-EGFP was cloned in a multi-copy expression vector bearing the strong constitutive *GPD* 1 (glyceraldehyde-3-phosophate-dehydrogenase) promoter. The respective yeast transfomants were only slightly impaired in growth and exhibited a needle-like green fluorescent pattern. Transmission electron microscopy (TEM) studies revealed the presence of closely packed cylindrical structures in the cytosol with regular symmetry comparable to those obtained after *in vitro* recrystallization. Similar structures are observed in HeLa cells expressing mSbsC-EGFP from the Cytomegalovirus (CMV IE) promoter.

Conclusion

The mSbsC-EGFP fusion protein is stably expressed both in the yeast, *Saccharomyces cerevisiae*, and in HeLa cells. Recombinant mSbsC-EGFP combines properties of both fusion partners: it assembles both *in vitro* and *in vivo* to cylindrical structures that show an intensive green fluorescence. Fusion of proteins to S-layer proteins may be a useful tool for high level expression in yeast and HeLa cells of otherwise instable proteins in their native conformation. In addition the self assembly properties of the fusion proteins allow their simple purification. Moreover the binding properties of the S-layer part can be used to immobilize the fusion proteins to various surfaces. Arrays of highly ordered and densely structured proteins either immobilized on surfaces or within living cells may be advantageous over the respective soluble variants with respect to stability and their potential interference with cellular metabolism.

BACKGROUND

Bacterial cell surface layers (S-layer) as the outermost cell envelope components are a common feature of many bacteria and archaea species (for review see [1, 2]). With few exceptions S-layers consist of a single species of subunits that occasionally is posttranslationally modified by phosphorylation [3], or glycosylation [4, 5]. S-layer monomers assemble to two-dimensional highly porous arrays with either oblique, square or hexagonal symmetry. The interactions between the S-layer subunits as well as between the S-layer and the supporting envelope can be disrupted in a reversible manner by cation substitution or high concentrations of chaotropic agents [6]. Upon removal of the denaturing agent the isolated S-layer subunits assemble *in vitro* into regular arrays exhibiting structural features of the authentic cell surface layer. Contrary to the situation *in vivo*, the *in vitro* self-assembly process of S-layer proteins in solution can also result in double layer sheets or in tube-like structures [7].

S-layers have been recognized as important structures for biotechnological applications [8]. However, large-scale preparation of S-layers from the authentic organisms is sometimes limited. For example, some bacterial strains have been reported to loose their ability to produce S-layers under laboratory conditions [9]. Due to alterations in the cultivation conditions expression of truncated forms of the S-layer protein may result in the loss of S-layer sheets [10]. To circumvent such difficulties recombinant S-layer proteins have been heterologously produced in prokaryotic systems, like *E. coli*, *Bacillus subtilis*, *Lactobacillus casei* [9], or *Lactococcus lactis* [11]. For example, high level expression of the S-layer protein SbsC from *Geobacillus stearothermophilus* ATCC 12980 has been reported in *E. coli* [12]. SbsC possesses an N-terminal secretion signal of 30 amino acids (aa) that is cleaved off during secretion, a secondary cell wall polymer (SCWP)-binding domain (aa 31–258), and a central portion responsible for formation of lattice symmetry and self-assembly [12]. Upon expression in *E. coli*, the mature S-layer protein mSbsC$_{(31-1099)}$ with an apparent molecular mass of 112 kDa forms monolayer cylinders and spirally wound sheets-like structures in the cytosol that can be recovered from the insoluble fraction upon cell lysis. Neither deletion of the C-terminal 179 aa (rSbsC$_{(31-920)}$) [7] nor fusion of this truncated form with major birch pollen antigen Bet v1 (rSbsC$_{(31-920)}$-Betv1) interferes with the assembly process and the oblique lattice symmetry. Interestingly, the Bet v1 epitope was accessible to antibodies demonstrating that the fused portion is protruding from the respective assembly structure [13]. Contrary to the situation of SbsC, recombinant S-layer protein SbpA from *B. sphaericus* CCM 2177 and its derivatives fail to assemble in the *E. coli* cytosol, but form insoluble inclusion bodies [14].

In the present study we addressed the question whether the cytosol of eukaryotic host cells can provide a suitable environment for the formation of S-layer self assembly products, despite the presence of numerous chaperons and proteases. Assembly of S-layer fusion proteins could provide a simple purification scheme for soluble proteins, especially if the formation of their native structures depends on the conditions of the eukaryotic cytosol. In addition, such fusion proteins could stabilize proteolytically sensitive proteins and thus increase their yield.

We used the yeast *Saccharomyces cerevisiae* and the HeLa cells as eukaryotic model organisms for expression of a bifunctional S-layer fusion protein, consisting of mature SbsC (aa 31–1099) and the enhanced green fluorescent protein (EGFP).

RESULTS

Expression of mSbsC-EGFP in *E. coli*

The 3207 bp open reading frame (ORF) encoding mSbsC$_{(31-1099)}$ was fused with the 720 bp ORF of EGFP and cloned into pET17b as described in Methods.

Both ORFs are separated by a two-aa-linker (Leu-Glu) that was introduced by the non-template-encoded *Xho* I site (CTCGAG). *E. coli* BL21(DE3), transformed with pET17b-mSbsC-EGFP, were cultivated at 30°C and 37°C and analysed by fluorescence microscopy. Expression at 30°C resulted in an enhanced fluorescence emission, indicating a higher yield of correctly folded EGFP [14]. 0, 1, 2, and 4 hours after induction by IPTG cell lysates were tested for the presence of the fusion protein by Western blot analysis with anti-GFP antibody. Already after 1 h the mSbsC-EGFP-fusion protein, with an expected molecular weight of 139 kDa, could be detected (Fig. 1). The protein was recovered from the pellet fraction, indicating the formation of inclusion bodies or of assembly products as recently reported in the case of mature SbsC [7]. In addition to the predominant 139 kDa band a number of low molecular weight protein bands are detected by the anti-GFP antibody, that likely reflect degradation products, although internal start of translation cannot be excluded.

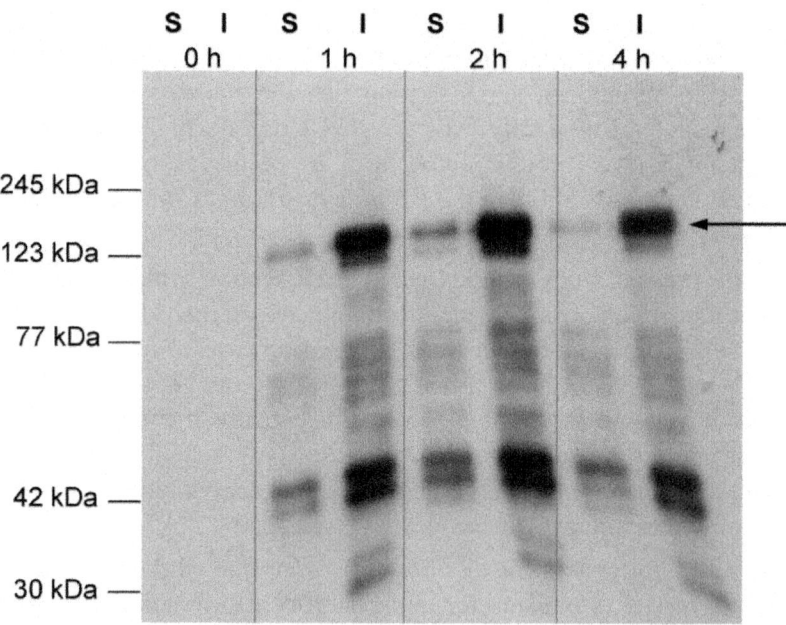

Figure 1: Expression of mSbsC-EGFP in *E. coli*. Expression of mSbsC-EGFP in *E. coli* transformants was induced by addition of IPTG as described in Material and methods. Immediately before induction (0 h), 1 h, 2 h and 4 h after induction samples of the cells were lysed and the soluble (S) and insoluble (I) fractions were prepared. 10 µg of each fraction were subjected to Western blot analysis with anti-GFP antibody. Sizes of marker proteins (Roti®-Mark prestained, Carl Roth GmbH; M) are indicated on the left hand side. Arrow indicates the full-size protein of 140 kDa.

Pure preparations of mSbsC-EGFP were obtained by FPLC as described in Methods. REM analysis of dialysed FPLC fractions of the purified full-size mSbsC-EGFP revealed predominantly tube-like structures. According to TEM analysis these tubes exhibit a regular symmetry (Fig. 2). The cylindrical structures had an average length of 10–20 μm and a diameter of approximately 60 nm. This partly resembles the situation of mSbsC which has been described to assemble into sheets and cylindric structures (albeit with a diameter of 70–110 nm) [7].

Figure 2: TEM analysis of recrystallized recombinant mSbsC-EGFP. Recombinant mSbsC-EGFP was isolated from *E. coli* transformants, denatured with 5 M Gua-HCl, dialysed against dH$_2$O, and the assembled structures were subjected to TEM-analysis. The cylindrical structures with a diameter of approximately 60 nm exhibit regular symmetry. bar = 1 μm.

When recrystallization was performed in the presence of *G. stearothermophilus* wild type strain ATCC 12980 devoid of its native S-layer due to Gua-HCl treatment, cells were covered with a green fluorescent layer indicating binding to the SCWP (data not shown). The same result was obtained with other *G. stearothermophilus* strains (DSM 297, DSM 13240 and DSM 1550), but not with the Gua-HCl treated yeast strain BY4741.

Expression of mSbsC-EGFP in *S. cerevisiae*

For expression in a eukaryotic system p426-GPD-mSbsC-EGFP was transformed into the *S. cerevisiae* strain BY4741. As a control BY4741 was transformed with p426-GPD-EGFP. Expression of the S-layer fusion protein has only a very moderate effect on growth of the transformants (Fig. 3). Despite of a slightly prolonged lag phase and a marginally elongated doubling time, cell densities in the stationary phase are identical. Western blot analysis of the 20,000 × g pellet from whole cell lysates with anti-GFP antibody showed a strong signal by a distinct protein band whose size is identical with that of mSbsC-EGFP expressed in *E. coli* (Fig. 4C). Contrary to the situation in *E. coli* no further signals were detected, indicating that expression mSbsC-EGFP in yeast is not accompanied by any obvious degradation. Fluorescence microscopy revealed that EGFP-expressing cells show a homogeneous distribution of green fluorescence (Fig. 4A), while mSbsC-EGFP expression results in the formation of green fluorescent needle-like structures in the cytosol (Fig. 4C). Contrary to the situation of EGFP-expressing cells which all exhibit a similar fluorescence intensity, yeast cells expressing the fusion protein vary widely in their green fluorescence from low to extremely strong signals.

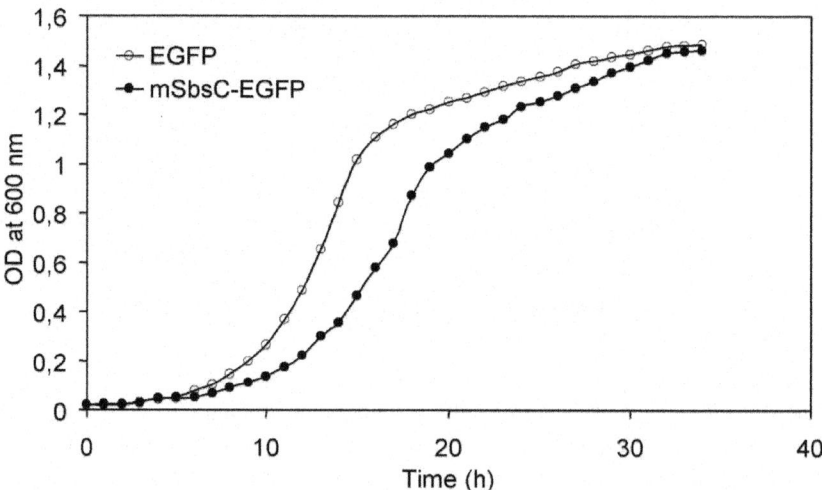

Figure 3: Growth curves of EGFP and mSbsC-EGFP expressing yeast transformants. Batch cultures of *S. cerevisiae* strain BY4741 expressing EGFP (open circles) or mSbsC-EGFP (filled circles) were inoculated in selective minimal medium with an OD_{600} of 0,02 and grown for 30 h at 30°C. Growth was followed by the increase of OD_{600}.

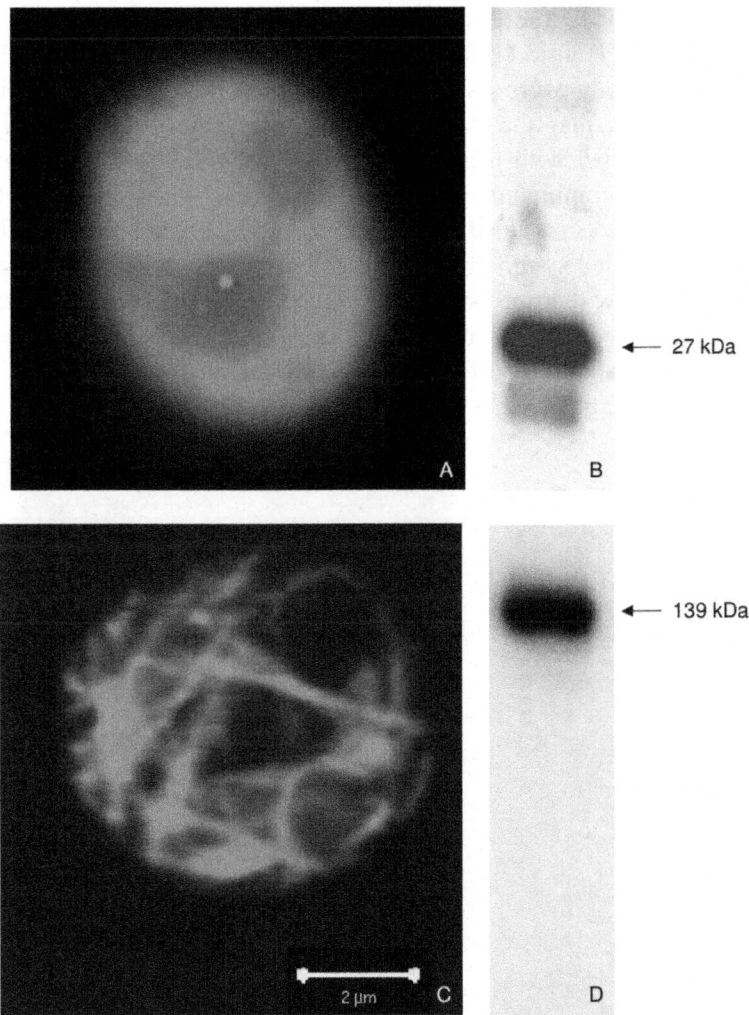

Figure 4: Fluorescence microscopy of EGFP and mSbsC-EGFP expressing yeast transformants. Yeast cells expressing EGFP (A) or mSbsC-EGFP (C) were cultivated in selective minimal medium to exponential phase and analysed by fluorescence microscopy. Westernblot analysis of whole cell extract using anti-GFP antibodies reveal the presence of either EGFP (B) or mSbsC-EGFP (D). bar = 2 μm.

TEM Analysis of Recombinant mSbsC-EGFP Structures in Yeast

TEM analysis of ultrathin sections of whole cells reveals tube-like self assembly products, that are exclusively located in the cytosol and neither associated with organelles, the cytoskeleton, or the plasma membrane (Fig. 5A). To verify the identity of these structures as the product of mSbsC-EGFP self assembly, anti-GFP antibody in combination with a secondary colloidal gold-conjugated antibody were used for detection in thin-sectioned protoplasts. Figure 5B shows that the densely packed protein crystals are selectively immunogold-labeled with primary anti-GFP antibody, thus confirming their identity as the recombinant S-layer fusion protein. TEM analysis of gently disrupted protoplasts revealed cylindrical, closely packed structures (Fig. 5C) exhibiting a regular symmetry (insert, for a more detailed image see 5D). These results show that mSbsC-EGFP monomers possess the intrinsic ability to form highly ordered assembly structures comparable to those observed after *in vitro* crystallization of mSbsC-EGFP (see Fig. 2).

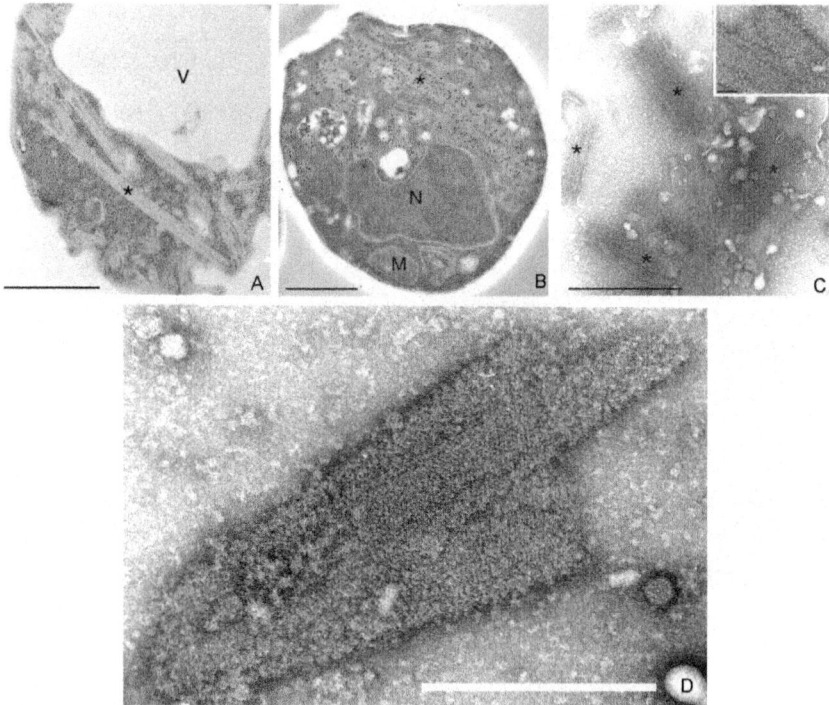

Figure 5: TEM-analysis of ultra-thin section of a yeast transformant expressing mSbsC-EGFP. (A) mSbsC-EGFP expressing yeast spheroplast showing rod like structures (*). (B) Immunocytochemistry of mSbsC-EGFP expressing yeast cell using antibodies against GFP. (C) Negative staining of osmotically shocked yeast protoplasts expressing mSbsC-EGFP showing *in vivo* formed assembly products with similar structures (*) as in Fig 2, the inset and image (D) show a higher magnification of these structures, a lattice is discernible. Key: M-mitochondrion, N-nucleus, V-vacuole. The bar represents 1 µm, for the inset 100 nm.

Expression of mSbsC-EGFP in HeLa-Cells

The ORFs of mSbsC and EGFP were fused in-frame within the vector pEGFP-N1. HeLa cells were liposome-transfected with the resulting plasmid pmSbsC-EGFP-N1. 16 h after transfection adherent HeLa cells exhibit green fluorescent filamentous structures. The fluorescence was exclusively restricted to these structures, other intracellular areas were competely devoid of green fluorescence. Obviously the fluorescent structures result from self assembly of mSbsC-EGFP within the cytosol. The morphology of these structures is very similar to that of the assembly products formed in mSbsC-EGFP producing yeast cells.

Transfected HeLa-cells were mechanically lysed with a dounce homogenizer, separated in a soluble and a pellet fraction by high speed centrifugation (20,000 × g), and the fractions analysed in a Western blot using anti-GFP antibodies. As shown in Figure 6, mSbsC-EGFP can be detected as an abundant protein band in the range of 139 kDa in the pellet fraction. Similar to the situation in yeast transformants we did not observe degradation products, indicating low or absent proteolytic activities.

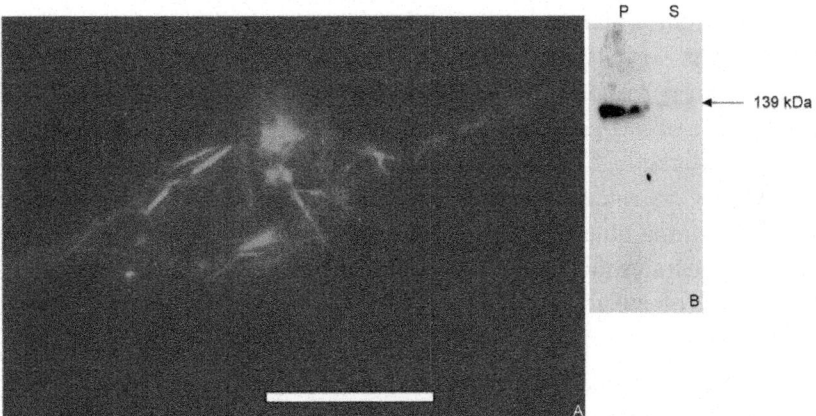

Figure 6: Fluoresence microscopy and Western blot analysis of HeLa cells expressing mSbsC-EGFP. (A) mSbsC-EGFP expressing HeLa cells 16 h after liposome transfection. Adherent cells show a filament-like green fluorescent pattern that resembles the structures obtained in yeast cells expressing mSbsC-EGFP. (B) Western blot analysis of lysate from mSbsC-EGFP expressing HeLa cells. Lysed cells were centrifuged at 20,000 × g and the resulting supernatant (S) and pellet (P) fractions were analysed by SDS-PAGE and Western blot with anti-GFP antibody. The mSbsC-EGFP fusion protein is mainly present in the insoluble fraction.

DISCUSSION

Despite of their potential for biotechnological application only few S-layers genes have been identified and were used for genetic modification and recombinant production. Usually prokaryotic systems like *Lactobacillus casei, B. subtilis, E. coli*(for review see: [9]) and *Lactococcus lactis* [11] were used for heterologous expression.

Our report describes for the first time the expression of a S-layer fusion protein in eukaryotic host cells. Because of the ability of mSbsC to form sheet-like and cylindrical structures in the cytosol of *E. coli* [7], we decided to use this S-layer protein for fusion with EGFP. Expression was first investigated in the yeast *S. cerevisiae*, one of the best-studied eukaryotic organism. We observed no alterations of the cell shape and only a slightly reduced vegetative growth, possibly reflecting the high energy demand due to the constitutive high level synthesis of 139 kDa mSbsC-EGFP. Western blot analysis of insoluble material with anti-GFP antibody gave no indication for N-terminal proteolytic degradation of the recombinant protein as observed in *E. coli*. Possibly the organization of mSbsC-EGFP in assembled structures confers protection from ubiquitin-mediated degradation in yeast. In respect to protein stability *S. cerevisiae* seems to be a superior expression system. Fluorescence microscopy revealed that yeast transformants exhibited an intense green fluorescence indicating proper folding of the EGFP part within mSbsC-EGFP. In line with the formation of self-assembly structures in the cytosol, the fluorescence pattern differs significantly from that of EGFP-expressing yeast cells. TEM studies revealed the presence of closely packed cylindrical structures with regular symmetry comparable with those obtained after *in vitro* recrystallization. Our study demonstrates that the assembly process of SbsC-EGFP is not affected by the presence of eukaryote-specific cytosolic chaperons, like members of the CCT family [15].

Recently it was reported that some S-layer monomers assemble on hydrophobic surfaces. S-layer proteins of the strains *G. stearothermophilus* PV72/p2 and NRS 2004/3a form protein monolayers on substrates like silicon wafers with a native oxide layer [16]. The self-assembly structures of SbsC-fusion proteins in yeast are not associated with the inner surface of the plasma membrane or cellular organelles despite their hydrophobic membranes. Obviously the hydrophobic membrane patches do not act in anchoring assembled S-layer structures.

As in yeast cells, mSbsC-EGFP monomers self assemble into filamentous structures within the cytosol of HeLa cells that are neither associated with subcellular structures nor affect the cell shape. We did not observe necrotic or apoptotic events in the transfected cells that expressed mSbsC-EGFP (data not shown). The detection of a single band of the expected molecular weight in Western blot analysis with anti-GFP antibodies indicates that the recombinant protein is stable and not subjected to N-terminal degradation.

A positively charged N-terminal domain of S-layer proteins directly interacts via electrostatic forces with the negatively charged SCWP whose chemical composition is identical in *G. stearothermophilus* wild-type strains [17]. In our study we show that fusion of EGFP to the mSbsC, resulting in an extension of the S-layer protein by ~240 aa, does not interfere with the binding.

CONCLUSION

Our results show that the mSbsC-EGFP fusion protein is efficiently synthesized, and self-assembles under the physiological conditions of the cytosol of eukaryotic cells. Both in yeast and human cells cytosolic accumulation of S-layer self assembly products is not accompanied by proteolytic degradation. The green fluorescence of mSbsC-EGFP suggests that both *in vitro* and *in vivo* folding of both protein portions within the fusion protein is independent, and not affected by each other. This observation opens the possibility to fuse other proteins, *e.g.* enzymes, to the S-layer part in order to obtain functional fusion constructs in a densely packed and stable structure.

Fusion of otherwise instable proteins to S-layer proteins may be a useful tool for their high level expression in eukaryotic cells. The self assembly properties of the fusion proteins can be exploited to purify them by simple centrifugation steps from cell lysate. By engineering a suitable protease cleavage site between the S-layer and the fused protein part, incubation of the assembly products with the respective proteases could allow separation and subsequent purification of the protein of interest. Arrays of highly ordered and densely structured S-layer fusion proteins can efficiently be immobilized on a variety of surfaces, either directly or upon treating with SCWP.

Besides technical applications, an ordered aggregation of S-layer fusion proteins within living cells may offer novel strategies for cell manipulation. For example, if expression of proteins would interfere with cellular functions, e.g. as a result of intracellular transport, fusion to an S-layer protein permits their retention in the cytosol. "*In vivo* affinity chromatography" may be another potential field for applying S-layer fusion proteins. Enrichment of soluble biomolecules within living cells may be achieved by assembly structures consisting of fusion proteins between the respective receptor and a suitable S-layer protein.

METHODS

Strains

Geobacillus (G.) stearothermophilus ATCC 12980

G. stearothermophilus DSM 1550 (German Collection of Microorganisms and Cell Cultures (DSMZ) GmbH, Braunschweig, Germany)

G. stearothermophilus DSM 13240 (DSMZ)

G. stearothermophilus DSM 297 (DSMZ)

Escherichia (E.) coli DH5á (BRL)

E. coli BL21(DE3) (Invitrogen)

Saccharomyces (S.) cerevisiae BY4741 (EUROSCARF)

HeLa cells (kind gift of Frank Pfennig, Institut für Zoologie, Technische Universität Dresden, Germany)

Primers

#1: 5' TATATATACATATGGCAACGGACGTGGCGACGGTC 3'

#2: 5' TATATATACTCGAGCGATGCTGATTTTGTACCAATTTG 3'

#3: 5' TATATATACTCGAGATGGTGAGCAAGGGCGAGGAG 3'

#4: 5' TATATATAGCTCAGCTTACTTGTACAGCTCGTCCATGC 3'

#5: 5' TATATATAGGATCCATGGCAACGGACGTGGCGACGGTC 3'

#6: 5' TATATATAGGTACCTCACTATTACTTGTACAGCTCGTCCATGC 3'

#7:   5'   TATATATACTCGAGATGGCAACGGACGTGGCG   3' #8:   5' TATATATACCGCGGCGATGCTGATTTTGTACCAATTTG 3'

Cultivation of *G. stearothermophilus* Strains, Isolation and Recrystallization of S-Layer Proteins

Batch cultures of the *G. stearothermophilus* strains were grown in tryptone-enriched LB-medium at 55°C (strain DSM 13240 at 65°C) to an OD_{600} of 0,6. Surface proteins attached to the cell wall layer via noncovalent forces were removed by Guanidine-HCl (Gua-HCl) treatment of cells from a 500 ml culture. Cells were washed 3 times in distilled water, suspended in 10 ml of 5 M Gua-HCl and incubated for 30 min on ice with occasional mixing. Cells devoid of their S-layers were recovered by centrifugation at 8,000 × g and washed twice with 5 M Gua-HCl. For preparation of S-layer proteins the supernatant obtained after centrifugation at 40,000 × g was dialyzed against distilled water for 18 h.

Preparation of Genomic DNA and DNA-Cloning

Genomic DNA was isolated with the QIAGEN Genomic-tip 100 kit according to the manufacturer's instruction. Plasmid DNA was prepared with the "Wizard® SV Gel and PCR Claen-up system kit» (Promega).

Cloning in pET17b

DNA encoding mSbsC$_{(31-1099)}$ was PCR-amplified with primers #1 and #2, using total DNA of G. stearothermophilus ATCC 12980 as a template. An ATG initiation codon was introduced 5'to the open reading frame by primer #1. The PCR-product was cut with Nde I and Xho I and ligated with plasmid pET17b (Novagen) to yield plasmid pET17b-mSbsC_oT.

The Xho I and Bpu 1102I sites of pET17b-mSbsC_oT were used to insert the EGFP reading frame, which was PCR-amplified from vector EGFP-N1 (Clontech) with primers #3 and #4. Primer #3 introduced a 5›-flanking Xho I-site in the PCR product, while primer #4 generated a 3'-flanking Bpu 1102I-site and introduced a TAA termination codon. The PCR-product was cut with Xho I and Bpu 1102I and ligated with pET17b-mSbsC_oT to yield plasmid pET17b-mSbsC_oT-EGFP.

Cloning in p426-GPD

A DNA fragment encoding mSbsC was PCR-amplified as described above with primers #5 and #2, cut with Bam HI and XhoI, and ligated with the S. cerevisiae expression vector p426-GPD, which bears the strong GPD promoter [18]. The resulting plasmid p426-GPD-mSbsC_oT was used for in frame fusion the EGFP orf, which was amplified with primers #3 and #6. The respective PCR product harbours a 5' Xho I site and at the 3'-end three consecutive termination codons followed by a Kpn I restriction site. Thereby plasmid p426-GPD-mSbsC_oT-EGFP was created.

Plasmid p426-GPD-EGFP for expression of EGFP in yeast was created by ligation of the PCR-fragment encoding the EGFP orf into p426-GPD via the flanking Xho I and Kpn I sites. Plasmids p426-GPD-mSbsC_oT-EGFP and p426-GPD-EGFP were propagated in E. coli DH5α and transformed into the S. cerevisiae wild type strain BY4741.

Cloning in pEGFP-N1

The reading frame coding for mSbsC was amplified with primers #7 and #8 and subsequently integrated within the Xho I and Sac II restriction sites in the vector pEGFP-N1 (Clontech) as an in frame fusion with the EGFP-ORF.

The resulting plasmid pmSbsC-EGFP-N1 was propagated in *E. coli* DH5α and transfected into HeLa cells.

Transfection of HeLa Cells

For transient transfection with plasmid pmSbsC-EGFP-N1, $2,5 \times 10^5$ cells were plated on chamber slides (Nunc) and incubated after one day with TFX™-20 liposome reagent (Promega) and 3 μg DNA for 1 h. Subsequently DMEM with high glucose and 10 % (v/v) fetal calf serum (PAA laboratories) was supplemented. After over night cultivation transfected cells were analysed by Western blot and fluorescence microscopy.

Isolation and Purification of Recombinant S-Layer Proteins from *E. coli*

E. coli BL21(DE3) was used as a prokaryotic host for the expression of mSbsC-EGFP. *E. coli* transformants bearing pET17b-mSbsC-EGFP were cultivated in LB medium containing 100 mg/l ampicillin at 30°C to an OD_{600} of 0,4. S-layer synthesis was initiated by adding IPTG to a final concentration of 0,5 mM.

8–16 h after induction cells obtained from a 500 ml culture were harvested by centrifugation at 10,000 × g for 10 min; 4°C. The cell pellet was washed 3 times with ice-cold distilled water and resuspended in 10 ml Tris-buffer (10 mM Tris, pH 7,5; 1 mM AEBSF hydrochloride (AppliChem)) containing 1% (v/v) Triton X-100. After disruption of cells by French press (25,000 PSI pressure, at 4°C), soluble and insoluble fractions were separated by centrifugation at 20,000 × g for 30 min at 4°C. Nucleic acids were removed from the pellet fraction by treatment with DNase I (1 mg/ml) and RNase A (1 mg/ml) in Tris-buffer. After three washing steps with Tris-buffer the insoluble material was denatured in 5 ml 5 M Gua-HCl (in 10 mM Tris pH 7,5). The suspension was stirred at room temperature (RT) for 30 min and centrifuged for 30 min at 40,000 × g (10°C). Renatured S-layer proteins were obtained by dialyzing the supernatant twice against 5 l of distilled water or Tris-buffer over night at 4–8°C using dialysis tubes with an exclusion size of 14 kDa.

For further purification the suspension containing renatured S-layer proteins was centrifuged at 12,000 × g for 30 min, the pellet washed 3 times with distilled water or Tris-buffer and dissolved in 3 M Gua-HCl in Tris-buffer. After a final centrifugation step (40,000 × g; 30 min) the clear supernatant was subjected to FPLC using a Sephacryl™ S-300 column (Pharmacia Biotech). Size exclusion chromatography was performed under denaturing conditions (3 M Gua-HCl in Tris-buffer) at RT with a flow rate of 0,5 ml/min. Fractions containing mSbsC-EGFP were dialyzed against dH_2O overnight and used for further analysis.

Isolation and Purification of Recombinant S-Layer Proteins from *S. cerevisiae*

Constitutive expression of S-layer protein coding genes was obtained by growing transformants of *S. cerevisiae* strain BY4741 bearing plasmid p426-GPD-mSbsC-EGFP in YNB medium (Invitrogen) supplemented with Leu, His, Met and glucose as carbon source. 500 ml of an exponential culture ($OD_{600} = 1$) were centrifuged at 5,000 × g for 3 min and 20°C. Cells were washed 3 times with Tris-buffer and subsequently lysed by osmotic swelling of protoplasts obtained by treatment with zymolyase-20T (ICN) as described [19]. S-layer self assembly products were pelleted from the suspension by centrifugation at 20,000 × g.

Isolation and Purification of Recombinant S-Layer Proteins from HeLa Cells

$4 × 10^6$ HeLa cells were transiently transfected with pmSbsC-EGFP-N1. After cultivation for 16 h cells were harvested by trypsin treatment followed by three washing steps with 1 × PBS w/o Ca^{2+}, Mg^{2+} (PAA-Laboratories). Cell lysis was performed by 30 strokes in a dounce homogenizer in the presence of 0,1 % (v/v) Triton X-100, 1 mM AEBSF and Pi-cocktail (protease inhibitor mix; Roche). The cell lysate was centrifuged at 20,000 × g, and the resulting supernatant and pellet fractions were analysed in a Western blot.

Protein Analysis

Sample preparation and SDS-polyacrylamide gelelectrophoresis were carried out as described by Laemmli [20]. Unless otherwise indicated 10 µg of protein were separated on 7,5% (w/v) acrylamide gels. For Westernblot analysis proteins were transferred onto a PVDF membrane (Millipore) and probed with monoclonal mouse antibody directed against the GFP epitope (Boehringer Mannheim). Detection of bound antibodies was performed with horseradish peroxidase (HRP)-conjugated secondary antibodies and the ECL-Plus Kit (Amersham Pharmacia Biotech). As positive control for immunreactivity of anti-GFP antibodies whole cell extract of EGFP expressing *S. cerevisiae* transformants was used (Fig. 4B).

Electron Microscopy

Whole yeast cells and spheroplasts were fixed and prepared for electron microscopy as described by Waterham *et al.* [21]. Immunolabeling was performed on ultrathin sections of Unicryl-embedded cells using polyclonal rabbit anti-GFP antibodies and 15 nm colloidal gold-conjugated goat anti-rabbit secondary antibodies (Amersham). Negative staining of osmotically swollen protoplasts and FPLC-purified recrystallized mSbsC-EGFP was done with 3% (w/v) ammonium-molybdate, pH 7,2 and examined in a Philips CM10 Transmission Electron Microscope.

Miscellaneous Procedures

Standard DNA techniques were as described [22]. Yeast cells were transformed by the lithium acetate method [23]. GENOMED™ columns were used for isolation of DNA fragments from agarose gels. The correct sequence of all constructs was confirmed by DNA sequencing with the dideoxy-chain termination method [24] using 5'-IRD800 labelled primers (MWG-BIOTECH). A Thermo Sequenase fluorescent labelled primer cycle sequencing kit with 7-deaza-dGTP (Amersham) was employed for sequencing with the LI-COR DNA sequencer 4000 (MWG-BIOTECH). Protein concentrations were determined by the Lowry method (BioRad).

ACKNOWLEDGEMENTS

This work was supported by a grant of the Saxonian Ministry of Science and the Fine Arts. We thank S. Selenska-Pobell, FZ Rossendorf for providing strain ATCC 12980 and S. Tokalov, Institut für Zoologie, Technische Universität Dresden for laser scan microscopy.

AUTHORS' CONTRIBUTIONS

AB carried out the molecular genetic studies, performed protein purification and drafted the manuscript.

KZ performed part of the DNA cloning and sequencing and carried out the transfection studies of HeLa cells.

KAS performed the immunogold labeling of yeast protoplasts and the TEM analysis. He provided the TEM images for publication.

MV contributed to the design of the study and conducted the electron microscopy analysis.

GR conceived of the study, participated in its design and coordination, and participated in drafting of the manuscript.

All authors read and approved the final manuscript.

REFERENCES

1. Sleytr UB, Beveridge TJ: Bacterial S-layers. Trends Microbiol. 1999, 7: 253-260. 10.1016/S0966-842X(99)01513-9.

2. Sleytr UB: Basic and applied S-layer research: an overview. FEMS Microbiol Rev. 1997, 20: 5-12. 10.1016/S0168-6445(97)00039-9.

3. Thomas SR, Trust TJ: Tyrosine phosphorylation of the tetragonal paracrystalline array of Aeromonas hydrophila: molecular cloning and high-level expression of the S-layer protein gene. J Mol Biol. 1995, 245: 568-581. 10.1006/jmbi.1994.0047.

4. Messner P: Prokaryotic glycoproteins: unexplored but important. J Bacteriol. 2004, 186: 2517-2519. 10.1128/JB.186.9.2517-2519.2004.

5. Schäffer C, Messner P: Glycobiology of surface layer proteins. Biochimie. 2001, 83: 591-599. 10.1016/S0300-9084(01)01299-8.

6. Sara M, Sleytr UB: S-Layer proteins. J Bacteriol. 2000, 182: 859-868. 10.1128/JB.182.4.859-868.2000.

7. Jarosch M, Egelseer EM, Huber C, Moll D, Mattanovich D, Sleytr UB, Sára M: Analysis of the structure-function relationship of the S-layer protein SbsC of Bacillus stearothermophilus ATCC 12980 by producing truncated forms. Microbiology. 2001, 147: 1353-1363.

8. Sleytr UB, Bayley H, Sara M, Breitwieser A, Kupcu S, Mader C, Weigert S, Unger FM, Messner P, Jahn-Schmid B, Schuster B, Pum D, Douglas K, Clark NA, Moore JT, Winningham TA, Levy S, Frithsen I, Pankovc J, Beale P, Gillis HP, Choutov DA, Martin KP: Applications of S-layers. FEMS Microbiol Rev. 1997, 20: 151-175. 10.1016/S0168-6445(97)00044-2.

9. Bahl H, Scholz H, Bayan N, Chami M, Leblon G, Gulik-Krzywicki T, Shechter E, Fouet A, Mesnage S, Tosi-Couture E, Gounon P, Mock M, Conway de Macario E, Macario AJ, Fernandez-Herrero LA, Olabarria G, Berenguer J, Blaser MJ, Kuen B, Lubitz W, Sára M, Pouwels PH, Kolen CP, Boot HJ, Resch S: Molecular biology of S-layers. FEMS Microbiol Rev. 1997, 20: 47-98. 10.1016/S0168-6445(97)00050-8.

10. Egelseer EM, Schocher I, Sleytr UB, Sára M: Evidence that an N-terminal S-layer protein fragment triggers the release of a cell-associated high-molecular-weight amylase in Bacillus stearothermophilus ATCC 12980. J Bacteriol. 1996, 178: 5602-5609.

11. Novotny R, Scheberl A, Giry-Laterriere M, Messner P, Schäffer C: Gene cloning, functional expression and secretion of the S-layer protein SgsE from Geobacillus stearothermophilus NRS 2004/3a in Lactococcus lactis. FEMS Microbiol Lett. 2005, 242: 27-35. 10.1016/j.femsle.2004.10.036.

12. Jarosch M, Egelseer EM, Mattanovich D, Sleytr UB, Sára M: S-layer gene sbsC of Bacillus stearothermophilus ATCC 12980: molecular characterization and heterologous expression in Escherichia coli. Microbiology. 2000, 146 (Pt 2): 273-281.

13. Breitwieser A, Egelseer EM, Moll D, Ilk N, Hotzy C, Bohle B, Ebner C, Sleytr UB, Sára M: A recombinant bacterial cell surface (S-layer)-major birch pollen allergen-fusion protein (rSbsC/Bet v1) maintains the ability to self-assemble into regularly structured monomolecular lattices and the functionality of the allergen. Protein Eng. 2002, 15: 243-249. 10.1093/protein/15.3.243.

14. Ilk N, Küpcü S, Moncayo G, Klimt S, Ecker RC, Hofer-Warbinek R, Egelseer EM, Sleytr UB, Sára M: A functional chimaeric S-layer-enhanced green fluorescent protein to follow the uptake of S-layer-coated liposomes into eukaryotic cells. Biochem J. 2004, 379: 441-448. 10.1042/BJ20031900.

15. Archibald JM, Logsdon JMJ, Doolittle WF: Origin and evolution of eukaryotic chaperonins: phylogenetic evidence for ancient duplications in CCT genes. Mol Biol Evol. 2000, 17: 1456-1466.

16. Pum D, Neubauer A, Györvary E, Sára M, Sleytr UB: S-layer proteins as basic building blocks in a biomolecular construction kit. Nanotechnology. 2000, 11: 100-107. 10.1088/0957-4484/11/2/310.

17. Schäffer C, Kahlig H, Christian R, Schulz G, Zayni S, Messner P: The diacetamidodideoxyuronic-acid-containing glycan chain of Bacillus stearothermophilus NRS 2004/3a represents the secondary cell-wall polymer of wild-type B. stearothermophilus strains. Microbiology. 1999, 145 (Pt 7): 1575-1583.

18. Mumberg D, Müller R, Funk M: Yeast vectors for the controlled expression of heterologous proteins in different genetic backgrounds. Gene. 1995, 156: 119-122. 10.1016/0378-1119(95)00037-7.

19. Daum G, Gasser SM, Schatz G: Import of proteins into mitochondria. Energy-dependent, two-step processing of the intermembrane space enzyme cytochrome b2 by isolated yeast mitochondria. J Biol Chem. 1982, 257: 13075-13080.

20. Laemmli UK: Cleavage of structural proteins during the assembly of the head of bacteriophage T4. Nature. 1970, 227: 680-685. 10.1038/227680a0.

21. Waterham HR, Titorenko VI, Haima P, Cregg JM, Harder W, Veenhuis M: The Hansenula polymorpha PER1 gene is essential for peroxisome biogenesis and encodes a peroxisomal matrix protein with both carboxy- and amino-terminal targeting signals. J Cell Biol. 1994, 127: 737-749. 10.1083/jcb.127.3.737.

22. Sambrook J, Fritsch EF, T M: Molecular Cloning. A Laboratory Manual. 1989, Cold Spring Harbor Laboratory Press: New York.

23. Schiestl RH, Gietz RD: High efficiency transformation of intact yeast cells using single stranded nucleic acids as a carrier. Curr Genet. 1989, 16: 339-346. 10.1007/BF00340712.

24. Sanger F, Nicklen S, Coulson AR: DNA sequencing with chain-terminating inhibitors. Proc Natl Acad Sci U S A. 1977, 74: 5463-5467.

Chapter 6

NEW METHOD FOR SELECTION OF HYDROGEN PEROXIDE ADAPTED BIFIDOBACTERIA CELLS USING CONTINUOUS CULTURE AND IMMOBILIZED CELL TECHNOLOGY

Valeria Mozzetti [1] , Franck Grattepanche[1] , Déborah Moine [2] , Bernard Berger [2] , Enea Rezzonico [2] , Leo Meile [1] , Fabrizio Arigoni [2] , Christophe Lacroix[1]

[1] Laboratory of Food Biotechnology, Institute of Food Science and Nutrition, Schmelzbergstrasse 7, ETH-Zurich, 8092 Zürich, Switzerland

[2] Nestlé Research Center, Vers-chez-les-Blanc, 1000 Lausanne 26, Switzerland

ABSTRACT

Background

Oxidative stress can severely compromise viability of bifidobacteria. Exposure of *Bifidobacterium* cells to oxygen causes accumulation of reactive oxygen species, mainly hydrogen peroxide, leading to cell death. In this study, we tested the suitability of continuous culture under increasing selective pressure combined with immobilized cell technology for the selection of hydrogen peroxide adapted *Bifidobacterium* cells. Cells of *B. longum* NCC2705 were immobilized in gellan-xanthan gum gel beads and used to continuously ferment MRS medium containing increasing concentration of H_2O_2 from 0 to 130 ppm.

Results

At the beginning of the culture, high cell density of 10^{13} CFU per litre of reactor was tested. The continuous culture gradually adapted to increasing H_2O_2 concentrations. However, after increasing the H_2O_2 concentration to 130 ppm the OD of the culture decreased to 0. Full wash out was prevented by the immobilization of the cells in gel matrix. Hence after stopping the stress, it was possible to re-grow the cells that survived the highest lethal dose of

H_2O_2 and to select two adapted colonies (HPR1 and HPR2) after plating of the culture effluent. In contrast to HPR1, HPR2 showed stable characteristics over at least 70 generations and exhibited also higher tolerance to O_2 than non-adapted wild type cells. Preliminary characterization of HPR2 was carried out by global genome expression profile analysis. Two genes coding for a protein with unknown function and possessing trans-membrane domains and an ABC-type transporter protein were overexpressed in HPR2 cells compared to wild type cells.

Conclusions

Our study showed that continuous culture with cell immobilization is a valid approach for selecting cells adapted to hydrogen peroxide. Elucidation of H_2O_2 adaptation mechanisms in HPR2 could be helpful to develop oxygen resistant bifidobacteria.

BACKGROUND

According to FAO/WHO (2002) [1], probiotics are defined as "live microorganisms which when administered in adequate amounts confer a health benefit on the host". One of the crucial points in the production and distribution of probiotic foods is hence to deliver enough live probiotic cells to the consumers. The minimum daily intake of probiotics to obtain a beneficial effect is still under debate. However, a concentration of 10^6 live cells of probiotic bacteria per gram of product at the time of consumption is generally accepted and selected to provide bacterial concentrations that are attainable and cost-effective for probiotic food products [2]. During production and storage of food, microorganisms experience a wide range of stresses, including oxidative stress, which can severely compromise cell viability of sensitive strains as well as their incorporation into food products.

Bacterial strains belonging to *Bifidobacterium* and *Lactobacillus* genera are the most widely used microorganisms in probiotic food products and supplements [3]. Bifidobacteria are obligate anaerobes. However, some species can tolerate oxygen, such as *B. psychroaerophilum*, *B. indicum* and *B. asteroides*, the latter two possessing catalase [4, 5]. Exposure to oxygen induces accumulation of reactive oxygen species, mainly hydrogen peroxide, which cause oxidative damage to vital cellular components, resulting in cell death of sensitive cells [6, 7].

Although oxygen tolerant bifidobacteria can be isolated from the environment [8], these isolates do not necessarily exhibit relevant probiotic characteristics. Mutagenic agents have also been applied to obtain oxygen resistant bifidobacteria [9]. However, random mutations affecting the probiotic characteristics of the strain may occur. Another method to isolate stress resistant bacterial strains consists in culturing cells in presence of a selective agent. Plating and cultivation in batch cultures with varying concentration of a selective agent are fairly simple procedures, but the number of generations over which selection can occur is limited. Therefore repeated cycles of subculturing may be required. In contrast to batch, continuous culture can be performed over an unlimited number of generations under strictly controlled conditions. Moreover, it can ensure continuous presence of unstable selective agents, such as hydrogen peroxide that can be broken up into nascent oxygen and water in contact with organic matter of rich media like MRS broth [10]. However, the use of continuous culture combined with selective pressure is limited because the resistance level of cells can greatly vary with fermentation time. In addition, an over-dosage of the selective agent leads generally to a wash-out of cells from the reactor. This major drawback of continuous cultures can be prevented using immobilized cell technology, which allows retaining cell in reactor even if dilution rate exceed growth rate of the culture [2, 11].

The aim of this study was to test the application of continuous culture with immobilized cells for selecting hydrogen peroxide adapted populations of *B. longum* NCC2705.

METHODS

Bacterial Strain

B. longum NCC2705 was obtained from the Nestlé Culture Collection (Lausanne, Switzerland) and cultivated in MRS [12] medium (Biolife, Milano, Italy). Two successive pre-cultures, inoculated at 1% from a frozen stock at -80°C in MRS with 10% glycerol (Sigma-Aldrich, Buchs, Switzerland), were performed for 16 h under anaerobic (AnaeroGen, Oxoid, Basingstoke, United Kingdom) conditions at 37°C before use.

Cell Immobilization

Cell immobilization was based on a two-phase dispersion process as previously described [13]. A mixed gel of 2.5% (w/v) gelrite gellan gum and 0.25% (w/v) xanthan gum (both Sigma-Aldrich) was inoculated at 2% (v/v) with a pre-culture of *B. longum* NCC2705, containing ca. 10^9 CFU ml^{-1}. Beads with diameters in the range of 1.0-2.0 mm were selected by wet sieving and used for fermentation. The entire process was completed within 1 h.

Continuous Culture

A volume of 70 ml of inoculated gel beads was transferred into a 1-l stirred tank reactor (Multifors, Infors-HT, Bottmingen, Switzerland) containing 630 ml MRS. The reactor was stirred at 100 rpm by an inclined blade impeller. Nitrogen was aseptically injected into the headspace of the reactor to maintain anaerobic conditions. Temperature was set at 37°C and pH was controlled and maintained at 6.0 by adding 5 M NaOH. Culture was started in batch mode for the first 24 h, followed by 24 h in continuous mode with feeding of MRS at 2.6 ml min^{-1}, using a peristaltic pump (Infors-HT), to allow colonization of gel beads. During this colonization step, immobilized population increased from 10^7 to 10^{11} CFU g^{-1} of gel beads. Afterwards, feeding of H$_2$O$_2$ solutions was started, using a calibrated peristaltic pump (Infors-HT), and the culture was carried out in continuous mode for 23 days. Inflow rate of MRS, initially set to 2.6 ml min^{-1} was decreased to 0.9 ml min^{-1} at day 9. H$_2$O$_2$ was continuously added to the reactor at a flow rate 10 fold smaller than that of the MRS and using concentrated solutions ranging from 50 to 1,300 ppm depending on the applied H$_2$O$_2$ level. H$_2$O$_2$ solutions were prepared using 30% H$_2$O$_2$ (VWR, Dietikon, Switzerland) in sterile water. H$_2$O$_2$ solution was kept on ice, protected from light and replaced daily, to avoid H$_2$O$_2$ breakdown during the experiment. Effluent samples were collected from the reactor at different time intervals for optical density (600 nm) measurements using sterile MRS medium as reference, and cell enumeration using plate counts. Aliquots of 2 ml effluent samples were centrifuged (6,000 g for 2 min) and pellets suspended in equal volume of fresh MRS containing 10% glycerol and stored at -80°C for further analyses. Gel bead samples of 1 g were placed into 1 ml of MRS with 10% glycerol and stored at -80°C until analysis.

Viable Cell Enumeration in Culture Effluent and Gel Beads

Samples from fermented effluent were serially diluted in phosphate buffered saline (pH 7.7) supplemented with 0.05% L-cysteine hydrochloride monohydrate (Sigma-Aldrich) (C-PBS). Appropriate dilutions were plated in duplicate on MRS agar (DIFCO, Becton Dickinson AG, Allschwil, Switzerland) and incubated anaerobically at 37°C for 48 h. The immobilized cell population was also monitored by plate counts after adding ca. 0.5 g gel beads to 20 ml 1% EDTA (Sigma-Aldrich) and treatment in a stomacher (Seward, Norfolk, UK) for 3 min for bead dissolution before dilution in C-PBS and plating.

H$_2$O$_2$ Resistance Level of Cells from Culture Effluent

Resistance to H$_2$O$_2$ over fermentation time of cells from culture effluent was tested as follows. Cells from frozen samples were twice subcultured in MRS inoculated at 1% (v/v) and incubated anaerobically at 37°C for 24 h. Aliquot of 1 ml containing ca. $1.11 \pm 0.67 \times 10^9$ CFU ml^{-1} was centrifuged at 6,000 g for 2 min and the pellet was suspended in 10 ml 400 ppm H_2O_2 solution. After 1.5 h incubation at room temperature, the cell suspension was diluted in C-PBS and plated on MRS agar and incubated anaerobically at 37°C for 48 h. Results were expressed as survival rate in percent before and after treatment. The test was performed in duplicate.

Isolation of Cells Adapted to H_2O_2

Frozen bead samples (1 g) collected at day 18 of continuous culture, were dissolved in 40 ml 1% EDTA, treated in a stomacher for 3 min and centrifuged at 6,000 g for 5 min. Cell pellets were suspended in 10 ml 40 ppm H_2O_2 and incubated for 60 min at room temperature to recover fractions of populations adapted to H_2O_2 stress. 1 ml of cell suspension was subsequently plated on MRS agar. Colonies visible within 48 h were twice subcultured anaerobically for 16 h in MRS broth at 37°C and frozen at -80°C in MRS with 10% glycerol until further analyses.

Characterization of H_2O_2 Adapted Isolates

H_2O_2 Resistance Level of Isolates

Frozen samples of wild type and H_2O_2 adapted isolate cultures containing approximately $1.03 \pm 0.33 \times 10^9$ CFU ml^{-1} were thawed at room temperature. 1 ml of sample was centrifuged at 6,000 g for 2 min. The pellet was suspended in 10 ml 200 ppm H_2O_2 solution. After 2 h incubation at room temperature cell suspensions were diluted in C-PBS and plated on MRS agar and incubated anaerobically at 37°C for 48 h. Results were expressed as survival rate in percent before and after treatment. The test was performed in triplicate.

Stability of H_2O_2 Adapted Phenotype

Stability of H_2O_2 adapted phenotype of isolates was tested by subculturing cells without selective pressure in MRS at 37°C under anaerobic conditions for 24 h. After each subculture, containing ca. $1.76 \pm 1.07 \times 10^9$ CFU ml^{-1}, resistance level to H_2O_2 of cells was tested using 400 ppm H_2O_2 solution as described above. The test was performed in duplicate.

Growth in Presence of Oxygen in Liquid Shaking Cultures

Ability of cells to grow in presence of oxygen was tested according to Meile et al.[8] with some modifications. The headspace of 500 ml serum flasks containing 400 ml MRS, after creating vacuum conditions was flushed with N_2 or CO_2 until normal atmosphere pressure was achieved; then 7.5 or 12.5% (v/v) sterile oxygen were added. Afterwards, the medium was inoculated at 2% with an overnight culture of wild type or H_2O_2 resistant cells and the flasks were incubated at 37°C in a shaker at 160 rpm (Kühner AG, Basel, Switzerland) for 24 h. Samples were taken every 2 h during the first 12 h of the culture to measure optical density at 600 nm. Two repetitions were carried out.

Growth in Reactor with and without H_2O_2

Cells were grown in an 800-ml working volume reactor (1-l reactor, Infors-HT) containing MRS inoculated at 5% with an overnight culture of wild type or H_2O_2 resistant cells. Temperature was controlled at 37°C and agitation set at 200 rpm. Anaerobic conditions were maintained by continuously sparging either CO_2 or N_2 in the medium, starting overnight prior inoculation. Cells were cultivated with and without addition of 42 ppm H_2O_2 in mid-exponential growth phase corresponding to an OD of 0.6. This H_2O_2 concentration of 42 ppm causes a cessation in growth for at least 40 min of exponentially wild type cells of *Bifidobacterium longum* NCC2705 without a decrease in viable cell counts during this growth arrest [14]. Growth was monitored by measuring OD at 600 nm. Samples for global transcriptional profiling were taken in mid-exponential growth phase after 3-3.5 h of culture at an OD of 0.7-0.8 in reactors sparged with CO_2 and without H_2O_2. Aliquots of 2 ml were centrifuged (4,000 g for 1 min), supernatants discarded and cell pellets snap frozen in liquid nitrogen and stored at -80°C until RNA-extraction. Fermentations were performed in triplicate.

Microarray Analysis

Microarray Design and RNA Extraction

DNA based arrays, produced by Agilent Technologies http://www.agilent.com, were obtained by *in situ* synthesis of 60 mer oligonucleotides on glass slides [15]. For each gene, 3 to 6 different probes were randomly distributed on the array. Total RNA was extracted with the macaloïd method and purified as previously described [16].

Array Hybridization

For each hybridization, cDNA was synthesized starting from 4 µg of total RNA and subsequently labeled using the Array 900MPX Genisphere kit (Genisphere Inc., Hatfield, PE, USA), following the protocol provided by the supplier.

Luciferase and kanamycin control mRNA (Promega, Zürich, Switzerland) at 1 and 10 ng, respectively, were mixed with total RNA before labeling to allow balancing of the two channels during scanning. After the hybridization procedure, array slides were scanned at 10 μm using a Scanarray 4000 (Packard Biochip Technologies, Billerica, MA, USA). Laser power and photomultiplier tube gain were set in order to prevent saturation of any spot, except the probes corresponding to rRNA.

Array Analysis

Data extracted with Imagene 5.6 (Biodiscovery, El Segundo, CA, USA) were treated with homemade scripts in Python language http://www.python.org and a local installation of the ArrayPipe web server [17]. Probes showing a signal smaller than twice the standard deviation of the local background were considered without signal. Probes showing no signal or saturated signals in both channels were discarded from the analysis. Assuming an intensity-dependent variation in dye signal, (limma) loess global normalization was applied on signal ratios. To calculate average gene expression values, data from 3 biological replicates were combined as follows. Within each hybridization data set, gene fold changes were calculated from the median of the corresponding probes values. The expression value of a gene was retained if a signal was detected in at least 50% of its probes. Genes were considered to be differentially expressed if their log_2-transformed signal ratios were higher than 1.5 or smaller than -1.5. Statistical analysis of the 3 biological array replicates of the hybridization between wild type and H_2O_2 resistant cells were performed using the statistical software R version 2.6.1 [18]. Bayes statistics for differential expression [19] was used to rank genes in order of evidence for differential expression. The data have been deposited in NCBI's Gene Expression Omnibus and are accessible through GEO Series accession number GSE16039. TMHMM-prediction server version 2.0 was used to identify transmembrane domains of predicted proteins [20].

RESULTS

Continuous Culture Monitoring

The concentration of H_2O_2 in the continuous culture with immobilized cells of *B. longum* NCC2705 was increased stepwise to 130 ppm in order to select for cells resistant to oxidative stress (Figure 1). At day 2, optical density of the fermented effluent reached 8.4, corresponding to 1.26×10^9 CFU ml^{-1}. Subsequently, OD decreased and increased after each increase in H_2O_2 concentration. Viable cells in effluent samples before changing the level of H_2O_2 concentration ranged from 6.31×10^8 to 2.00×10^9 CFU ml^{-1} until day 8 (Figure 1). Following H_2O_2 concentration increase to 130 ppm at day 9, the OD decreased to 0. To enrich potentially hydrogen peroxide resistant survivors after this harsh treatment, the addition of H_2O_2 was stopped for 2 days and the flow rate of the medium was decreased to 0.9 ml min^{-1}, resulting in a new phase of growth with an increase of OD to 3.2 at day 11 in effluent samples, corresponding to 6.31×10^5 CFU ml^{-1}. Concentration of H_2O_2 was then set again at 100 ppm and OD decreased again to 0 with viable cell counts in the effluent gradually decreasing from 3.98×10^5 at day 14 to 2.51×10^4 CFU ml^{-1} at day 20 (Figure 1). Cell counts in gel beads were generally 1 to 2 log higher than in effluent samples, except at the end of culture where both population reached similar levels (Figure 1).

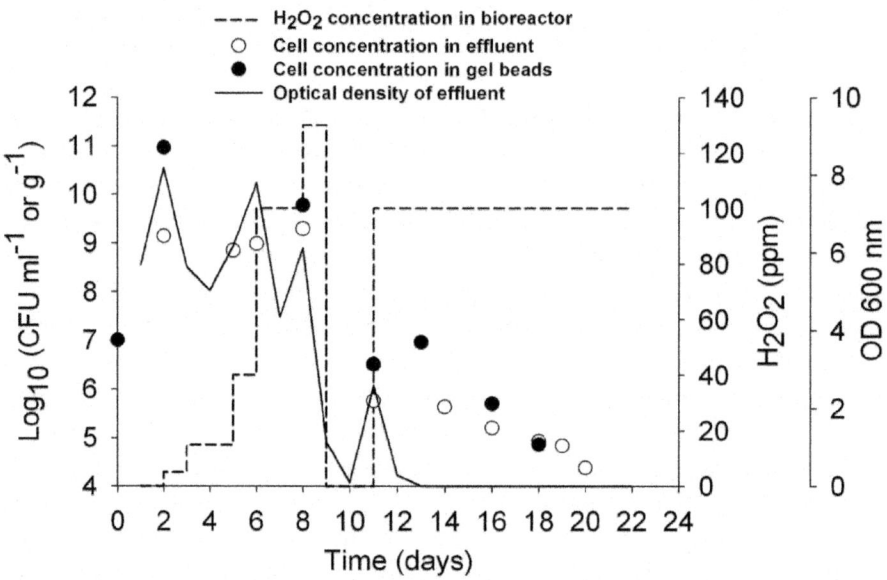

Figure 1: Continuous culture monitoring. The figure shows the optical density of fermented effluent, H_2O_2 concentration in bioreactor, viable cells counts in gel beads and in the effluent during continuous culture with immobilized cells of **B. longum** NCC2705 and H_2O_2 selective pressure.

H_2O_2 resistance of free cells in the culture effluent remained stable during the first 8 days with a survival rate of $0.00010 \pm 0.00004\%$ (Figure 2). At day 14 and 18 the survival rate was 760 and 16 folds higher than for day 1, respectively (Figure 2).

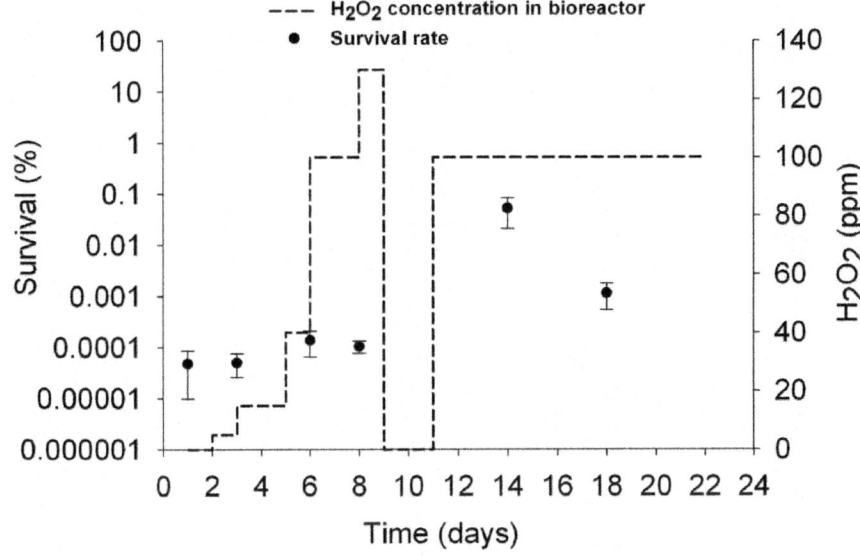

Figure 2: Resistance of cells collected from fermented effluent to H_2O_2. Resistance of cells from fermented effluent to H_2O_2 as a function of fermentation time was assessed after incubation in 400 ppm H_2O_2 solution for 1.5 h and expressed in % survival. Applied concentrations of H_2O_2 in the bioreactor are also indicated. Data are means of two replicates ± standard deviation.

Selection of Resistant Cells and Stability of H_2O_2 Resistance Phenotype

Because a large fraction of cell biomass of the system was in the bead matrix (ca. 10^{11} CFU g^{-1} beads after colonization) at the beginning of the fermentation, a higher number of adapted cells was expected in the gel beads, which were physically retained in the reactor, than in effluent at the end of culture. Cells from gel beads collected at day 18 (6.31×10^4 CFU g^{-1}) were subjected to an isolation step, and tested for their resistance to H_2O_2. Two colonies, namely HPR1 and HPR2, were detected after this isolation step. Tolerance to H_2O_2 of cells from these two isolates was 20-and 30-folds higher, respectively, than that of wild type cells (Figure 3). A stability test was performed with these two isolates using a hydrogen peroxide concentration of 400 ppm and incubation time of 2 hours. After three successive subcultures, HPR1 cells exhibited similar resistance level to wild type cells, whereas HPR2 maintained its phenotype over at least 11 subsequent cultures (Figure 4). The survival tests showed large variations depending on the testing day; however a repetition of this test showed similar trends (data not shown).

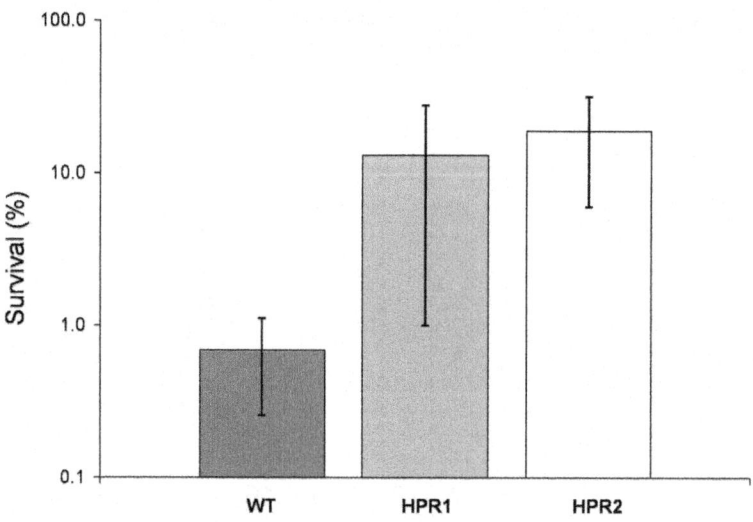

Figure 3: Resistance level of isolates HPR1 and HPR2 and of wild type (WT) *B. longum* NCC2705 to H_2O_2. Resistance was assessed after incubation of cells in 200 ppm H_2O_2 solution for 2 h and expressed in % survival. Data are means of three replicates ± standard deviation.

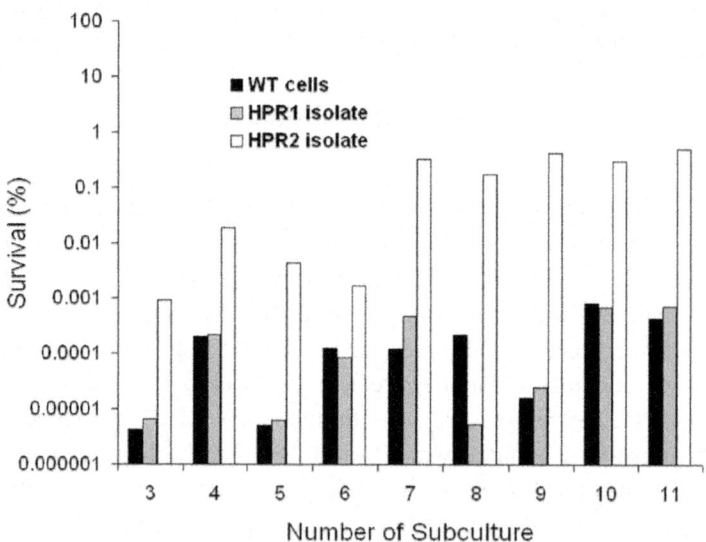

Figure 4: Stability of H_2O_2 adapted phenotype of isolates HPR1 and HPR2. The figure shows the survival (in %) to 2 h, 400 ppm H_2O_2 solution of wild type, and of H_2O_2 resistant phenotype HPR1 and HPR2 cells after successive subcultures without selective pressure.

Growth in Presence of Oxygen in Liquid Shaking Cultures

HPR2 and wild type cells grew at similar rate of 0.47 ± 0.03 h^{-1} and reached the same optical density of 5.22 ± 0.36 after 24 h culture in presence of 100% nitrogen or carbon dioxide in the headspace (Figures 5 and 6). When the atmosphere of the headspace was composed of 7.5 and 92.5% of oxygen and nitrogen, respectively, HPR2 cells grew at a lower rate of 0.09 ± 0.00 h^{-1} and reached an OD of 0.60 ± 0.02 after 24 h culture while growth of wild type cells was negligible with a growth rate of 0.01 ± 0.00 h^{-1} and a final OD of 0.15 ± 0.02 (Figure 5). Using a headspace of 12.5 and 87.5% of oxygen and carbon dioxide, respectively, the growth rate during the first 11.5 h was 0.12 ± 0.01 and 0.04 ± 0.01 h^{-1} for HPR2 and wild type cells, respectively (Figure 6). From 11.5 to 24 h of culture, both strains grew at similar rate of 0.15 ± 0.01 h^{-1} and reached a final OD of 1.44 ± 0.25 and 0.81 ± 0.19 for HPR2 and wild type cells, respectively (Figure 6).

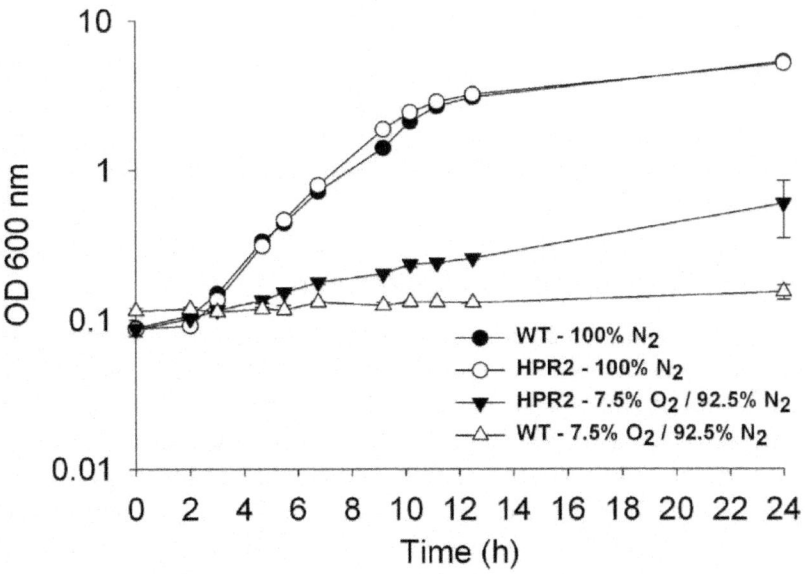

Figure 5: Growth of HPR2 and wild type *B. longum* NCC2705 in presence of 100% N_2 and of 7.5% O_2-92.5% N_2. Cells were grown in liquid shaking cultures. Growth was monitored using optical density at 600 nm. Data are means of two replicates ± standard deviation.

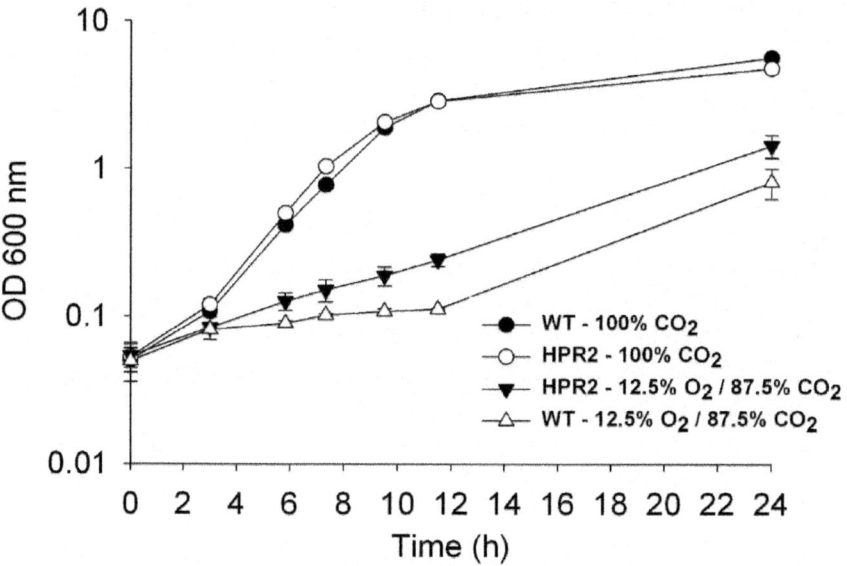

Figure 6: Growth of HPR2 and wild type _B. longum_ NCC2705 in presence of 100% CO_2 and of 12.5% O_2-87.5% CO_2. Cells were grown in liquid shaking cultures. Growth was monitored using optical density at 600 nm. Data are means of two replicates ± standard deviation.

Growth in Presence of H_2O_2 in Reactors

Both wild type and HPR2 cells recovered rapidly and in a similar manner after addition of H_2O_2 in medium sparged with carbon dioxide, and reached the same OD of 6.06 ± 0.60 than non treated cells after 24 h of culture (Figure 7). In presence of nitrogen, HPR2 strain started to recover earlier (1.63 vs. 2.51 h after addition of hydrogen peroxide) and grew at a higher rate (0.24 ± 0.02 vs. 0.13 ± 0.03 h^{-1}) than wild type cells after H_2O_2 addition, reaching an optical density of 1.41 ± 0.02 and 0.70 ± 0.18, respectively, after 8 h culture (Figure 8). Growth of HPR2 and wild type cells, treated or not with H_2O_2, were negatively affected by sparging nitrogen compared to carbon dioxide in the medium (Figures 7 and 8).

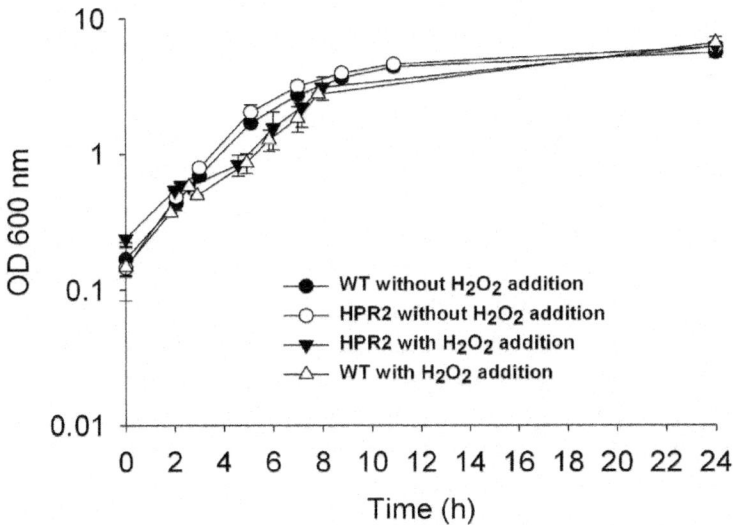

Figure 7: Growth of HPR2 and wild type *B. longum* NCC2705 in bioreactors under CO_2 atmosphere, with and without addition of H_2O_2. Cells were cultivated under CO_2 atmosphere and with or without addition of 42 ppm H_2O_2 in mid-exponential growth phase corresponding to an OD of 0.6. Growth was monitored using optical density at 600 nm. Data are means of three replicates ± standard deviation.

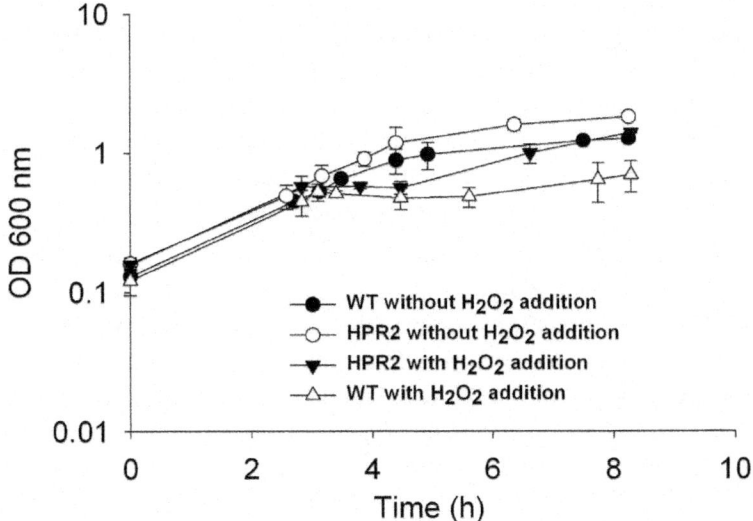

Figure 8: Growth of HPR2 and wild type *B. longum* NCC2705 in bioreactors under N_2 atmosphere, with and without addition of H_2O_2. Cells were cultivated under N_2 atmosphere and with or without addition of 42 ppm H_2O_2 in mid-exponential

growth phase corresponding to an OD of 0.6. Growth was monitored using optical density at 600 nm. Data are means of three replicates ± standard deviation.

Genome-Wide Transcription Analysis

Transcriptome analysis of exponentially growing cells without selective pressure showed that in HPR2 compared to wild type cells, the genes *BL1404* and *BL0931* were overexpressed. *BL1404* was over-expressed with an average $\log_2$ ratio of 3.1 (2.3-3.8, 95% confidence interval) and 1.5 (1.6-1.7, 95% CI) for *BL0931*. The start codon of *BL0931* gene is 45 nucleotide downstream of *BL0932* stop codon, a gene encoding an ABC-transporter ATP-binding protein. Both genes are predicted to be on the same operon, a prediction which is supported by the similar average gene expression profile displayed by the *BL0932* gene (1.4; 0.9-1.9, 95% CI). All remaining genes had $\log_2$ transformed ratios lower than 1.5 and non-significant p-values.

DISCUSSION

In this study, we tested the suitability of continuous culture under increasing selective pressure combined with immobilized cell technology for the selection of hydrogen peroxide adapted bifidobacteria cells. Continuous system allows cultivation of cells over an unlimited number of generations under strictly controlled conditions. In addition continuously cultured cells may undergo a number of consecutive mutational events, each contributing to improve adaptation of cells to their environment [21, 22]. Evolutionary engineering approaches using chemostats or repeated batch cultures, combined with constant or gradually increasing selective pressure have been successfully exploited to increase resistance of industrially relevant microorganisms, mainly yeast cells, to environmental stresses [23–25].

A very high cell concentration was reached in beads (10^{11} CFU g^{-1} beads) leading to a high total population, including immobilized and free cells, of 10^{13} CFU per litre of reactor similar to that reported by Doleyres *et al.*[26] with immobilized *B. longum* cells in gellan gum beads during continuous culture operated at a dilution rate of 0.5 h^{-1}. As a comparison, batch cultures with free cells of *B. longum* NCC2705 generally reach 10^9 CFU ml^{-1}, corresponding to 10^{12} CFU per litre of reactor at the end of the culture (data not shown). This high cell density could favor the occurrence of mutations since the rate of appearance of mutational events is proportional to the amount of biomass in the culture [27].

The continuous culture gradually adapted to increasing H_2O_2 concentrations, as shown by the culture oscillating OD. However, the *B. longum* population tested in the effluent reached an upper limit of its adaptive capabilities at day 9 with H_2O_2 concentration of 130 ppm. Immobilization of the cells in gel matrix has prevented full wash-out, which would have very likely occurred with a free cell system. Hence after stopping the stress, it was possible to re-grow the cells that survived this high H_2O_2 concentration. After this enrichment step and under 100 ppm H_2O_2, resistance level of continuously produced cells was higher than those tested at the beginning of the fermentation but decreased to an intermediate level at day 18. Heterogeneity in the immobilized and free populations which can exhibit different resistance level and/or adaptive mechanisms to the selective pressure, as discussed below for cells isolated from gel beads, could explain this change in resistance to H_2O_2 with fermentation time and using the same conditions. Çakar et al.[25] reported a heterogeneous resistance levels to cobalt for single cells within an evolved population of *Saccharomyces cerevisiae* subjected to continuous increasing levels of cobalt stress.

HPR1 and HPR2, isolated from gel beads at day 18, showed a 20-and 30-fold higher resistance to H_2O_2 than wild type cells, respectively. Two different mechanisms can be proposed to explain the adaptive response of these two isolates in regard to the stability of their phenotype. HPR1 isolate lost rapidly its hydrogen peroxide adapted phenotype which could be the result of a transient adaptation caused by the stressing conditions encountered in gel beads and not necessarily by H_2O_2 selective pressure. Indeed, it was already reported that immobilized *B. longum* cells, continuously cultivated without specific selective pressure adapt to H_2O_2 stress [28]. This phenotype was reversible after subculturing, and could be related to a non specific stress adaptation caused by diffusional limitations of both substrates and inhibitory products in the gel beads [29]. Resistance phenotype of HPR2 isolate to H_2O_2 was much more stable (at least over 70 generations) than that of HPR1 indicating a probable stable mutation. To our knowledge, this is the first description of a mutant strain of bifidobacteria resistant to H_2O_2. HPR2 cells were then further characterized.

HPR2 could tolerate higher O_2 level than wild type cells of *B. longum* NCC2705 and the moderately oxygen-tolerant *B. thermophilum* RBL67 as reported by von Ah et al.[29]. The ability of HPR2 isolate to grow in presence of oxygen can be associated with its H_2O_2 adapted phenotype. Indeed, accumulation of H_2O_2 during culture of sensitive bifidobacteria in presence of oxygen is generally considered as the primary reason for growth inhibition [7, 30, 31]. Additionally, HPR2 and wild type cells seem to tolerate more O_2 in presence of CO_2 than with N_2 in the headspace of the culture. This result is different from that of Kawasaki et al.[30] who reported that colony development of different *Bifidobacterium* species under an atmosphere composed of 5 and 95% O_2 and N_2, respectively, was not improved by addition of CO_2. This difference can be explained by the methodology and strains used in our studies. The better growth of HPR2 and wild type cells observed in reactor with medium sparged with CO_2 instead of N_2 are in agreement with others studies reporting the essential role of CO_2, even at low level, to stimulate growth of bifidobacteria [30, 32]. Presence of residual dissolved CO_2 in MRS medium for liquid shaking culture, can also explain the lack of difference between growth of HPR2 and wild type cells in presence of 100% N_2 or CO_2 in the headspace, in contrast to cultures in reactor where these gases were directly sparged into the medium.

A known mechanism influencing sensitivity of Bifidobacteria to oxygen is the type of NADH oxidase activity. In O_2 sensitive species, NADH oxidase exhibits H_2O_2 forming activity while H_2O is produced in microaerophilic species [6, 31]. *B. longum*NCC2705 possess a gene (*BL1266*), which codes for a putative NADH oxidase with an active site for the four electron reduction of O_2 to H_2O and could therefore be implied in the detoxification of H_2O_2[14]. However, *BL1266* is not differentially expressed in HPR2 compared to wild type cells. Constitutive overproduction of oxidative stress related proteins can also protect cells to H_2O_2 as observed in a mutant strain of *Bacteroides fragilis* resistant to H_2O_2[33]. In HPR2 another mechanism seems to be involved in the resistance to H_2O_2. Two genes, *BL1404* and *BL0931*, were constitutively overexpressed in HPR2 compared to wild type cells. These two genes were not differentially expressed in wild type cells exposed to H_2O_2[14]. BLAST search [34] showed that BL1404, which is annotated as hypothetical protein [35], is highly homologous, 99% on nucleotide level, to the integral membrane protein BLD_0271 from *B. longum* DJ010A. The BLAST search could not identify homologous genes in other bacteria, indicating that *BL1404* encodes a protein specific to bifidobacteria. The BL1404 protein possesses three predicted transmembrane domains and 2 outer and 2 inner domains. BL0931 is annotated as possible ABC-type transport system involved in lipoprotein release. The functional predictions of the two differentially expressed genes do not suggest any mechanistic explanation for the resistance phenotype of the HPR2 isolate. For this purpose further experiments are required. Transcriptional profiling in the presence of H_2O_2 might be more informative for characterization of the resistance phenotype. Moreover, transcriptional profiling analysis is known to be useless for mutations that do not impact gene expression. Therefore, for future experiments, whole genome sequencing using next-generation sequencing technologies will most likely provide better mechanistic insight.

CONCLUSIONS

Our study showed that continuous culture with cell immobilization is a valid approach for selecting cells adapted to hydrogen peroxide. Cell immobilization allowed maintaining high cell numbers in the reactor, even when high selective pressure was applied. This enabled controlled application of stress at high levels on the culture over a long time. Additionally, preliminary characterization of HPR2 revealed the constitutive induction of two genes associated with the cell membrane. However, their function needs further characterization. Elucidation of H_2O_2 resistance mechanisms in HPR2 could be helpful to improve resistance of bifidobacteria to oxidative stress.

ACKNOWLEDGEMENTS

This study was carried out thanks to the financial support of the Commission of Technology and Innovation of Switzerland (CTI-Project Nr. 75272 LSPP-LS) and Nestle (Switzerland).

AUTHORS' CONTRIBUTIONS

VM carried out the experimental part and contributed to draft the manuscript. FG participated to the design of the experiments and drafted the manuscript. DM performed all microarray experiments. BB and ER provided essential inputs for the microarray analysis and together with CL, FA and FG, conceived the initial approaches. LM participated to the analysis of the genomic data and to draft the manuscript. CL acted as overall supervisor and corresponding author of the work. All authors have read and approved the final version of the manuscript.

REFERENCES

1. FAO/WHO: Joint FAO/WHO Working Group Report on Drafting Guidelines for the Evaluation of Probiotics in Food: April 30 and May 1, London, Ontario, Canada. 2002

2. Lacroix C, Yidirim S: Fermentation technologies for the production of probiotics with high viability and functionality. Curr Opin Biotechnol. 2007, 18: 176-183. 10.1016/j.copbio.2007.02.002.

3. Vasijevic T, Shah NP: Probiotics - From Metchnikoff to bioactives. Int Dairy J. 2008, 18: 714-728. 10.1016/j.idairyj.2008.03.004.

4. Simpson PJ, Ross RP, Fitzgerald GF, Stanton C: *Bifidobacterium psychraerophilum* sp. nov. and *Aeriscardovia aeriphila* gen. nov., sp.

nov., isolated from a porcine caecum. Int J Syst Evol Microbiol. 2004, 54: 401-406. 10.1099/ijs.0.02667-0.

5.   Biavati B, Mattarelli P: The family Bifidobacteriaceae. The Prokaryotes. Edited by: Falkow S, Rosenberg E, Schleifer KH, Stackebrandt E, Dworkin M. 2001, 322-382. New-York: Springer, 3

6.   Kawasaki S, Mimura T, Satoh T, Takeda K, Niimura Y: Response of the microaerophilic *Bifidobacterium* species, *B. boum* and *B. thermophilum*, to oxygen. Appl Environ Microbiol. 2006, 72: 6854-6858. 10.1128/AEM.01216-06.

7.   Talwalkar A, Kailasapathy K: The role of oxygen in the viability of probiotic bacteria with reference to *L. acidophilus* and *Bifidobacterium* spp. Curr Issues Intest Microbiol. 2004, 5: 1-8.

8.   Meile L, Ludwig W, Rueger U, Gut C, Kaufmann P, Dasen G, Wenger S, Teuber M: *Bifidobacterium lactis* sp. nov, a moderately oxygen tolerant species isolated from fermented milk. Syst Appl Microbiol. 1997, 20: 57-64.

9.   Shiuann YS, Chen MC, Liao CC: Bifidobacteria strains with acid, bile salt and oxygen tolerance and their culture method. US Patent 5, 711, 977. 1998

10.   Narendranath NV, Thomas KC, Ingledew WM: Urea hydrogen peroxide reduces the numbers of lactobacilli, nourishes yeast, and leaves no residues in the ethanol fermentation. Appl Environ Microbiol. 2000, 66: 4187-4192. 10.1128/AEM.66.10.4187-4192.2000.

11.   Doleyres Y, Lacroix C: Technologies with free and immobilised cells for probiotic bifidobacteria production and protection. Int Dairy J. 2005, 15: 973-988. 10.1016/j.idairyj.2004.11.014.

12.   De Man JC, Rogosa M, Sharpe EM: A medium for the cultivation of lactobacilli. J Appl Bacteriol. 1960, 23: 130-135.

13.   Cinquin C, Le Blay G, Fliss I, Lacroix C: Immobilization of infant fecal microbiota and utilization in an in vitro colonic fermentation model. Microb Ecol. 2004, 48: 128-138. 10.1007/s00248-003-2022-7.

14.   Klijn A: Physiological and molecular characterization of stress responses in *Bifidobacterium longum* NCC2705. PhD thesis. 2005, University College Cork

15.   Wolber PK, Collins PJ, Lucas AB, De Witte A, Shannon KW: The Agilent in situ-synthesized microarray platform. Methods Enzymol. 2006, 410: 28-57. 10.1016/S0076-6879(06)10002-6.

16. Parche S, Beleut M, Rezzonico E, Jacobs D, Arigoni F, Titgemeyer F, Jankovic I: Lactose-over-glucose preference in *Bifidobacterium longum* NCC2705: *glcP*, encoding a glucose transporter, is subject to lactose repression. J Bacteriol. 2006, 188: 1260-1265. 10.1128/JB.188.4.1260-1265.2006.

17. Hokamp K, Roche FM, Acab M, Rousseau ME, Kuo B, Goode D, Aeschliman D, Bryan J, Babiuk LA, Hancock RE, : ArrayPipe: a flexible processing pipeline for microarray data. Nucleic Acids Res. 2004, 32: W457-459. 10.1093/nar/gkh446.

18. Team RDC: R: A language and environment for statistical computing. 2007, Vienna, Austria: R Foundation for Statistical Computing

19. Smyth GK: Linear models and empirical bayes methods for assessing differential expression in microarray experiments. Stat Appl Genet Mol Biol. 2004, 3: Article3

20. Krogh A, Larsson B, von Heijne G, Sonnhammer ELL: Predicting transmembrane protein topology with a hidden Markov model: Application to complete genomes. J Mol Biol. 2001, 305: 567-580. 10.1006/jmbi.2000.4315.

21. Butler PR, Brown M, Oliver SG: Improvement of antibiotic titers from Streptomyces bacteria by interactive continuous selection. Biotechnol Bioeng. 1996, 49: 185-196. 10.1002/(SICI)1097-0290(19960120)49:2<185::AID-BIT7>3.0.CO;2-M.

22. Novick A, Szilard L: Experiments with the chemostat on spontaneous mutations of bacteria. PNAS. 1950, 36: 708-719. 10.1073/pnas.36.12.708.

23. Sauer U: Evolutionary engineering of industrially important microbial phenotypes. Adv Biochem Eng Biotechnol. 2001, 73: 129-169.

24. Çakar ZP, Seker UOS, Tamerler C, Sonderegger M, Sauer U: Evolutionary engineering of multiple-stress resistant *Saccharomyces cerevisiae*. FEMS Yeast Res. 2005, 5: 569-578. 10.1016/j.femsyr.2004.10.010.

25. Çakar ZP, Alkim C, Turanli B, Tokamn N, Akman S, Sarikaya M, Tamerler C, Benbadis L, FranÇois JM: Isolation of cobalt hyper-resistant mutants of *Saccharomyces cerevisiae* by *in vivo* evolutionary engineering approach. J Biotechnol. 2009, 143: 130-138. 10.1016/j.jbiotec.2009.06.024.

26. Doleyres Y, Paquin C, LeRoy M, Lacroix C: *Bifidobacterium longum* ATCC 15707 cell production during free-and immobilized-cell cultures in MRS-whey permeate medium. Appl Microbiol Biotechnol. 2002, 60: 168-73. 10.1007/s00253-002-1103-8.

27. Lane PG, Oliver SG, Butler PR: Analysis of a continuous-culture technique for the selection of mutants tolerant to extreme environmental stress. Biotechnol Bioeng. 1999, 65: 397-406. 10.1002/(SICI)1097-0290(19991120)65:4<397::AID-BIT4>3.0.CO;2-X.

28. Doleyres Y, Fliss I, Lacroix C: Increased stress tolerance of *Bifidobacterium longum* and *Lactococcus lactis* produced during continuous mixed-strain immobilized-cell fermentation. J Appl Microbiol. 2004, 97: 527-539. 10.1111/j.1365-2672.2004.02326.x.

29. von Ah U, Mozzetti V, Lacroix C, Kheadr EE, Fliss I, Meile L: Classification of a moderately oxygen-tolerant isolate from baby faeces as *Bifidobacterium thermophilum*. BMC Microbiol. 2007, 7:

30. Kawasaki S, Nagasaku M, Mimura T, Katashima H, Ijyuin S, Satoh T, Niimura Y: Effect of CO2 on colony development by *Bifidobacterium* species. Appl Environ Microbiol. 2007, 73: 7796-7798. 10.1128/AEM.01163-07.

31. Kawasaki S, Satoh T, Todoroki M, Niimura Y: b-type dihydroorotate dehydrogenase is purified as a H_2O_2-forming NADH oxidase from *Bifidobacterium bifidum*. Appl Environ Microbiol. 2009, 75: 629-636. 10.1128/AEM.02111-08.

32. Ninomiya K, Matsuda K, Kawahata T, Kanaya T, Kohno M, Katakura Y, Asada M, Shioya S: Effect of CO_2 concentration on the growth and exopolysaccharide production of *Bifidobacterium longum* cultivated under anaerobic conditions. J Biosci Bioeng. 2009, 107: 535-537. 10.1016/j.jbiosc.2008.12.015.

33. Rocha ER, Smith CJ: Characterization of a peroxide-resistant mutant of the anaerobic bacterium *Bacteroides fragilis*. J Bacteriol. 1998, 180: 5906-5912.

34. Zhang Z, Schwartz S, Wagner L, Miller W: A greedy algorithm for aligning DNA sequences. J Comput Biol. 2000, 7: 203-214. 10.1089/10665270050081478.

35. Schell MA, Karmirantzou M, Snel B, Vilanova D, Berger B, Pessi G, Zwahlen MC, Desiere F, Bork P, Delley M, : The genome sequence of *Bifidobacterium longum* reflects its adaptation to the human gastrointestinal tract. PNAS. 2002, 99: 14422-14427. 10.1073/pnas.212527599.

Chapter 7

COMPARATIVE STUDIES FOR THE BIOTECHNOLOGICAL PRODUCTION OF L-LYSINE BY IMMOBILIZED CELLS OF WILD-TYPE CORYNEBACTERIUM GLUTAMICUMATCC 13032 AND MUTANT MH 20-22 B

Meerza Abdul Razak[1] and Buddolla Viswanath[2]

[1]Natco Pharma Limited, Natco House, Road No. 2, Banjara Hills, Hyderabad 500 034, India

[2]Department of Virology, Sri Venkateswara University, Tirupati 517502, A. P, India

ABSTRACT

Establishing a cost and time efficient approach for bioprocess optimization is desired but is challenging. In the present work, we have addressed the effectiveness of using immobilized cells for aerobic processes, behaviour of immobilized cells, optimization and upstream bioprocess analysis for the production of lysine by immobilized cells of *Corynebacterium glutamicum* ATCC 13032 and MH 20-22 B in stirred tank bioreactor. Optimized operational conditions for maximal yield and productivity were determined with six parameters i.e., pH, temperature, fermentation time, airflow rate, glucose concentration and aeration rate. With the obtained results, it was evident that the optimum values for the upstream parameters viz., fermentation time, pH, temperature, glucose concentration, air flow rate and agitation rate are 96 h, 7.5, 30 °C, 90 g/l, 1.0 vvm and 200 rpm for both immobilized cells of *C. glutamicum* ATCC 13032 and MH 20-22 B. Immobilized cells of *C. glutamicum* MH 20-22 B, which is a leucine auxotroph has yielded more l-lysine compare to the immobilized cells of wild type strain *C. glutamicum* ATCC 13032.

INTRODUCTION

Amino acids have been produced with the help of microorganisms for almost 50 years. The economic significance of these cellular building blocks is significant and hence demand is consistently growing and constant efforts to enhance production performance are directed towards the microorganisms, as well as towards technological improvements of the relevant processes. The l-glutamic acid is in first place of highest produced amino acid, followed by l-lysine and dl-methionine whereas the other amino acids follow behind (Anastassiadis 2007). The cause for the increased demand for amino acids stems from their consumption as food preservative, feed supplements, therapeutic agents and precursors for the production of peptides or agrochemicals (Leuchtenberger et al. 2005). l-lysine is essential as a feed additive for poultry and pig breeding.

Gram-positive, rod-shaped bacteria *Corynebacterium glutamicum* has usually occupied a special position within the amino acid producing microorganisms with industrial significance. The essential amino acid l-lysine is one of the most important amino acids applied as supplement in animal feed (Wendisch and Bott2005; Bercovici and Fuller 2007). The supplementation of such feed materials with a lysine rich source leads to optimized growth of pigs or chicken (Wendisch and Bott 2005). The direct addition of lysine hereby has proven especially valuable. It does not cause an extra uptake and metabolization of other aminoacids beyond their requirement so that superfluous formation of ammonia and environmental burden by increased nitrogen loads in the manure is avoided. The progressing development of an increased utilization of white meat in different countries of the western as well as the eastern world has led to a vast market growth for lysine during the past decades (Tryfona and Bustard 2005). The high importance of lysine in nutrition has inspired intensive research on the lysine, bioprocess optimization, biosynthetic pathways and their regulation and the search for microorganisms capable of over-producing this amino acid. The first steps towards industrial production of lysine were done in Japan in the 1950s when Kyowa Hakko Co., Ltd., Tokyo initiated a research program targeted at finding a microorganim capable to produce glutamate. One of the results from this was the isolation of a microorganism *Micrococcus glutamicus*, later on renamed to *C. glutamicum*, which was able to produce glutamate (Kinoshita et al. 1957; Udaka 1960). During mutagenesis and screening program lysine producing mutants were discovered (Kinoshita et al. 1958) and the basis for lysine production was made. Within a few years the first industrial scale lysine manufacturing facility was developed. Since then lysine fermentation processes have been employed for large scale production. As the research interest for a cost effective production has increased, l-Lysine product prices have decreased. l-Lysine is produced by aerobic fermentation process using the bacterium *C. glutamicum*, it is an aerobic, non-sporulating, gram-positive, bacterium with GRAS (Generally Regarded As Safe) status that has been extensively used for the industrial production of a number of food grade amino acids, feed, and pharmaceutical products for many decades based on classical metabolic engineering. *C. glutamicum* abilities to produce other amino acids likel-threonine (Shiio 1990; Kase and Nakayama 1974; Shiio et al. 1991), l-methionine (Nakayama and Araki1973; Kalinowski et al. 2003), l-serine (Eggeling 2007), l-histidine (Araki et al. 1974), l-valine (Ruklisha et al.2007), l-tryptophan (Ikeda 2006), l-phenylalanine and l-tyrosine (Ikeda and Katsumata 1992), l-leucine (Patek 2007) and l-isoleucine (Guillout et al. 2002) has made it major work horse of industrial biotechnology. The biotechnological production of l-lysine by *C. glutamicum* requires a continuous improvement of the lysine production process with a special attention on optimization of the production process and strains engineering (Pfefferle et al. 2003). Therefore, the main objective of this paper is to analyze upstream bioprocess and comparative studies of l-Lysine production by free cells of *Corynebacterium glutamicum*ATCC 13032 and MH 20-22 B in stirred tank bioreactor.

MATERIALS AND METHODS

The wild-type *C. glutamicum* ATCC 13032 was acquired from American Type and Culture Collection, Manassas, USA and *C. glutamicum* MH 20-22 B was donated by Professor Eggeling, Biotechnology Institute, Julich, Germany. Both these strains were cultured on agar slopes containing peptone (5 g), beef extract (3 g), NaCl (5 g), agar (15 g), distilled water (1,000 ml) was maintained at pH 7.

Immobilization Method

Growth Medium Composition

The composition of growth medium is as follows: Glucose (2 g), beef extract (1 g), bacto peptone (1 g), NaCl (0.25 g), agar (2 g), distilled water (100 ml) and the pH 7.

Agar slants of *C. glutamicum* cells which were grown for 24 h were used to inoculate 50 ml of growth media and kept on shaker for 48 h (150 rpm) at 30 °C. Hundred milliliter of 72 h culture was used to prepare immobilized beads of calcium alginate. Fifteen percent volume of beads were employed throughout this study.

Immobilization of *C. glutamicum* cells was done in strict aseptic conditions. Gluteraldehyde entrapment method using cross linked calcium alginate was used to immobilize *C. glutamicum* cells (Jetty et al. 2005; Marek et al. 1985). Hundred microliter gluteraldehyde and 3 % sodium alginate were thoroughly mixed with 0.06 % cells on dry cell weight basis (DCW) (w/v), to get uniform suspension. This uniform suspension was transferred into 0.2 M CaCl$_2$ solution using peristaltic pump through a cut micropipette tip (or) orifice. The curing of the formed beads was done by incubating in 0.2 M CaCl$_2$ solution for 24 h and washed twice with sterile saline solution [0.9 % NaCl solution (w/v)] and preserved at 4 °C in saline solution for further use. The above immobilization method is followed for both strains of *C. gluatamicum* to get immobilized cells of them.

Fermentation Procedure for Lysine Production by Immobilized Cells of *C. glutamicum* ATCC 13032 and MH 20-22 B

Batch or fed-batch processes are employed for the commercial production of amino acids. In batch operations all of the nutrients are added at the beginning. And moreover in batch fermentations microorganisms grows until one or more of essential nutrients get exhausted or until fermentation conditions like oxygen limitation, pH decrease and product inhibition become unfavourable. In the present work fermentation experiments with immobilized cells have been conducted in batch mode using a sterilized stirred tank bioreactor. The immobilized *C. glutamicum* cells were used to inoculate the fermentation medium in the bioreactor and this batch fermentation was carried out for 200 h. The above batch fermentation procedure was carried out at different parameters separately for immobilized cells of both strains, to optimize the upstream parameters. New Brunswick Bioreactor of 5 L was used for the present batch fermentation experiments and was carried out for 120 h. Initially all reactor parts are separated. All parts are washed with distilled water thoroughly and again with acetone before sterilization. After washing with distilled water all the above parts are wrapped with aluminium foil before placing into the autoclave. All these accessories are kept in the autoclave and the autoclave is operated at a temperature of 120 °C and a pressure of 15 psi for a period of 20 min. The aeration rate for the batch fermentation was set to different vvm by the integrated gas flow controller. A pH electrode (Mettler Toledo, Giessen, Germany) and automated addition of 25 % NH$_4$OH was used to maintain the pH at 7 in the stirred tank bioreactor. The added volume was determined gravimetrically (Lab Balance Cupis, Sartorius, Gottingen, Germany). Dissolved oxygen was determined using a pO2 (partial pressures of oxygen) electrode (Mettler Toledo, Giessen, Germany) and by variation of the stirrer speed. This was controlled by the process control software Base Lab (BASFSE, Ludwigshafen, Germany).Temperature was maintained at 30 °C using jacket cooling. CO$_2$ and O$_2$ in the exhaust gas were analyzed by a mass spectrometer. All processes data were monitored online and recorded by BaseLab. The composition of media used for fermentation process in this work is as follow:

CaCl$_2$ 2H$_2$O (1 g), (NH$_4$)2SO$_4$ (30 g), MgSO$_4$ 7H$_2$O (0.4 g), NaCl (0.05 g), MnSO$_4$ H$_2$O (0.0076 g), FeSO$_4$7H$_2$O (0.001 g), KH$_2$PO$_4$ (1 g), K$_2$HPO$_4$ (1 g), Urea (2 g), Yeast extract (1.5 g), Peptone (2 g), d-Glucose (150 g), Thiamine (0.2 mg), d-Biotin (0.5 mg), l-Serine (0.1 mg) and distilled water (1.0 L). And the pH was maintained at 7.0.

Analytical Estimation of l-Lysine and Substrates

Cell concentration was determined photometrically at 660 nm (UV–Visible spectrophotometer, Thermo Electron Corporation). Supernatant from batch fermentation was used for quantification of substrates and products which were obtained by separation of the biomass by centrifugation at 8,500 g for 5 min at 4 °C. l-lysine was estimated by the method of Chinard (Chinard 1952). Glucose concentration was determined by anthrone method (Morris 1948; Neish 1952). Residual sugar was determined as glucose in the supernatant fluid by the colorimetric DNS Method (Miller 1959). Biomass in the broth was estimated in 1 ml of the sample by centrifugation, and dried in an oven at 105 °C until constant cell weight obtained.

RESULTS AND DISCUSSION

Earlier we have carried out many batch fermentations to improve the production rate of l-Lysine by free and immobilized cells of *C. glutamicum* under various process conditions. It was reported from one of our former investigation that l-Lysine can be produced more efficiently by immobilized growing *C. glutamicum* cells compared to free cells in stirred tank bioreactor (Razak and viswanath 2014). In addition, many investigators have stated about the multistage continuous fermentation processes and their difficulties to run continuousl-Lysine production systems (Becker 1982; Michalski et al. 1984). So in the present investigation we have focused on the upstream bioprocess analysis, comparative studies and optimization of l-Lysine production by immobilized cells of wild type *C. glutamicum* ATCC 13032 and *C.glutamicum mutant* MH 20-22 B.

Effect of fermentation time on l-lysine production by immobilized cells of *C. glutamicum* ATCC 13032 and MH 20-22 B

The investigation on effect of fermentation time on l-lysine production by immobilized cells of *Corynebacterium glutamicum* ATCC 13032 and MH 20-22 B cells was carried out under operating fermentation conditions like temperature 28 °C, pH 7.0, air flow rate 1.5 vvm, agitation rate of 300 rpm and glucose concentaration of 100 (g/l). It is observed that as a fermentation time increases the concentration of the residual glucose decreases between 40 and 120 h. Similar type of studies was carried out by Amin and Al-Talhi for production of l-Glutamic acid by using immobilized cells of *C. glutamicum* (Amin and Al-Talhi2007). l-lysine concentration was relatively lower for the first 70 h and the cell were multiplied with rapid growth of biomass. As the fermentation proceeds the glucose consumption and l-lysine production increased at 96 h with high sugar utilization and maximum l-lysine concentration. After 96 h, maximum product achieved a downward trend, l-lysine concentration and yields were found to be decreased; in-spite of an increase in fermentation time due to depletion of nutrients. Figure 1 results describe that the maximum lysine concentration is 22.83 ± 0.21 g/l and biomass concentration is 14.26 ± 0.18 g/l at 96 h for *C. glutamicum* ATCC 13032 immobilized cells.

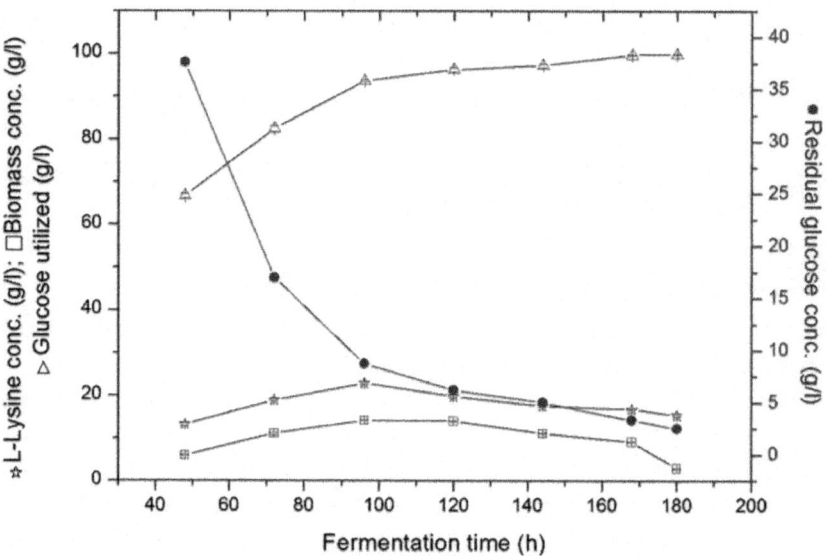

Figure. 1: Effect of fermentation time on l-lysine production by immobilized cells of *C. glutamicum* ATCC 13032.

The results presented in the Fig. 2 shows that l-lysine production was more at 96 h, it is the best fermentation time for the production of l-lysine by immobilized cells of *C. glutamicum* MH 20-22 B. The maximal yield of the l-lysine by immobilized cells of *C. glutamicum* MH 20-22 B at 96 h is 23.44 ± 0.27 g/l and biomass concentration is 15.28 ± 0.22 g/l The results presented in the Figs. 1 and 2 shows that l-lysine production was good at 96 h, which is the best fermentation time for the production of l-lysine by immobilized cells of *C. glutamicum* MH 20-22 B.

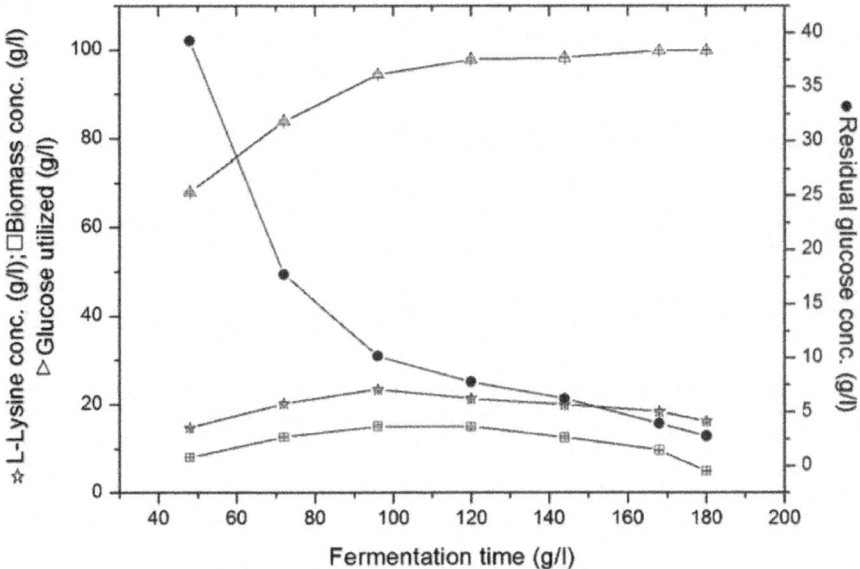

Figure. 2: Effect of fermentation time on l-lysine production by immobilized cells of *C. glutamicum* MH 20-22 B.

Effect of Temperature on l-Lysine Production by Immobilized Cells of *C. glutamicum* ATCC 13032 and MH 20-22 B

The growth rate of bacterial microorganism is basically dependent on temperature during fermentation which makes changes in whole metabolism. Any change in the temperature can alter the substrate utilization rate of microorganism which leads to unbalanced nutrients in the medium with respect to growth rate of the *C. glutamicum* cells. If any of the crucial nutrients is exhausted soon or unused this can make the culture from balanced to unbalanced growth and results in its performance change. The effect of temperature on l-Lysine production by immobilized cells of *C. glutamicum* ATCC 13032 and MH 20-22 B was studied under different operating conditions like fermentation time of 96 h, pH of 7.0, air flow rate of 1.5 vvm, agitation rate of 300 rpm and glucose concentration of 100 g/l. Glucose utilization was high at 30 °C. The down trend of residual glucose as the temperature increases can be seen after 30 °C the l-lysine biomass concentration was decreased. It was observed that *C. glutamicum* ATCC 13032 immobilized cells had shown maximum l-lysine concentration of 23.03 ± 0.17 (g/l) along with 18.34 ± 0.18 g/l of maximum biomass production at 30 °C (Table 1). Hilliger et al. (1984) observed temperature as one of the key fermentation parameters which has good influence on growth and l-Lysine formation by *C. glutamicum* (Hilliger et al. 1984). An increase in temperature resulted in decreased productivity, which suggested that a little increase in temperature has profound effect on cellular activities, which might be because of repression of metabolic enzymes. *C. glutamicum* MH 20-22 B immobilized cells shows maximum l-lysine production of 24.45 ± 0.19(g/l) and biomass concentration 19.46 ± 0.18 (g/l) at 30 °C (Table 2). The production of l-Lysine by immobilized cells of *C. glutamicum* MH 20-22 B is more when compared with *C. glutamicum* ATCC 13032 immobilized cells.

Table 1: Effect of temperature on l-lysine production by immobilized cells of *C. glutamicum* ATCC 13032

Sl. no.	Temperature (°C)	Lysine conc. (p) (g/l)	Biomass (x) (g/l)	Residual glucose conc. (g/l)	Glucose utilized (s) (g/l)
1	27	16.78	13.49	11.12	92.55
2	28	20.22	15.67	7.85	95.88
3	29	21.91	16.00	7.95	96.77
4	30	23.03	18.34	6.17	98.44
5	31	22.56	17.98	3.48	97.65
6	32	19.87	15.90	2.19	95.48

Table 2: Effect of temperature on l-lysine production by immobilized cells of *C. glutamicum* MH 20-22 B

Sl. no.	Temperature (°C)	Lysine conc. (p) (g/l)	Biomass (x) (g/l)	Residual glucose conc. (g/l)	Glucose utilized (s) (g/l)
1	27	17.89	15.15	12.33	89.76
2	28	21.74	15.99	9.64	92.45
3	29	22.33	16.79	8.61	93.48
4	30	24.45	19.45	6.42	95.67
5	31	23.05	18.82	4.66	97.43
6	32	21.74	16.44	2.68	99.41

Effect of pH on l-Lysine Production by Immobilized Cells *C. glutamicum* ATCC 13032 and MH 20-22 B

The pH is very important process parameter strongly influences the microbial fermentation. In fermentation process the pH of the broth decreases due to accumulation of other byproducts. As a result the bacterial growth was cease with affiliated decrease in the yield. Basic compounds such as sodium hydroxide, potassium hydroxide, ammonium hydroxide, calcium carbonate, urea, ammonia and inorganic acid compounds such as phosphoric sulphuric acid are used in controlling pH in l-lysine producing cultures ranging from 5.0 to 8.0. The effect of pH on l-lysine yield was found to be very major and crucial parameter. In the present study, effect of pH on l-lysine production was carried out at the fermentation time of 96 h, temperature of 30 °C, air flow rate of 1.5 vvm, agitation rate of 300 rpm and glucose concentration rate of 100 g/l. Figures 1 and 2 shows the effect of pH on l-Lysine production by immobilized cells of *C. glutamicum* ATCC 13032 and MH 20-22 B. pH of the fermentation processes can also be maintained by adding ammonia water along with small amount of $CaCO_3$. (Wang et al. 1991). The optimum pH for *C. glutamicum* ATCC 13032 immbolized cells is 7.5 at which the maximum lysine concentration of 23.98 ± 0.15 (g/l) and biomass concentrarion of 19.19 ± 0.15 (g/l) was observed (Table 3). To maintain optimum pH, reagents like calcium carbonate must be added to the culture medium at the beginning of the fermentation, thus calcium carbonate was used as internal neutralizing agent. The maximum l-lysine concentration of *C. glutamicum* MH 20-22 B immobilized cells is 25.63 ± 0.18 (g/l) and biomass concentration is 20.43 ± 0.21 (g/l), it was obtained at 7.5 pH (Table 4). Above results prove that 7.5 pH is the best suited for more lysine production by immobilized cells of *C. glutamicum* ATCC 13032 and MH 20-22 B.

Table 3: Effect of pH on l-Lysine production by immobilized cells of *C. glutamicum* ATCC 13032

Sl. no.	pH	Lysine conc. (p) (g/l)	Biomass (x) (g/l)	Residual glucose conc. (g/l)	Glucose utilized (s) (g/l)
1	6	13.16	13.33	11.93	92.02
2	6.5	18.73	17.29	9.32	95.64
3	7	20.91	18.66	6.10	98.65
4	7.5	23.98	19.19	4.79	99.72
5	8	20.19	16.59	3.89	97.08
6	8.5	18.70	14.40	2.59	96.65

Table 4: Effect of pH on l-lysine production by immobilized cells of *C. glutamicum* MH 20-22 B

Sl. no.	pH	Lysine conc. (p) (g/l)	Biomass (x) (g/l)	Residual glucose conc. (g/l)	Glucose utilized (s) (g/l)
1	6	14.69	15.47	12.86	89.23
2	6.5	19.9	18.68	10.44	91.65
3	7	21.78	19.33	7.34	94.75
4	7.5	25.62	20.42	5.60	96.49
5	8	21.71	17.53	4.41	97.68
6	8.5	19.7	15.62	2.68	99.41

Effect of Glucose Concentration on l-Lysine Production by Immobilized Cells of *C. glutamicum* ATCC 13032 and MH 20-22 B

In bioreactor the cell concentration and glucose concentration plays significant task in the performance of a bioprocess for the production of l-lysine. The immobilized cells use available glucose instantly if the process conditions in the reactor are encouraging. The capability and growth of the cells are dependent on the glucose availability in the reactor. *C. glutamicum* can utilize a variety of carbon sources, such as glucose, fructose, sucrose, and maltose for lysine production. The effect of glucose concentration on the l-lysine yield was expressed on the basis of lysine produced per unit substrate utilized. Effect of glucose concentration on l-Lysine production by immobilized cells of *C. glutamicum* ATCC 13032 and MH 20-22 B was studied by carrying fermentation under conditions like pH 7.5, fermentation time of 96 h, temperature 30 °C, air flow rate of 1.5 vvm, agitation rate of 300 rpm and different ranges of glucose concentrations are 70, 80, 90, 100, 110 and 120 (g/l). From running different batches, it was clearly confirmed and concluded that lysine production is cell-growth associated and the growth of the cells is influenced by glucose concentration. The immobilized cells of *C. glutamicum* ATCC 13032 grown at 90 g/l. of glucose concentration have produced maximum lysine with 25.42 ± 0.23 (g/l) and biomass concentration of 18.34 ± 0.14 (g/l). Glucose concentration on l-lysine production was investigated by Hirose and Shibai (1985). and it was described that higher concentration of glucose inhibited bacterial growth along with low yield Hirose and Shibai 1985. Effect of glucose concentration on l-lysine production by immobilized cells of *C. glutamicum* ATCC is summarized in Table 5. Table 6 clearly illustrates that lysine produced by *C. glutamicum* MH 20-22 B immoblized cells is 26.60 ± 0.17 (g/l) and biomass concentration is of 19.68 ± 0.11 at glucose concentration of 90 g/l. As the glucose concentration is increased simultaneously biomass concentration is also increased. It was also seen in the present experiments any excessive substrate concentration present in the fermentation broth leads to decrease in the product concentration. Hadj Sassi et al. (1988) reported that the early concentration of glucose have impact on the production of l-lysine by *Corynebacterium* Sp. in batch culture and found that the specific production rate was obtained at or above 65 g/l of glucose (Hadj Sassi et al. 1988). It was also observed that pH and substrate concentration has significant effect in comparison to temperature.

Table 5: Effect of glucose concentration on l-lysine production by immobilized cells of *C. glutamicum* ATCC 13032

Sl. no.	Glucose concentration (g/l)	Lysine conc. (p) (g/l)	Biomass (x) (g/l)	Residual glucose conc. (g/l)	Glucose utilized (s) (g/l)
1	70	16.59	5.37	3.97	67.74
2	80	20.11	10.76	5.93	76.55
3	90	25.42	18.34	6.22	84.92
4	100	22.21	17.91	10.61	90.25
5	110	19.07	16.60	15.49	96.54
6	120	15.99	15.05	18.98	97.48

Table 6: Effect of glucose concentration on l-lysine production by immobilized cells of *C. glutamicum* MH 20-22 B

Sl. no.	Substrate conc. (g/l)	Lysine conc. (p) (g/l)	Biomass (x) (g/l)	Residual glucose conc. (g/l)	Glucose utilized (s) (g/l)
1	70	18.5	7.93	4.61	67.48
2	80	21.10	11.38	6.66	75.43
3	90	26.59	19.68	7.47	84.62
4	100	23.7	18.64	12.33	89.76
5	110	20.10	17.46	16.85	97.24
6	120	17.5	16.64	20.66	99.34

Effect of Airflow Rate on l-Lysine Production by Immobilized Cells of *C. glutamicum* ATCC 13032 and MH 20-22 B

In submerged cultures, the oxygen availability influences microbial production of amino acids. Therefore the oxygen plays crucial task in regulation of both intermediately metabolism and biomass formation coupled with alteration of l-lysine synthesis. If excess of oxygen is supplied, it may lead to formation of other metabolites like succinic acids and lactic acid. While the low availability of oxygen decrease the targeted product production. To investigate the optimum value of air flow rate for immobilized cells of *C. glutamicum* ATCC 13032 and MH 20-22 B experimental studies were conducted in a stirred tank bioreactor under conditions pH 7.5, fermentation time of 96 h, temperature 30°C, agitation rate of 300 rpm, glucose concentration of 90 g/l and the aeration rate was maintained at different ranges from 0.25 to 1.5. Tables 7 and 8 showed the characteristics batches of l-lysine fermentation at a range of 0.25–1.5 volume air per volume of medium per minute (vvm). The best air flow rate for l-lysine production by *C. glutamicum* ATCC 13032 immobilized cells is 1.0 vvm at which the lysine concentration is 25.21 ± 0.15 g/l and the biomass productivity was is 19.11 ± 0.19 (g/l) (Table 7). Glucose utililized was 85.56 ± 0.27 g/l at 1.0 vvm. So 1.0 vvm was considered as the optimum airflow rate for *C. glutamicum* ATCC 13032. Wang et al. (1991) worked at 200 rpm for fermentation of l-lysine on rotary shaker and suggested optimum air flow rate of l-lysine production is 1.0 vvm or above 1.0 vvm approximately. Table 8 states that maximum lysine production is 26.82 ± 0.17 (g/l), biomass concentration is 20.93 ± 0.20 (g/l) and glucose utililized is 86.48 ± 0.35 (g/l) for *C. glutamicum* MH 20-22 B immobilized cells. Maximum product concentration, biomass and glucose utilized are good at the air flow rate of 1.0 vvm. Therefore optimum air flow rate for l-lysine production by *C. glutamicum* MH 20-22 B immobilized cells is found to be 1.0 vvm.

Table 7: Effect of airflow rate on l-lysine production by immobilized cells of *C. glutamicum* ATCC 13032

Sl. no.	Air flow rate (vvm)	Lysine conc. (p) (g/l)	Biomass (x) (g/l)	Residual glucose conc. (g/l)	Glucose utilized (s) (g/l)
1	0.25	14.37	9.94	10.68	76.23
2	0.5	17.97	13.17	9.11	79.34
3	0.75	20.13	18.07	5.69	84.32
4	1	25.21	19.11	4.87	85.56
5	1.25	24.12	16.54	3.04	87.34
6	1.5	23.41	14.76	2.79	87.00

Table 8: Effect of airflow rate on l-lysine production by immobilized cells of *C. glutamicum* MH 20-22 B

Sl. no.	Aeration rate (vvm)	Lysine conc. (p) (g/l)	Biomass (x) (g/l)	Residual glucose conc. (g/l)	Glucose utilized (s) (g/l)
1	0.25	16.97	11.39	12.41	78.68
2	0.5	19.68	14.95	10.35	81.56
3	0.75	21.56	19.09	6.96	85.04
4	1	26.85	20.93	5.26	86.47
5	1.25	25.97	18.37	3.53	88.56
6	1.5	24.44	16.78	3.02	88.97

Effect of Agitation Rate on l-Lysine Production by Immobilized Cells of *C. glutamicum* ATCC 13032 and MH 20-22 B

In bioreactor, fermentation medium was agitated to provide homogeneity across the vessel. Agitation and aeration in a stirred tank bioreactor always causes foaming. Excess foaming drives the broth out of the bioreactor and contaminates the system rapidly. So optimum aeration and agitation must be operated which minimizes foaming and maximizes the lysine production rate. l-lysine production at different agitation rates of 100, 150, 200, 300, 350, 400 rpm under different parameters like pH 7.5, fermentation time of 96 h, temperature of 30 °C, aeration rate of 1.0 vvm and glucose concentration of 90 g/l were investigated to find out the optimum agitation rate by immobilized cells of *C. glutamicum* ATCC and MH 20-22 B. From Table 9 we can observe that maximum l-lysine concentration is 30.17 ± 0.19 g/l and biomass is 16.97 ± 0.19 g/l at 200 rpm. Therefore optimum agitation rate for high l-lysine production is 200 rpm for immobilized cells of *C. glutamicum* ATCC 13032. Shah et al. (2002) investigated about influence of agitation rate and reported that the optimum range is between 50 and 300 rpm (Shah et al. 2002). Similar studies were carried out by Razak and Viswanath for biotechnological production of l-lysine by stirred tank bioreactor and they reported that optimum agitation rate of l-lysine production is 200 rpm for *C. glutamicum* MH 20-22 B (Razak and viswanath 2014). Table 10 shows that the highest lysine yield is 31.58 ± 0.24 (g/l) and maximum biomass concentration is 17.73 ± 0.18 (g/l) for immobilized cells of *C. glutamicum* MH 20-22 B at 200 rpm.

Table 9: Effect of agitation rate on l-lysine production by by immobilized cells of *C. glutamicum* ATCC 13032

Sl. no.	Agitation rate (rpm)	Lysine conc. (p) (g/l)	Biomass (x) (g/l)	Residual glucose conc. (g/l)	Glucose utilized (s) (g/l)
1	100	21.45	11.87	9.37	93.55
2	150	25.10	14.54	7.19	94.95
3	200	30.17	16.97	3.99	96.68
4	250	27.32	15.11	6.14	97.35
5	300	25.98	16.40	5.43	98.85
6	350	23.41	15.31	8.97	97.16
7	400	22.72	13.96	10.20	95.33

Table 10: Effect of agitation rate on l-lysine production by by immobilized cells of *C. glutamicum* MH 20-22 B

Sl. no.	Agitation rate (rpm)	Lysine conc. (p) (g/l)	Biomass (x) (g/l)	Residual glucose conc. (g/l)	Glucose utilized (s) (g/l)
1	100	23.63	13.45	10.22	80.87
2	150	26.09	15.79	8.41	83.65
3	200	31.58	17.72	4.07	86.03
4	250	28.49	16.77	7.48	84.16
5	300	27.94	17.39	6.82	83.27
6	350	25.78	16.94	10.67	81.42
7	400	23.33	15.09	11.45	80.64

CONCLUSIONS

In order to improve the production rate of l-lysine, immobilized cell of *C. glutamicum* ATCC 13032 and MH 20-22 B were analyzed under various physical and chemical process parameters. Taken together, the results of this study indicated that the optimum values of fermentation time, pH, temperature, glucose concentration, airflow rate and aeration rate were 96 h, 7.5, 30 °C, 90 g/l, 1.0 vvm and 200 rpm respectively for immobilized cells (Table 11). From the results obtained from this investigation we conclude that the immobilized cells of the *C. glutamicum* mutant MH 20-22 B produced more lysine compared to the wild type *C. glutamicum* ATCC 13032. The present study reveals that the bioreactor studies for the optimization of different physical and chemical parameters for the maximum production of l-lysine can also be extended to the fluidized bed bioreactor. The advantage of the bioprocess based on immobilized cells include enhancing microbial cell stability, allowing continuous process operation and avoiding the biomass— liquid separation requirement. Immobilizing *C. glutamicum* cells is one of the bioprocess engineering approaches for improving biotechnological production of l-Lysine. Based on the present studies further studies may be carried out by genetically modified strains to evaluate physical and chemical process parameters for l-lysine production.

Table 11: Optimized fermentation parameters for immobilized cells of *C. glutamicum* *C. glutamicum* ATCC 1303 and MH 20-22 B

Fermentation conditions	Immobilized cells of *C. glutamicum* ATCC 13032	Immobilized cells of *C. glutamicum* MH 20-22 B
Fermentation time (h)	96	96
pH	7.5	7.5
Temperature (°C)	30	30
Glucose concentration (g/l)	90	90
Airflow rate (vvm)	1.0	1.0
Aeration rate (rpm)	200	200

ACKNOWLEDGMENTS

Authors are grateful to NATCO for all their encouragement in fermentation studies. Authors want to thank anonymous reviewers for their valuable suggestions to improve this paper.

REFERENCES

1. Amin GA, Al-Talhi A (2007) Production of l-Glutamic acid by immobilized cell reactor of the bacterium*Corynebacterium glutamicum* entrapped into carrageenan gel beads. World Appl Sci J 2(1):62–67

2. Anastassiadis S (2007) l-Lysine fermentation. Recent Pat Biotechnol 1(1):11–24

3. Araki K, Kato F, Aral Y, Nakayama K (1974) Histidine production by auxotrophic histidine analog-resistant mutants of *Corynebacterium glutamicum*. Agri Biol Chem 38:837

4. Becker MJ (1982) Product biosynthesis in continuous fermentation. Folia Microbiol 27:315–318

5. Bercovici D, Fuller MF (2007) Industrial amino acids in non ruminant animal nutrition. In: Wallace RJ, Chesson A (eds) Biotechnology in animal feeds and animal feeding. Wiley-VCH, Weinheim. doi:10.1002/9783527615353.ch6

6. Chinard FP (1952) Photometric estimation of proline and ormithine. J Biol Chem 199:91–95

7. Eggeling L (2007) l-Serine and glycine. In: Wendisch VF (ed) Amino acid biosynthesis pathway, regulation and metabolic engineering, vol 5. Springer, Berlin, p 259–272

8. Guillout S, Rodal AA, Lessard PA, Sinskey AJ (2002) Methods for producing l-isoleucine. USA patent (US 6451564)

9. Hadj Sassi A, Fauvart L, Deschamps AM, Lebeault JM (1988) Fed batch production of l-Lysine by*Corynebacterium glutamicum*. Biochem Eng J 1:85–90

10. Hilliger M, Haenel F, Menz J (1984) Influence of temperature on growth and l-Lysine formation in *Corynebacterium glutamicum*. J Appl Microbiol 24:437–441

11. Hirose Y, Shibai H (1985) l-Glutamic acid fermentation. Compreh Biotechnol 3:595–600

12. Ikeda M (2006) Towards bacterial strains overproducing l-tryptophan and other aromatics by metabolic engineering. Appl Microbiol Biotechnol 69:615–626

13. Ikeda M, Katsumata R (1992) Metabolic engineering to produce tyrosine or phenylalanine in a tryptophan producing *Corynebacterium glutamicum* strain. Appl Env Microbiol 58:781–785

14. Jetty A, Gangagni Rao A, Sarva Rao B, Madhavi G, Ramakrishna SV (2005) Comparative studies of ALR and FBR for Streptomycin production by immobilized cells of *Streptomyces* Sp. Chem Biochem Eng 19:179–184

15. Kalinowski J, Bathe B, Bartels D, Bischoff N, Bott M, Burkovski A, Dusch N, Eggeling L, Eikmanns BJ, Gaigalat L (2003) The complete *Corynebacterium glutamicum* ATCC 13032 genome sequence and its impact on the production of aspartate derived amino acids and vitamins. J Biotechnol 104:5–25

16. Kase H, Nakayama K (1974) Studies on l-threonine fermentation, mechanism of l-threonine and l-lysine production by analog-resistant mutants of *Corynebacterium glutamicum*. Agric Biol Chem 38:993–1000

17. Kinoshita S, Udaka S, Shimono M (1957) Amino acid fermentation I. Production of l-glutamic acid by various microorganism. J Gen Appl Microbiol 3:193–205

18. Kinoshita S, Nakayama K, Akita S (1958) Taxonomical study of glutamic acid accumulating bacteria, *Micrococcus glutamicus*. Agri Chem Soc Jpn 22:176

19. Leuchtenberger W, Huthmacher K, Drauz K (2005) Biotechnological production of amino acids and derivatives: current status and prospects. Appl Microbiol Biotechnol 69(1):1–8

20. Marek PJ, Kierstan M, Coughlan MP (1985) Immobilized cells and enzymes. Practical approaches UK, Ireland press limited 43–44

21. Michalski HJ, Krzystek L, Blazczyk R, Jamroz T, Wieczorek A (1984) The effect of mean residence time and aeration intensity on the l-Lysine production in a continuous culture. In: Third Europian congress on biotechnology, Verlag Chemie, Weinheim, 2, p 527–532

22. Miller GL (1959) Use of dinitrosalicylic acid reagent for detection of reducing sugar. Ann Chem 31:427–431

23. Morris DL (1948) Microbial physiology. Wiley, New York 161

24. Nakayama HV, Araki K (1973). Process for producing l-lysine, US patent (3708395)

25. Neish AC (1952) Analytical methods for bacterial fermentation. Manual, published by the National Research Council of Canada

26. Patek M (2007) Branched chain amino acids biosynthesis pathways, regulation and metabolic engineering. Springer, Heidelberg 778

27. Pfefferle W, Moeckel B, Bathe B, Marx A (2003) Biotechnological manufacture of l-Lysine. In Scheper T (ed) Advances in biochemical engineering/biotechnology, vol 79. Springer, Berlin, p 59

28. Razak MA, Viswanath B (2014) Optimization of fermentation upstream parameters and immobilization of*Corynebacterium glutamicum* MH 20-22 B cells to enhance the production of l-Lysine. 3 Biotech. doi 10.1007/s13205-014-0252-7

29. Ruklisha M, Paegle L, Denina I (2007) L-Valine biosynthesis during batch and fed batch cultivations of*Corynebacterium glutamicum*: relationship between changes and bacterial growth rate and intracellular metabolism. Prog Biochem 42:634–640

30. Shah AH, Hameed A, Khan GM (2002) Fermentative production of l-lysine, bacterial fermentation. J Med Sci 2:152–157

31. Shiio I (1990) Threonine production by dihydrodipicolinate synthase-defective mutants of *Brevibacterium flavum*. Biotechnol Adv 8:97–103

32. Shiio I, Toride Y, Yokota A, Sugimoto S, Kawamura K (1991) Process for the production of l-threonine by fermentation, USA Patent (5077207)

33. Tryfona T, Bustard MT (2005) Fermentation production of lysine by *Corynebacterium glutamicum*: transmembrane transport and metabolite flux analysis. Process Biochem 40(2):499–508

34. Udaka S (1960) Screening method for microorganisms accumulating metabolites and its use in the isolation of*Micrococcus glutamicus*. J Bacteriol 79:754–755

35. Wang JS, Kuo YC, Chang CC, Liu YT (1991) Optimization of culture conditions of l-lysine fermentation by*Brevibacterium species*. Nat Biotechnol 68:154–159

36. Wendisch VF, Bott M (2005) In Handbook of *Corynebacterium glutamicum, CRC Press*. Taylor & Francis, Boca Raton, pp 377–396

Chapter 8

IMMOBILIZATION OF ANODE-ATTACHED MICROBES IN A MICROBIAL FUEL CELL

Rachel C Wagner, Sikandar Porter-Gill and Bruce E Logan

Department of Civil and Environmental Engineering, 212 Sackett Building, The Pennsylvania State University, University Park, PA 16802, USA

ABSTRACT

Current-generating (exoelectrogenic) bacteria in bioelectrochemical systems (BESs) may not be culturable using standard *in vitro* agar-plating techniques, making isolation of new microbes a challenge. More *in vivo* like conditions are needed where bacteria can be grown and directly isolated on an electrode. While colonies can be developed from single cells on an electrode, the cells must be immobilized after being placed on the surface. Here we present a proof-of-concept immobilization approach that allows exoelectrogenic activity of cells on an electrode based on applying a layer of latex to hold bacteria on surfaces. The effectiveness of this procedure to immobilize particles was first demonstrated using fluorescent microspheres as bacterial analogs. The latex coating was then shown to not substantially affect the exoelectrogenic activity of well-developed anode biofilms in two different systems. A single layer of airbrushed coating did not reduce the voltage produced by a biofilm in a microbial fuel cell (MFC), and more easily applied dip-and-blot coating reduced voltage by only 11% in a microbial electrolysis cell (MEC). This latex immobilization procedure will enable future testing of single cells for exoelectrogenic activity on electrodes in BESs.

INTRODUCTION

Bioelectrochemical systems (BESs) are based on electron transfer between microbes and an electrode surface. Most investigations into the mechanisms of electron transfer from a microbe to an anode have focused on two microorganisms, *Geobacter sulfurreducens* (Marsili et al. 2008; Holmes et al. 2006; Strycharz et al. 2010; Inoue et al. 2010; Nevin et al. 2009; Srikanth et al. 2008) and *Shewanella oneidensis*(Bretschger et al. 2007; Gorby et al. 2006), where it has been shown that specific genes and proteins are involved in exogenous electron transfer. Further study of current-generating (exoelectrogenic) bacteria and biofilms will benefit from isolating and identifying other microorganisms that are capable of electron transfer to an electrode.

Isolation techniques to identify novel exoelectrogens have typically involved dilution-to-extinction in BESs, or isolation on ferric iron agar plates. A U-tube reactor was developed (Zuo et al. 2008) that would allow a single microbe, obtained by serial dilutions, to deposit by sedimentation onto a flat anode surface. This technique was used to identify novel exoelectrogens *Ochrobactrum anthropi* YZ-1 (Zuo et al. 2008) and *Enterobacter cloacae* FR (Rezaei et al. 2009). However, the cumbersome process required many serial transfers to obtain these isolates. A microbe related to *Clostridium butyricum* was isolated from a microbial fuel cell (MFC) using ferric iron agar plates (Park et al. 2001), but this method of isolation does not target all exoelectrogens as some microbes have been isolated that can generate current but not reduce iron (Kim et al. 2004; Zuo et al. 2008).

In addition to spread-plating techniques, screening of arrays of microorganisms on ferric iron agar plates is possible through printer technology (Ringeisen et al. 2009). This approach can be used to print very small droplets of a cell suspension diluted to contain single microbes. To take advantage of this technology, for example by printing single cells in a grid pattern onto an electrode for isolation, a robust immobilization layer is required to bind the cells to the electrode so that they do not move after application to the electrode surface. This layer should not interfere with the ability of microbes to transfer electrons to an electrode surface, or with the diffusion of substrate to the cells. Latex films were evaluated here to see if they could be used to fulfill these requirements. Latex films have previously been used to entrap microbes on non-conducting surfaces, producing a high density of organisms in a thin film that survived freezing and drying (Gosse et al. 2007; Lyngberg et al. 1999; Flickinger et al. 2007). We show here effective entrapment of bacteria-sized particles using fluorescent microspheres, and demonstrate that latex entrapped anode biofilms allow exoelectrogenic activity.

MATERIALS AND METHODS

Latex was applied to two different types of anodes, carbon paper (without wet proofing; E-Tek) or graphite blocks (Grade GM-10; GraphiteStore.com Inc.), in two different types of BESs in order to evaluate the immobilization method under different conditions. Carbon paper was used as the anode in a single-chamber 28-mL microbial fuel cell (MFC) reactor with a platinum-catalyzed air cathode (Cheng et al. 2006; Liu and Logan 2004) (both electrodes with projected surface area of 7 cm^2). Graphite blocks (projected surface area of 4.6 cm^2) were used as anodes for a single-chamber 5-mL microbial electrolysis cells (MECs) with a 1.0×1.5 cm^2 304 stainless steel 90×90 mesh cathode (Call and Logan 2011). Carbon paper (projected surface area of 3.0 cm^2) was also used as anode material in some 5-mL MECs. All reactors were inoculated using cell suspensions from pre-acclimated MFCs that were originally inoculated with domestic wastewater and acetate. A multimeter (2700, Keithley Instruments, Inc.) was used to monitor the voltage across an external resistor ($R_{ex} = 10$ Ω, MEC; 1000 Ω, MFC). A power source (3645A, Circuit Specialists, Inc.) was connected to the MEC circuit to add -0.7 V to the cathode. All BESs were maintained at 30°C.

MFC medium was 100 mM phosphate buffer with 17 mM acetate as the substrate (per L: 0.62 g NH$_4$Cl, 4.9 g NaH$_2$PO$_4$·H$_2$O, 9.15 g Na$_2$HPO$_4$, 0.26 g KCl, 1.4 g sodium acetate, and Wolfe's vitamins and minerals) (Lovley and Phillips 1988). MEC medium was 30-mM bicarbonate buffer with 10-mM acetate as the substrate, based on the ATCC recipe for *G. sulfurreducens*, #1957 (per L: 1.5 g NH$_4$Cl, 0.6 g NaH$_2$PO$_4$, 0.1 g KCl, 2.5 g NaHCO$_3$, 0.82 g sodium acetate, and Wolfe's vitamins and minerals), without the addition of the electron acceptor. MFC and MEC reactors were operated in fed-batch mode until they successively produced at least 3 equivalent batch cycles, indicating a well-established anodic biofilm.

A monodisperse latex emulsion (SF-091; Rohm & Haas) was amended with 5% glycerol to optimize the degree of coalescence and subsequent diffusivity of the film to the substrate (Lyngberg et al. 2001; Gosse et al. 2007). This solution was applied in two different ways to well-established biofilms in the different BESs by removing the anodes temporarily from the reactors. Glycerol-amended latex (referred to simply as "latex") was applied to the carbon paper biofilm from the MFC using an air brush (Paache, BearAir, S. Easton, MA; 4.5 L/min of airflow). One, three, or five layers were applied, allowing 15 minutes between each layer, and one hour after the final layer, for drying at room temperature. For the graphite blocks and carbon paper anodes from the MEC a simpler application procedure was used, where the latex was applied by dipping the blocks or paper into the latex, and excess solution was drawn off the anode with a laboratory wipe. In other experiments, the glycerol-amended latex was diluted in water to 30% to see if performance improved with a thinner layer of latex.

The effectiveness of the latex to immobilize bacteria on the anode materials was examined using several different techniques. Direct observation of individual bacteria on an electrode, when bacteria were stained using acridine orange, was not possible due to high levels of background fluorescence. Therefore, application of individual microbes on an electrode was simulated by applying droplets of fluorescent microspheres (Fluoresbrite spheres, 4.1-μm diameter, Invitrogen) to graphite electrodes. Latex was applied by the dipping method described above. After drying, the latex-coated electrode was immersed in MFC media to simulate the electrode in a BES. The droplets were observed with fluorescence microscopy before and after latex application and MFC simulation.

For SEM visualization, small sections of carbon paper anodes with exoelectrogenic biofilms with and without latex coating were mounted in cryo-matrix and frozen. Thin slices were removed from the cross-section with a microtome until a smooth surface was obtained. The surface was etched with the cryo-SEM electron beam to remove ice crystals before viewing.

RESULTS

Latex Preparation

Application of glycerol-amended latex with the airbrush resulted in ~2.1 mg dry weight of latex per cm^2 anode area per layer. Application by dipping and blotting of the glycerol-amended latex onto graphite block resulted in ~5.3 mg/cm^2/layer for 100% latex-glycerol, and 0.67 mg/cm^2/layer for 30% latex-glycerol. On carbon paper, ~8.1 mg/cm^2/layer was applied for 100% latex-glycerol, and 2.5 mg/cm^2/layer for 30% latex-glycerol.

Immobilization of Microspheres and Microbes

Fluorescent microspheres are often used as analogs for microorganisms (Smith and McKay 2005; Solomon and Matthews 2005). The location and shape of a droplet of microspheres (4.1-μm diameter) on an electrode were retained after latex application and drying, and after submersion in standard MFC media.

The latex film applied with an airbrush to an exoelectrogenic biofilm on a carbon paper anode remained completely intact, without dissolving or cracking, after 6 cycles in an MFC (Figure 1). The layers of latex coalesced into one continuous overlay. The biofilm was not visible in SEM images due to preparation requirements for the latex; however, the presence of the biofilm was confirmed by the exoelectrogenic activity through current production in the MFC. The latex layer applied to the MEC carbon paper anode using the dip-and-blot method also remained visibly intact throughout the experiment. The latex layer applied to the graphite block with the dip-and-blot method had variable performance. The layer made using the 30% dilution remained intact.

However, at full strength, the latex layer did not consistently remain adhered to the block, and in some reactors, the latex began to peel off after ten days.

Figure 1: SEM image of 3 layers of latex ("latex overlay"; approximately 165 **μm** thick) on a carbon paper anode with exoelectrogenic biofilm after 6 cycles in an MFC. The biofilm is not visible due to SEM preparation techniques necessary to maintain the latex layer.

Latex coatings on Anode Biofilms

When one layer of glycerol-amended latex was applied with the airbrush to a biofilm on carbon paper in an MFC, the reactor recovered immediately to its pre-latex voltage. When three layers were applied, the reactor returned to its original performance in 6 cycles. However, when five layers were applied, the MFC only reached 45% of its original voltage even after 6 cycles (Figure 2).

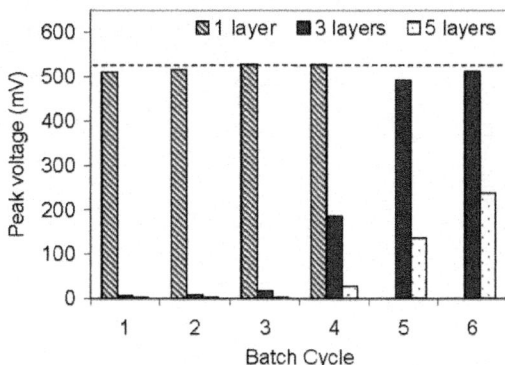

Figure 2: An exoelectrogenic biofilm on a carbon paper anode in an MFC with 1, 3, or 5 layers of latex applied to a carbon paper anode using an airbrush, compared to a reactor with no latex (dashed line). Representative reactors are shown.

Using undiluted glycerol-amended latex for immobilization of microbes on a graphite block, the MEC with graphite block anode returned to 42% (± 8%) of its original current within three cycles of latex application by dipping and blotting. However, after three cycles, which took approximately 10 days, the overlay had started to delaminate from the graphite block, so testing was discontinued. Using a 30% dilution of the latex-glycerol, the current recovery in the MECs improved, reaching 85% (± 9%) of the original current within 3 cycles of latex application with consistent results over 3 additional cycles (Figure 3). In addition, the latex remained adhered to the anode.

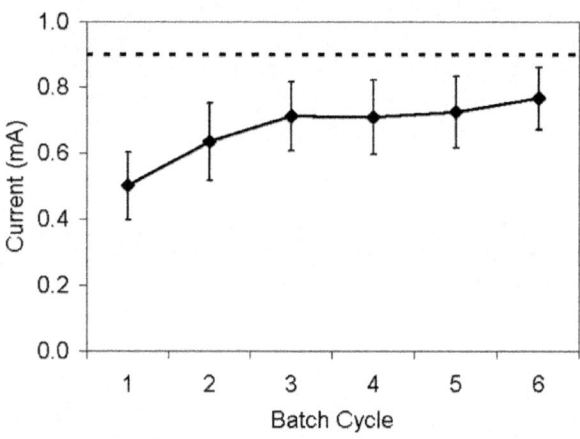

Figure 3: An exoelectrogenic biofilm on a graphite block anode in an MEC immobilized with glycerol-amended latex diluted to 30% strength, compared to the biofilm with no overlay (dashed line).

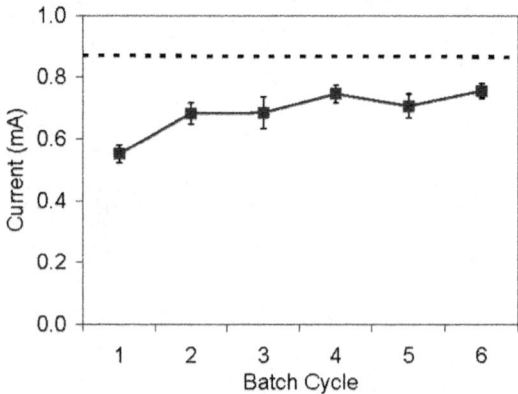

Figure 4: An exoelectrogenic biofilm on a carbon paper anode in an MEC immobilized with glycerol-amended latex diluted to 30% strength, compared to the biofilm with no overlay (dashed line).

When the undiluted overlay was applied to carbon paper anodes in the MECs, current returned to 43% of the original level within 3 cycles and was maintained in further batches. With the thinner, 30% diluted layer, current returned to 89% (± 6%) of the original current within 3 cycles of application and remained consistent in subsequent batches (Figure 4).

DISCUSSION

Latex films were shown be effective in holding individual particles (fluorescent microspheres) or active biofilms on electrically conductive surfaces. Microbes trapped on two different surfaces (carbon paper and graphite block) using different application methods (airbrushing and dip-and-blot) retained most of their exoelectrogenic capability. On both surfaces, and in both MFC and MEC reactors, increasing the amount of latex applied onto the biofilm adversely affected the ability of the anode to recover exoelectrogenic activity to pre-application current levels (Lyngberg et al. (2001)). found that effective diffusivity through the latex was highly dependent on layer thickness. Therefore, this decrease in activity was likely due to a reduction in mass transfer to (substrate) and from (protons) the biofilm with thicker layers of latex.

The latex coating thickness, measured by dry weight, on the graphite block was less than that of the graphite paper, and the full strength latex coating did not stick well to the block. The coating on the carbon paper when applied by the air brush to the MFC anode or the dip-and-blot method (at 30% strength) to the MEC anode was similar (slightly more than 2 mg/cm^2/layer). While the MFC regained 100% of its pre-application performance, the MEC was limited to about 89% of its pre-application performance. It is unlikely that there was any decrease in the performance of the MEC in these experiments due to exposure of the biofilm to oxygen during the latex application, as MEC biofilms are routinely exposed to air when it they are refilled (often intentionally to reduce methanogenesis) without adverse affects to current production (Call and Logan 2008). In addition, the biofilm in an MFC is routinely exposed to oxygen in air due to oxygen diffusion through the cathode and into the anode chamber without apparent adverse effects. If desired, the latex film could be applied under strictly anoxic conditions in an anaerobic glove box. Previous work with bio-catalytic films used for hydrogen gas production has shown that the coating itself is not adversely affected by the presence or absence of air, nor is the performance of that biofilm (Gosse et al. 2007). However, it is possible that some strict anaerobes might be affected by oxygen during this procedure, so anaerobic application of the latex biofilm may be of interest in future studies.

The ability to immobilize microbes on an electrode using a latex film has two valuable applications for BESs, but for successful application in BESs, immobilization of microbes on electrodes must not interfere with the ability of cells to transfer electrons. Bioelectrochemical features seen in cyclic voltammograms of pectin-entrapped *Geobacter* biofilms have been shown to be similar to naturally-grown *Geobacter* biofilms (Srikanth et al. 2007). This suggests that entrapment by itself is not changing the electrical capability of the cells, although they found current was somewhat decreased as observed here as well. One application of an immobilization layer for cells on a BES electrode is isolation of microbes directly on an electrode. This requires immobilization of an array of single cells, without greatly compromising current generation, which our latex overlay achieves. In addition, a biofilm of specific microbes can be developed on an electrode in a controlled setting, immobilized and protected under a latex coating, and then introduced to a more complex, non-sterile environment. Under the coating, these organisms would not have to compete with other microbes for the electron-accepting surface. Exoelectrogenic biofilm activity under a glycerol-amended latex film can be restored to nearly the same levels as pre-application activity, making it a suitable immobilization layer for these applications.

ACKNOWLEDGEMENTS

This material is based upon work supported under National Science Foundation Graduate Research Fellowships (RCW), and award KUS-I1-003-13 by King Abdullah University of Science and Technology (KAUST).

REFERENCES

1. Bretschger O, Obraztsova A, Sturm CA, Chang IS, Gorby YA, Reed SB, Culley DE, Reardon CL, Barua S, Romine MF, Zhou J, Beliaev AS, Bouhenni R, Saffarini D, Mansfeld F, Kim B-H, Fredrickson JK, Nealson KH: Current production and metal oxide reduction by*Shewanella oneidensis* MR-1 wild type and mutants. *Appl Environ Microbiol* 2007,73(21):7003–7012.

2. Call DF, Logan BE: A method for high throughput bioelectrochemical research based on small scale microbial electrolysis cells.*Biosens Bioelectron* 2011,26(11):4526–4531.

3. Call D, Logan BE: Hydrogen production in a single chamber microbial electrolysis cell (MEC) lacking a membrane. *Environ Sci Technol* 2008,42(9):3401–3406.

4. Cheng S, Liu H, Logan BE: Increased performance of single-chamber microbial fuel cells using an improved cathode structure.*Electrochemistry Communications* 2006, 8:489–494.

5. Flickinger MC, Schottel JL, Bond DR, Aksan A, Scriven LE: Painting and printing living bacteria: engineering nanoporous biocatalytic coatings to preserve microbial viability and intensify reactivity. *Biotechnol Prog* 2007,23(1):2–17.

6. Gorby YA, Yanina S, McLean JS, Rosso KM, Moyles D, Dohnalkova A, Beveridge TJ, Chang IS, Kim BH, Kim KS, Culley DE, Reed SB, Romine MF, Saffarini DA, Hill EA, Shi L, Elias DA, Kennedy DW, Pinchuk G, Watanabe K, Ishii S, Logan BE, Nealson KA, Fredrickson JK:Electrically conductive bacterial nanowires produced by *Shewanella oneidensis* strain MR-1 and other microorganisms. *Proc Natl Acad Sci* 2006,103(30):11358–11363.

7. Gosse JL, Engel BJ, Rey FE, Harwood CS, Scriven LE, Flickinger MC: Hydrogen production by photoreactive nanoporous latex coatings of nongrowing *Rhodopseudomonas palustris* CGA009. *Biotechnol Prog* 2007,23(1):124–130.

8. Kim BH, Park HS, Kim Kj, Kim GT, Chang IS, Lee J, Phung NT: Enrichment of microbial community generating electricity using a fuel-cell-type electrochemical cell. *Appl Microbiol Biotechnol* 2004,63(6):672–681.

9. Liu H, Logan BE: Electricity generation using an air-cathode single chamber microbial fuel cell in the presence and absence of a proton exchange membrane. *Environ Sci Technol* 2004,38(14):4040–4046.

10. Lovley DR, Phillips EJP: Novel Mode of Microbial Energy Metabolism: Organic Carbon Oxidation Coupled to Dissimilatory Reduction of Iron or Manganese. *Appl Environ Microbiol* 1988,54(6):1472–1480.

11. Lyngberg OK, Ng CP, Thiagarajan V, Scriven LE, Flickinger MC: Engineering the microstructure and permeability of thin multilayer latex biocatalytic coatings containing *E. coli* . *Biotechnol Prog* 2001,17(6):1169–1179.

12. Lyngberg OK, Thiagarajan V, Stemke DJ, Schottel JL, Scriven LE, Flickinger MC: A patch coating method for preparing biocatalytic films of *Escherichia coli* . *Biotechnol Bioeng* 1999,62(1):44–55.

13. Park HS, Kim BH, Kim HS, Kim HJ, Kim GT, Kim M, Chang IS, Park YK, Chang HI: A novel electrochemically active and Fe(III)-reducing bacterium phylogenetically related to *Clostridium butyricum* isolated from a microbial fuel cell. *Anaerobe*2001,7(6):297–306.

14. Rezaei F, Xing D, Wagner RC, Regan JM, Richard TL, Logan BE: Simultaneous cellulose degradation and electricity production by*Enterobacter cloacae* in an MFC. *Appl Environ Microbiol* 2009,75(11):3673–3678.

15. Ringeisen BR, Lizewski SE, Fitzgerald LA, Biffinger JC, Knight CL, Crookes-Goodson WJ, Wu PK: Single cell isolation of bacteria from microbial fuel cells and Potomac River sediment. *Electroanalysis* 2009,22(7–8):875–882.

16. Smith HD, McKay CP: Drilling in ancient permafrost on Mars for evidence of a second genesis of life. *Planetary and Space Science*2005,53(12):1302–1308.

17. Solomon EB, Matthews KR: Use of fluorescent microspheres as a tool to investigate bacterial interactions with growing plants. *J Food Prot* 2005,68(4):870–873.

18. Srikanth S, Marsili E, Flickinger MC, Bond DR: Electrochemical characterization of *Geobacter sulfurreducens* cells immobilized on graphite paper electrodes. *Biotechnol Bioeng* 2007,99(5):1065–1073.

19. Zuo Y, Xing D, Regan JM, Logan BE: Isolation of the exoelectrogenic bacterium *Ochrobactrum anthropi* YZ-1 by using a U-tube microbial fuel cell. *Appl Environ Microbiol* 2008,74(10):3130–3137.

Chapter 9

DESIGN AND INVESTIGATION OF POLYFERMS IN VITRO CONTINUOUS FERMENTATION MODELS INOCULATED WITH IMMOBILIZED FECAL MICROBIOTA MIMICKING THE ELDERLY COLON

Sophie Fehlbaum[1], Christophe Chassard[1], Martina C. Haug[1], Candice Fourmestraux[2], Muriel Derrien[2], Christophe Lacroix[1]

[1] Laboratory of Food Biotechnology, Institute of Food, Nutrition and Health, ETH Zurich, Zurich, Switzerland

[2] Danone Nutricia Research, Palaiseau, France

ABSTRACT

In vitro gut modeling is a useful approach to investigate some factors and mechanisms of the gut microbiota independent of the effects of the host. This study tested the use of immobilized fecal microbiota to develop different designs of continuous colonic fermentation models mimicking elderly gut fermentation. Model 1 was a three-stage fermentation mimicking the proximal, transverse and distal colon. Models 2 and 3 were based on the new PolyFermS platform composed of an inoculum reactor seeded with immobilized fecal microbiota and used to continuously inoculate with the same microbiota different second-stage reactors mounted in parallel. The main gut bacterial groups, microbial diversity and metabolite production were monitored in effluents of all reactors using quantitative PCR, 16S rRNA gene 454-pyrosequencing, and HPLC, respectively. In all models, a diverse microbiota resembling the one tested in donor's fecal sample was established. Metabolic stability in inoculum reactors seeded with immobilized fecal microbiota was shown for operation times of up to 80 days. A high microbial and metabolic reproducibility was demonstrated for downstream control and experimental reactors of a PolyFermS model. The PolyFermS models tested here are particularly suited to investigate the effects of environmental factors, such as diet and drugs, in a controlled setting with the same microbiota source.

INTRODUCTION

The human colon harbors a large number of microbes forming a complex ecosystem responsible for various processes in the host. Under normal conditions, the gut microbiota acts as a barrier against enteropathogens, contributes to the development of the immune system and exerts important metabolic functions; which includes the production of short chain fatty acids (SCFA; such as acetate, propionate and butyrate) by breaking down complex carbohydrates that provide energy to epithelial cells and to the host [1, 2]. Each human harbors a unique gut microbiota composition consisting of bacteria belonging mainly to the phyla Firmicutes or Bacteroidetes and, to a lesser extent, to Actinobacteria, Proteobacteria and Verrucomicrobia [3, 4]. Colonization of the gut occurs first during birth, and throughout the first 2–3 years of life the microbial composition becomes established towards an adult-like microbiota. Recent studies indicate that the gut microbiota remains stable in adulthood, except for temporary alterations due to diet, disease and antibiotic treatment. However, an important shift in the microbial composition occurs during old age that is associated with a reduction in stability and often in biodiversity [3–7]. Additionally, a large inter-individual variability of the gut microbiota composition was reported for elderly Irish subjects of a large-scale *in vivo* study, with pyrosequencing reads assigned to the phyla *Bacteroidetes* and *Firmicutes* ranging from 3 to 94% [8]. To date, no common core microbiota for the elderly was defined, partly due to the various physiological factors, including lifestyle, diet and need for medications that change in old age. Thus, establishing the changes in the composition with ageing still require further investigations [4, 7].

Intestinal fermentation models allow the *in vitro* cultivation of gut microbiota to study their composition and function, uncoupled from the host. As such, models provide greater control, easier manipulation, and no ethical restrictions relative to *in vivo* studies, and are very complementary to *in vivo* strategies for elucidating mechanisms of gut microbiota [9]. Intestinal models have developed from batch for short-term fermentation studies to single or multistage continuous models that allow long-term studies due to substrate replenishment and toxic product removal [10]. However, one of the main challenges of the continuous culture models is the reproduction of the biofilm-associated microbes of the gut that is important to prevent washout of the less competitive bacteria. Immobilization of gut microbiota in gellan-xanthan gel beads has shown to reproduce the free and biofilm associated states of bacterial populations and to maintain the bacterial diversity at high cell densities in continuous intestinal reactors over periods of up to 71 days [10–12]. Furthermore, reproducibility and biological replication of continuous intestinal models was recently improved by the introduction of the PolyFermS model that allows the parallel testing of treatments with the same gut microbiota, and which has been validated for the child and the swine proximal colon [13, 14].

In vitro intestinal fermentations models have been developed and validated [10] to investigate factors of microbiota composition and metabolism of infants to adults while the elderly gut microbiota was only scarcely analyzed. Several studies were performed in continuous three-stage models for investigating the effects of antibiotics on *Clostridium difficile* infection. For these studies *C. difficile* was inoculated with mixed fecal samples from multiple elder donors and the system was challenged with antibiotics to promote the germination and growth of the sporulated bacteria, while microbiota analysis was only done with cultivation [15–18]. In a recent study batch cultures and continuous three-stage models inoculated with microbiota from single fecal samples of elder donors were used to investigate probiotics, prebiotics and synbiotics. Fluorescent *in situ* hybridization methods were used to monitor gut microbiota composition [19]. To date, no study has reported a detailed analysis of gut microbiota establishment and diversity in *in vitro* fermentation models reproducing the gut of aged (over 65 years) people.

The aim of this study was to investigate the use of immobilized fecal microbiota to develop different designs of continuous colonic fermentation models mimicking elderly gut fermentation. Immobilization of fecal microbiota obtained from three different donors was performed independently. Fecal beads were used to inoculate an immobilized cell reactor (IR) operated with conditions selected to mimic the proximal colon section of an elder. Three model designs, all starting with an IR used to generate a constant gut microbiota composition in proximal colon conditions, were tested for different experimental questions. These models were set in sequential order with the aim to investigate colonization of *Clostridium difficile* (data not shown in this paper). Model 1 was based on the three-stage design, with immobilized microbiota inoculated in a first proximal colon reactor connected to a transverse and a distal colon reactor, previously validated for infant and child microbiota fermentation [12, 20, 21]. Models 2 and 3 were developed based on the PolyFermS platform, which recently has been validated with child [13] and swine [14] microbiota, with adjusted conditions for the elderly microbiota. In model 2 IR containing immobilized fecal microbiota was used to inoculate (10% v/v) two parallel sets of 2-stage reactors mimicking the proximal and distal colon. Because *C. difficile* growth was only detected in distal colon reactors, in model 3 IR was used to feed (100% v/v) five reactors mounted in parallel, and mimicking conditions of the distal colon. This design allowed to test in parallel four treatments compared to a control in distal colon reactors.

The microbiota composition in reactor effluents was monitored and compared to that of the corresponding fecal donor, and temporal stability of the models and reproducibility of downstream reactors within a PolyFermS model were demonstrated. Microbiota composition, diversity (qPCR and pyrosequencing) and activity (HPLC) were monitored in reactor effluents over operation periods of up to 80 days.

MATERIALS AND METHODS

Ethics Statement

The Ethics Committee of ETH Zurich exempted this study from review because sample collection was not in terms of intervention. An informed written consent was, however, obtained from the fecal donors.

Fecal Inoculum and Immobilization

For each fermentation experiment a fresh fecal sample from a different donor was used for the immobilization procedure. Fecal samples were collected from three healthy women, aged 71 (fermentation 1), 72 (fermentation 2) and 78 years (fermentation 3), who did not receive antibiotic treatment for at least three months prior to sample collection, and who did not consume probiotics on a regular basis.

Immediately after defecating, the fecal sample was transferred to a tube containing 5 mL of sterile, pre-reduced peptone water (0.1%, pH 7), placed in an anaerobic jar (Anaerojar, Oxoid, Hampshire, England), and transported and processed within three hours. Handling and encapsulation of the fecal microbiota into 1–2 mm gel beads composed of gellan (2.5% w/v), xanthan (0.25% w/v), and sodium citrate (0.2% w/v, Sigma-Aldrich Chemie GmbH, Buchs, Switzerland) was performed in an anaerobic chamber as previously described [21].

Fermentation Medium

The fermentation medium was based on the composition described by MacFarlane et al. [22] for simulation of adult chyme entering the colon. It contained (g L^{-1} of distilled water): pectin from citrus (2), xylan from oat spelt (2), arabinogalactan from larch wood (2), guar gum (1), inulin (1), soluble potato starch (5), mucin (4), casein acid hydrolysate (3), peptone water (5), tryptone (5), yeast extract (4.5), cysteine (0.8), bile salts (0.4), KH_2PO_4 (0.5), $NaHCO_3$ (1.5), NaCl (4.5), KCl (4.5), $MgSO_4$ anhydrous (0.6), $CaCl_2$ x $2H_2O$ (0.1), $MnCl_2$ x $4H_2O$ (0.2), $FeSO_4$ x $7H_2O$ (0.005), hemin (0.05) and Tween 80

(1). One mL of a filter-sterilized (0.2 μm pore-size) vitamin solution [23] was added to 1 L of autoclaved (20 min, 120°C) and cooled medium. All components of the nutritive medium were purchased from Sigma-Aldrich Chemie, except for inulin (Orafti®, BENEO kindly provided by RPN Foodtechnology AG, Sursee, Switzerland), peptone water (Oxoid AG, Pratteln, Switzerland), bile salts (Oxoid AG), tryptone (Becton Dickinson AG, Allschwil, Switzerland) and KH_2PO_4 (VWR International AG).

Experimental Setup

Various reactor set-ups were applied for the three continuous intestinal fermentation experiments (Fig 1). The inoculum reactor (IR) seeded with donor's microbiota immobilized in polysaccharide gel beads, and operated with proximal colon conditions, was common to all models. Model 1 was a classical three-stage system consisting of three reactors placed in series and operated under conditions of the proximal (PC corresponding to IR), transverse (TC) and distal colon (DC) [21]. Model 2 was based on the recently developed PolyFermS platform [13, 14] and consisted of an IR with immobilized fecal microbiota in proximal colon conditions used to continuously inoculate two parallel systems (10% v/v of the feed), each composed of a proximal (PC1 and PC2) and a transverse-distal reactor (DC1 and DC2). For model 3, the chyme medium fermented in IR with immobilized microbiota and operated with proximal colon conditions was used to continuously feed (100% v/v of the feed) five reactors mounted in parallel and mimicking conditions of a transverse-distal colon. In models 2 and 3, one system or reactor downstream to IR was used as control while the other system (PC2-DC2) or reactors (TR1-TR4) were used to comparatively test treatments, respectively.

A

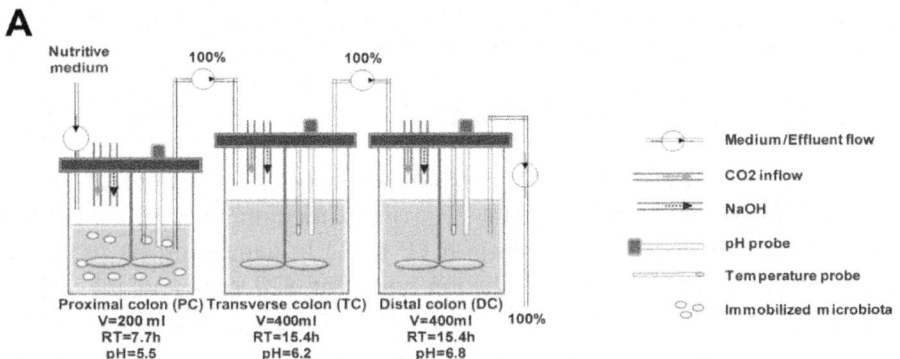

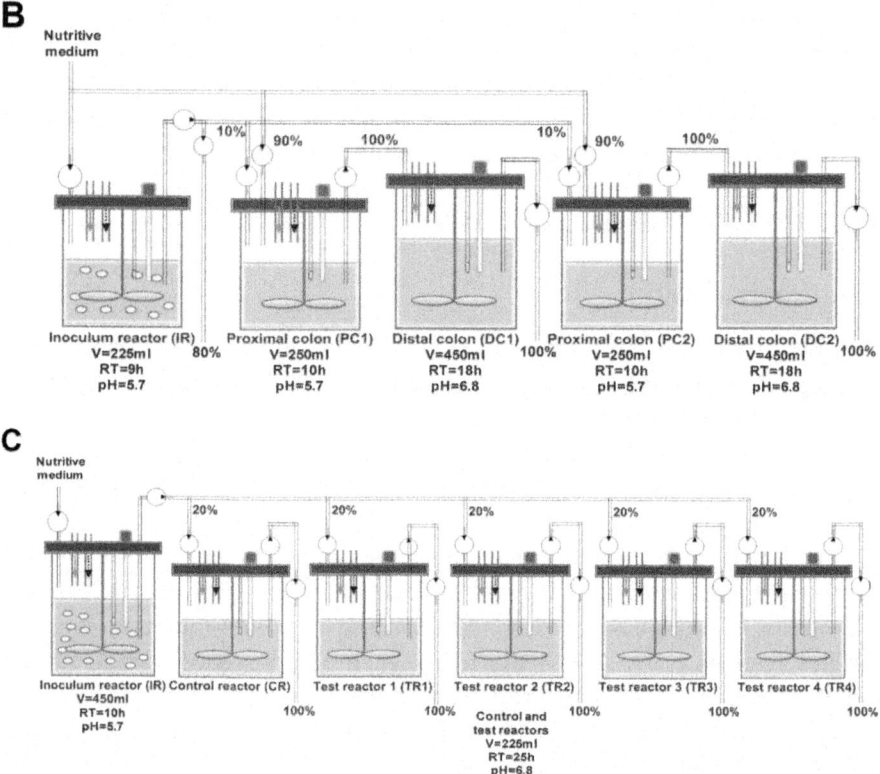

Figure 1. Set-up of the continuous fermentation models with immobilized gut microbiota. (A) Model 1. 3-stage model consisting of a proximal, transverse and distal colon reactor (B) Model 2. 2-stage model with an inoculum reactor connected to two parallel test systems consisting of a proximal and distal colon reactor (C) Model 3. 2-stage model with an inoculum reactor (proximal colon conditions) feeding 5 distal colon reactors connected in parallel; RT: Retention time, V: Volume.

Fermentation Procedures

The IR inoculated with 30% v/v gel beads was used to first colonize the beads with repeated-batch fermentations. The fresh medium was replaced every 12 h, for a total fermentation time of 60 to 72 h, depending on the model. Temperature was set at 37°C, stirring speed at 120 rpm and the pH was controlled at 5.5 or 5.7 by addition of 2.5 M NaOH. Sterile nutritive medium (4°C) was pumped continuously via a peristaltic pump (Reglo analog, Ismatec, Glattbrugg, Switzerland). Total mean retention times between 28 to 38.5 h were used in models 1–3, which is in the range of previously measured values for total colonic transit time in healthy elderly subjects that were between 25 and 66 h [24–28].

The reactors and nutritive media were continuously flushed with a low flow of CO_2 to maintain anaerobic conditions during fermentation. The conditions of the models along with the model specific design and trials are briefly presented below and summarized in Fig 1.

Model 1.

Six repeated-batch fermentations (72 h) were performed to colonize beads in PC (fermentation volume of 200 mL, pH of 5.5). Then two reactors mimicking conditions of transverse (TC) and distal colon (DC) (400 mL, pH 6.2 and 6.8, respectively) were connected in series to PC. The feed flow rate of the nutritive medium was set at 26 mL h^{-1}, giving mean retention times of 7.7 h for IR and 15.4 h for TC and DC, for a total system retention time of 38.5 h. The model was stabilized for 14 days.

Model 2.

Five repeated-batch fermentations (60 h) were performed to colonize beads in IR (225 mL and pH 5.7). The model was then switched to continuous mode with a medium flow rate of 25 mL h^{-1} for an additional five days. Two proximal colon reactors (PC1 and PC2, 250 mL and pH 5.7) were attached to IR, and each PC was connected to distal colon reactors (DC1 and DC2, 450 mL and pH 6.8) used to mimic transverse and distal colon conditions. Compared with model 1, the pH control set-point was increased from 5.5 to 5.7. This modification was implemented in order to better match the pH of the proximal colon section *in vivo*, which has been reported to be in the range from 5.5 to 5.9 [29] and to enhance metabolic activity in IR and PC's. PC1 and PC2 were continuously inoculated with 2.5 mL h^{-1} (10%) fermented medium from IR and 22.5 mL h^{-1} (90%) fresh nutritive medium, for total flow rate of 25 mL h^{-1}. Mean retention times in PC and DC reactors were 10 and 18 h, respectively, for a total retention time of 28 h (PC1 and DC1; PC2 and DC2). The model was continuously operated for an additional 18 days to reach stability and used for testing, for a total fermentation time of 55 days.

Model 3.

Beads colonization was carried out in five repeated-batch cultures (60 h) in IR (450 mL and pH 5.7). Five DC reactors (225 mL and pH 6.8) mounted in parallel and mimicking conditions of transverse-distal colon were connected to IR. The fresh nutritive medium flow rate in IR was set at 45 mL h^{-1}, while each DC reactors was fed with 9 mL h^{-1} medium fermented in IR. The mean retention times in IR and DC reactors of model 3 were 10 and 25 h, respectively, for a total retention time of 35 h. The model was stabilized for 14 days and used for testing, for a total fermentation of 80 days.

Sampling and Analysis

During continuous fermentations, effluent samples (10 mL) were collected daily from each reactor. Because the different models were also used for experimental trials with *Clostridium difficile* and antibiotic and probiotic treatments, only samples obtained during periods of control conditions are reported for model assessment. Analyses of microbial composition by quantitative polymerase chain reaction (qPCR) and pyrosequencing (model 2 and 3) were performed on samples from three days at the end of stabilization: days 9, 11 and 13 of model 1, days 16, 17 and 18 of model 2 and days 10, 11 and 12 of model 3. Metabolite concentrations in effluents of all reactors were tested daily during the entire fermentation by high-performance liquid chromatography (HPLC). Data corresponding to control condition periods (no treatment applied) are reported. Long term temporal stability was tested with IR data from models 2 and 3 since these reactors was not subjected to any manipulation over the entire culture period.

DNA Extraction

For qPCR and pyrosequencing analyses total microbial DNA of 200 mg feces and 2 mL effluent samples was extracted using the FastDNA® SPIN Kit for Soil (MP Biomedicals, Illkirch, France) and a final elution volume of 100 µL. DNA concentrations were determined using a Nanodrop® ND-1000 Spectrophotometer (Witec AG, Littau, Switzerland).

qPCR Analysis

Total bacteria and predominant bacterial groups were enumerated using specific primers (S1 Table). One µL of 10- or 100-fold diluted DNA was amplified in a total volume of 25 µL as described in [21], using 2 x SYBR Green PCR Master Mix (Applied Biosystems, Zug, Switzerland). Each reaction was run in duplicate on an ABI PRISM 7500-PCR sequence detection system (Applied Biosystems). For quantification, standard curves were produced by amplification of the DNA of the reference strain of the respective target group [30].

HPLC Analysis

SCFA (acetate, propionate, butyrate,valerate, isobutyrate and isovalerate) as well as formate and lactate concentrations) in fermentation effluent samples from all reactors were determined by HPLC analysis (Thermo Fisher Scientific Inc. Accela, Wohlen, Switzerland) in duplicate [31]. Effluent supernatants were 2-fold diluted with sterile ultra-pure water and filtered directly into vials through a 0.45 µm nylon HPLC filter (Infochroma AG, Zug, Switzerland). The analysis was run at a flow rate of 0.4 mL min^{-1} using an Aminex HPX-87H column (Bio-Rad Laboratories AG, Reinach, Switzerland) and 10 mM H_2SO_4 as eluent.

Microbiota Profiling by 454 Pyrosequencing

454-pyrosequencing analysis of total genomic DNA of fecal and effluent samples was carried out at DNAVision (Gosselies, Belgium). The V5-V6 hypervariable 16S RNA region was amplified using specific primers 784F (5'- AGGATTAGATACCCTKGTA-3') and 1061R (5'-CRRCACGAGCTGACGAC-3') [32]. The forward primer contained the sequence of the Titanium A adaptor and a unique barcode sequence. Pyrosequencing was carried out using primer A on a 454 Life Sciences Genome Sequencer FLX instrument (Roche Applied Science, Vilvoorde, Belgium) following Titanium chemistry. The data obtained was analyzed using the open source software package Quantitative Insights Into Microbial Ecology (QIIME), v1.7 [33]. Raw sequencing reads were filtered based on selected quality criteria such as: (1) no mismatch with the primer sequences and barcode tags; (2) no ambiguous bases (Ns); (3) read-lengths not shorter than 200 base pairs (bp) or longer than 1000 bp; (4) the average quality score in a sliding window of 50 bp not to fall below 25; (5) excluding homopolymer runs higher than 6 nt. Sequences that passed quality filtering were clustered into OTUs at 97% identity level using cd-hit [34]. Representative sequences (the most abundant) for each OTU were aligned using PyNAST and taxonomically assigned using Greengenes v_13_08 database. ChimeraSlayer was used to discard chimeric sequences, based on a reference data set of sequences [35]. This led to 8245 +/- 1924 (mean +/- SD) reads per sample. These phylogenies were combined with absence/presence or abundance information for each OTU to calculate unweighted or weighted UniFrac distances, respectively, using rarefaction of 7000 sequences per samples. Unifrac measures the phylogenetic distance between sets of taxa in a phylogenetic tree as the fraction of the branch length of the tree that leads to descendants from either one environment or the other, but not both [36, 37]. Weighted and unweighted Unifrac metrics were used to build phylogenetic distance matrices. Principal coordinates analysis

(PCoA) was applied to the distance matrices for visualization. Alpha diversity (diversity within sample) was calculated using Shannon (evenness) indexes. All 454-pyrosequencing files have been deposited to the National Center for Biotechnology Information (NCBI) Sequence Read Archive (SRA) under accession number SRP053000.

Statistical Analysis

Statistical analyses of HPLC and qPCR data (log10-transformed) were performed using JMP 8.0 (SAS Institute Inc., Cary, NC). Data are expressed as means ± SD of three days at the end of the stabilization period of each fermentation model. For every model the qPCR and HPLC data were compared between the reactors using the nonparametric Kruskal-Wallis test. *P* values < 0.05 were considered significant. Monte Carlo permutation procedure was used to determine difference between proximal and distal colon using 999 permutations. Correlation between genus-level phylotypes and metabolites (acetate, propionate, butyrate, isobutyrate, isovalerate and valerate) were done in fermentation models 2 and 3. Analysis was done using R package "Microbiome" [38] using Spearman correlation. P-values were corrected for multiple testing using Benjamini–Hochberg. Resulting q values < 0.05 were considered as significant.

RESULTS

Microbial Composition of Fecal Microbiota

The composition of dominant bacterial groups in fecal donor samples was assessed by analyzing the 16S rRNA gene copy numbers of total and selected bacterial groups using qPCR. All bacterial populations tested were detected in fecal samples, except for the *Roseburia* spp./*E. rectale* group and *Methanobacteriales* that were below the detection limit in the fecal inoculum of model 1 (Table 1). Predominant bacterial groups of all three fecal samples were *Bacteroides* spp. and *Clostridium* Cluster IV within which *Faecalibacterium prausnitzii* was dominant. *Enterobacteriaceae*, *Lactobacillus* spp. and *Bifidobacterium* spp. belonged to the subdominant populations in all three fecal inocula.

Design and Investigation of PolyFermS In Vitro Continuous Fermentation...   157

Table 1. qPCR enumeration of bacterial groups in fecal inocula and effluent samples of models' reactors at the end of the stabilization period

	Total 16S rRNA gene	Bacteroides spp.	Entero-bacteriaceae	Lactobacillus spp.	Bifido-bacterium spp.	F. prausnitzii	Clostridium Cluster IV	Roseburia spp./ E. rectale	Methano-bacteriales
Model 1									
Donor[1]	11.4	9.3	7.7	7.6	6.6	8.7	8.7	ND	ND
PC[2]	10.6 ± 0.02[A]	7.1 ± 0.1[A]	9.4 ± 0.1[A]	7.5 ± 0.2[A]	8.2 ± 0.1[A]	5.4 ± 0.2[A]	6.7 ± 0.4[A]	ND	ND
DC[2]	10.5 ± 0.1[A]	9.5 ± 0.01[B]	8.2 ± 0.1[B]	8.0 ± 0.2[B]	8.3 ± 0.1[A]	7.4 ± 0.1[B]	8.1 ± 0.1[B]	ND	ND
Model 2									
Donor[1]	11.1	9.6	7.2	8.3	8.5	10.1	10.3	9.4	8.6
IR[2]	10.2 ± 0.2[A]	10.0 ± 0.2[A]	8.9 ± 0.1[A]	6.1 ± 0.1[A]	6.7 ± 0.1[A]	9.1 ± 0.6[A]	10.1 ± 0.2[A]	8.6 ± 0.3[A]	8.8 ± 0.1[A]
PC1[2]	10.3 ± 0.05[A]	10.0 ± 0.2[A]	9.1 ± 0.3[A]	6.1 ± 0.02[A]	7.7 ± 0.1[B]	9.2 ± 0.4[A]	10.1 ± 0.1[A]	8.6 ± 0.3[A]	7.0 ± 0.4[B]
DC1[2]	10.3 ± 0.2[A]	9.9 ± 0.1[A]	8.9 ± 0.1[AB]	6.2 ± 0.1[AB]	7.6 ± 0.1[B]	9.2 ± 0.2[A]	9.9 ± 0.1[A]	8.6 ± 0.2[A]	8.9 ± 0.04[A]
PC2[2]	10.3 ± 0.1[A]	10.0 ± 0.1[A]	8.8 ± 0.3[AB]	6.2 ± 0.1[A]	6.6 ± 0.1[A]	9.0 ± 0.6[A]	10.0 ± 0.2[A]	8.4 ± 0.3[A]	7.3 ± 0.3[B]
DC2[2]	10.1 ± 0.2[A]	9.8 ± 0.2[A]	8.7 ± 0.1[B]	6.3 ± 0.001[B]	6.8 ± 0.2[A]	9.0 ± 0.2[A]	9.9 ± 0.1[A]	8.3 ± 0.2[A]	8.9 ± 0.1[A]
Model 3									
Donor[1]	11.5	10.6	6.3	7.5	6.2	10.6	10.3	9.7	7.7
IR[2]	11.2 ± 0.03[A]	9.9 ± 0.1[A]	9.6 ± 0.1[A]	6.7 ± 0.1[A]	7.8 ± 0.2[AB]	10.2 ± 0.1[A]	9.9 ± 0.1[A]	8.8 ± 0.1[A]	8.8 ± 0.04[A]
CR[2]	11.1 ± 0.1[A]	10.0 ± 0.1[AB]	9.5 ± 0.1[A]	6.9 ± 0.1[AB]	7.6 ± 0.1[AB]	10.1 ± 0.1[A]	9.8 ± 0.1[A]	8.5 ± 0.2[B]	8.3 ± 0.04[A]
TR1[2]	11.2 ± 0.2[A]	10.1 ± 0.03[B]	9.5 ± 0.1[A]	7.0 ± 0.01[B]	7.5 ± 0.1[A]	10.1 ± 0.04[A]	9.8 ± 0.1[A]	8.5 ± 0.2[AB]	8.3 ± 0.04[A]
TR2[2]	11.2 ± 0.05[A]	10.1 ± 0.03[B]	9.5 ± 0.1[A]	7.0 ± 0.2[AB]	7.6 ± 0.2[AB]	10.2 ± 0.02[A]	9.9 ± 0.02[A]	8.6 ± 0.1[AB]	8.3 ± 0.2[A]
TR3[2]	11.2 ± 0.1[A]	10 ± 0.2[AB]	9.5 ± 0.1[A]	7.0 ± 0.2[AB]	7.5 ± 0.2[AB]	10.1 ± 0.1[A]	9.8 ± 0.1[A]	8.5 ± 0.2[B]	8.3 ± 0.2[A]
TR4[2]	11.2 ± 0.1[A]	10.0 ± 0.1[AB]	9.5 ± 0.1[A]	6.8 ± 0.4[AB]	7.7 ± 0.2[B]	10.2 ± 0.1[A]	9.9 ± 0.2[A]	8.6 ± 0.3[AB]	7.9 ± 1.0[A]

PC, proximal colon reactor; DC, distal colon reactor; IR, inoculum reactor; CR, control reactor; TR, test reactor; ND, not detected

[1]Data are mean $\log_{10}$ copies 16S rRNA gene g[-1] feces; samples were analyzed in duplicate.

[2]Data are mean $\log_{10}$ copies 16S rRNA gene g[-1] fermentation effluent ± SD of three last days at the end of the stabilization period; samples were analyzed in duplicate. Values with different letters are significantly different within one model (P < 0.05)

doi:10.1371/journal.pone.0142793.t001

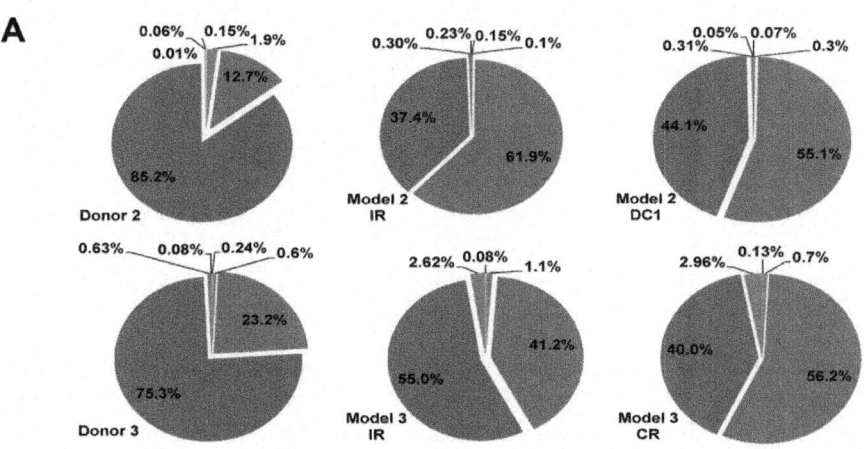

Other ▪ Actinobacteria ▪ Bacteroidetes ▪ Firmicutes ▪ Proteobacteria ▪ Tenericutes

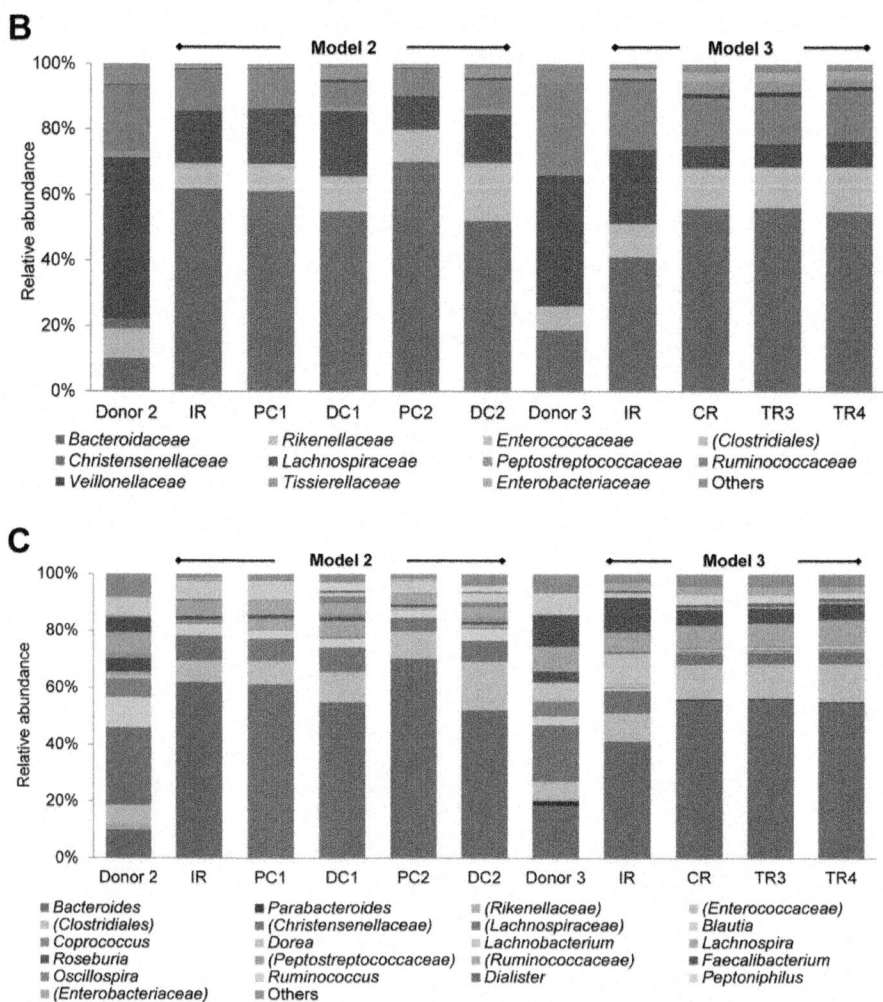

Figure 2. Microbial composition of fecal samples and reactors of PolyFermS models measured by 454 pyrosequencing. Relative abundance at **(A)** phylum level of fecal samples of donors 2 and 3, IR and DCI of model 2 and IR and CR of model 3 **(B)** family level and **(C)** genus level of fecal samples of donors 2 and 3, all reactors of model 2 and reactors IR, CR, TR3 and TR4 of model 3 identified by pyrosequencing of the V5-V6 hypervariable regions of the 16S rRNA gene. Effluent samples are average values of three last days at the end of the stabilization period. Parentheses indicate an unknown family belonging to an order or an unknown genus belonging to a family or order. Values < 1% are summarized in the group "others".

The microbial profile of fecal inocula of model 2 and 3 was additionally analyzed by pyrosequencing (Fig 2). Phyla of both fecal samples were mainly assigned to Firmicutes and Bacteroidetes, with a Bacteroidetes:Firmicutes ratio of 0.15 and 0.30 for donor 2 and 3, respectively, and followed by Actinobacteria, Proteobacteria and Tenericutes (Fig 2A). The dominant families, *Lachnospiraceae*, *Bacteroidaceae* and *Ruminococcaceae* were similar for both microbiota (Fig 2B). At the genus level an unassigned genus belonging to the family of*Lachnospiraceae* was most abundant in fecal sample 2, followed by *Blautia* and *Bacteroides*(Fig 2C). In fecal sample 3 the same unassigned genus at similar abundance ($\sim$ 20%) to fecal sample 2 was most abundant, closely followed by *Bacteroides*.

Microbial Composition of Effluents Determined by qPCR

Reactor effluents at the end of stabilization of all models were analyzed by qPCR to compare the microbial composition to the fecal donor and between reactors of a model (Table 1). Total bacterial numbers were, in general, high ($> 10 \log_{10}$ gene copies mL^{-1}) but between 0.3 to 1.0 $\log_{10}$ lower compared to the corresponding donor's fecal sample while no differences were observed between reactors within a model. In effluent samples of model 1, the *Roseburia*spp./*E. rectale* group and *Methanobacteriales* were not detected which was consistent with the lack of these groups in the corresponding fecal donor. *Bacteroides* spp. and*Enterobacteriaceae* were predominant in the PC and DC reactors of model 1, respectively. However copy numbers of specific population groups were significantly different between PC and DC effluents of model 1, with the exception of the total 16S rRNA gene and*Bifidobacterium* spp. The microbial composition of the DC reactor was more similar to the fecal donor than for the PC reactor, except for total bacteria, *Lactobacillus* spp. and *Bifidobacterium*spp.

Similar to the corresponding fecal samples, the predominant bacterial groups in effluents from model 2 and 3 comprised *Bacteroides* spp., *Clostridium* Cluster IV and *Faecalibacterium prausnitzii*. In model 2, no significant difference among all reactors was found for copy numbers of total 16S rRNA gene, *Bacteroides* spp., *Faecalibacterium prausnitzii*, *Clostridium* Cluster IV and *Roseburia* spp./*E. rectale* group. Only small (less than 0.4 $\log_{10}$) but significant differences were detected in DC2 for *Enterobacteriaceae* compared to IR and PC1, and for *Lactobacillus*spp. compared to IR, PC1 and PC2. *Bifidobacterium* spp. gene copy numbers were approx. 1 log higher in test system 1 compared to test system 2 and IR of model 2. *Methanobacteriales*numbers were significantly lower at PC conditions compared to IR and DC reactors.

In model 3, no significant difference was observed between reactors for most tested populations. Only small ($\leq 0.3 \log_{10}$ gene copies mL^{-1}), but significant differences were measured for *Bifidobacterium* spp. numbers in

TR4 compared to TR1, for *Bacteroides* and *Lactobacillus* spp. between IR and test reactors or CR, and for *Roseburia* spp. in CR and TR3 compared to IR.

Microbiota Profile and Diversity in Effluents Determined by Pyrosequencing

To assess the microbial diversity sequencing of the V5-V6 region of 16S rRNA gene was performed by 454 FLX pyrosequencing of all reactors effluent samples of model 2 and selected reactors (IR, CR, TR3 and TR4) of model 3 (Fig 2) at the end of stabilization phase and compared to diversity of corresponding feces. The main phyla in IR of model 2 and 3 were Firmicutes and Bacteroidetes followed by Actinobacteria, Proteobacteria and Tenericutes (Fig 2A). The ratio Bacteroidetes:Firmicutes was increased in IR and distal colon reactors of model 2 (ratios of 1.7 in IR and 1.2 in DC1) and model 3 (ratios of 0.8 in IR and 1.4 in CR) relative to the corresponding fecal inoculum (ratios 0.2 and 0.3, respectively). Bacteroidetes:Firmicutes ratios in DC2 of model 2 and TR3 and TR4 of model 3 were similar to DC1 and CR, respectively (data not shown). On the family and genus levels *Bacteroidaceae* and *Bacteroides* were dominant in all reactors of both models 2 and 3 (Fig 2B and 2C). *Bacteroidaceae* abundances increased from 10% to approx. 60% and from 36% to approx. 53% in the effluent samples of model 2 and 3, respectively, compared to the fecal donor samples. In general, very similar microbial patterns (family and genus level) were obtained for all reactors within a model. In both DC reactors of model 2 the abundance of *Bacteroidaceae* (54.8% for DC1 and 52.0% for DC2, Fig 2) decreased compared to PC reactors (61.2% for PC1 and 70.1% for PC2). Other small differences at family and genus levels between the PC and DC reactors of each test system were observed. In model 3, minor differences between composition in IR and DC reactors were observed while microbial patterns were highly comparable between the DC reactors.

Beta diversity (that measures the diversity between samples) of bacterial populations at the end of stabilization phase of model 2 and 3 was analyzed using Principal coordinate analysis (PCoA) (Fig 3). Significant differences between DC and PC were observed using both Unifrac distances ($p<0.005$). In models 2 and 3 a clear separation of reactors operated with proximal colon conditions (IRs, PC1 and PC2) and distal colon conditions (DC1- DC2, CR-TR3-TR4, respectively) was observed.

Weighted Unifrac

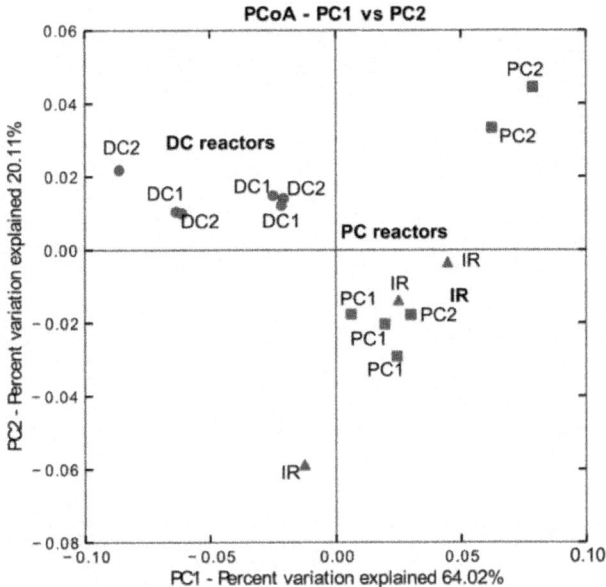

Unweighted Unifrac

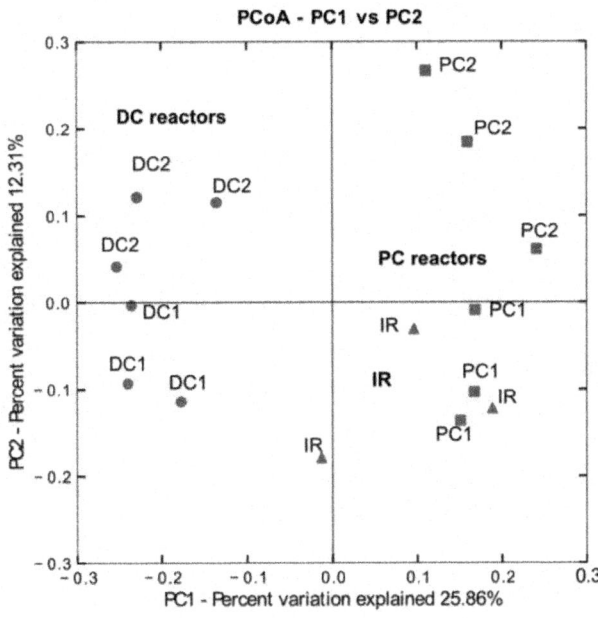

(A)

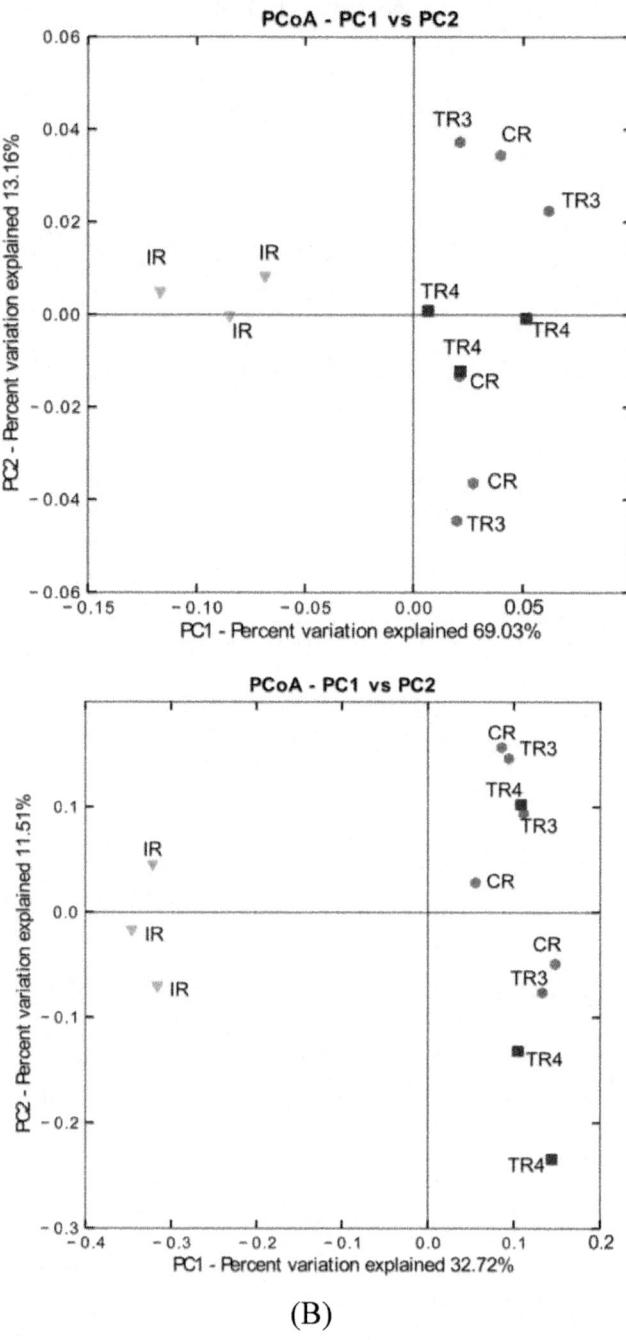

Figure 3. PCoA analysis of PolyFermS models based on weighted and unweighted UniFrac analysis. Each symbol is representing a different reactor. (A) Three last days

at the end of the stabilization period of all reactors of model 2 (IR, PC1, DC1, PC2 and DC2) and **(B)**three last days at the end of the stabilization period of model 3 (IR, CR, TR3 and TR4).

The Shannon diversity index was assessed for fecal donor samples and effluent samples of models 2 and 3 (Fig 4). A lower diversity was measured in model 2 (mean Shannon index of 5.3 ± 0.4 calculated for all reactors) and model 3 reactors (mean Shannon index of 6.4 ± 0.1 calculated for IR, CR, TR3 and TR4) compared to that of the corresponding fecal samples (Shannon index of 7.5 and 7.4 for model 2 and 3, respectively). In model 2, a higher diversity was obtained for DC (5.6 ± 0.2) compared to PC reactors (4.9 ± 0.2); while in model 3 the Shannon diversity was similar for all tested reactors with values between 6.2 and 6.6.

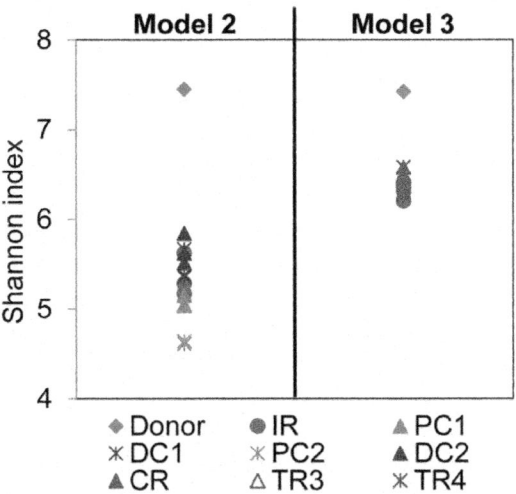

Figure 4. Shannon diversity index of fecal samples and reactors of PolyFermS models. The Shannon index was assessed in fecal donors 2 and 3, all reactors of model 2 and reactors IR, CR, TR3 and TR4 of model 3 of three last days at the end of stabilization phase. A higher Shannon index reflects a more diverse community (in abundance and evenness).

Metabolic Activity

SCFA were measured by HPLC in fermentation effluents at the end of the stabilization period of all reactors of models 1, 2 and 3 to assess the metabolic activity and intra model stability (Table 2). After the initial stabilization periods, high and stable metabolic activities were measured over the entire fermentation in IR's of models 2 and 3 which were operated under constant conditions and used to demonstrate temporal stability of the PolyFermS models (Fig 5).

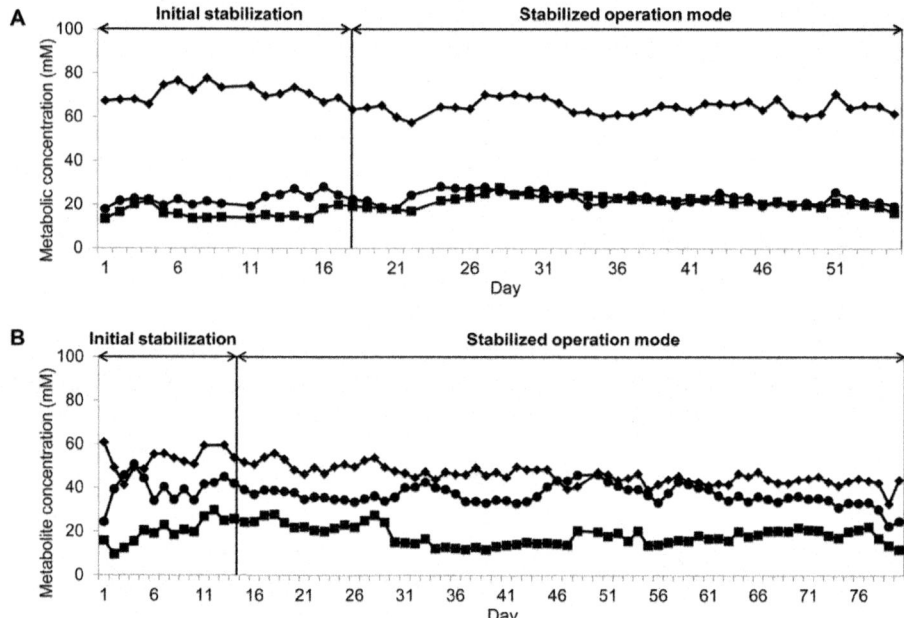

Figure 5. Daily mean SCFA concentrations in fermentation effluents of IR of PolyFermS models measured by HPLC. Initial stabilization: stabilization period in continuous mode to reach pseudo steady-state.; stabilized operation mode: continuous operation mode during pseudo steady-state conditions. (A) Model 2 and (B) model 3; (♦) acetate, (●) butyrate, and (■) propionate.

Table 2. Metabolites concentration (mM) and ratios (%) measured by HPLC in effluent samples of models' reactors at the end of the stabilization period

	Concentrations (mM)							Ratios (%)				
	Acetate	Butyrate	Propionate	Valerate	Isobutyrate	Isovalerate	Total SCFA	Acetate	Butyrate	Propionate	Valerate	BCFA
Model 1												
PC	74.1 ± 2.4^A	5.4 ± 1.0^A	5.4 ± 0.7^A	ND	ND	ND	84.9 ± 2.7^A	87.3	6.4	6.4		-
TC	108.6 ± 7.1^B	15.4 ± 2.1^B	15.3 ± 0.3^B	ND	1.9 ± 2.1^A	4.9 ± 1.2^A	146.1 ± 7.4^B	74.3	10.5	10.5		4.7
DC	110.7 ± 7.8^B	15.4 ± 0.8^B	15.3 ± 2.0^B	ND	1.6 ± 0.3^A	4.6 ± 1.5^A	147.6 ± 8.1^B	75.0	10.4	10.4		4.2
Model 2												
IR	66.3 ± 2.7^A	25 ± 2.9^A	19.1 ± 0.8^A	ND	ND	ND	110.4 ± 4.0^A	60.1	22.6	17.3		
PC1	69.1 ± 2.1^A	20.8 ± 1.4^AB	19.9 ± 0.6^A	ND	ND	ND	109.8 ± 2.6^A	62.9	18.9	18.1		
DC1	84.7 ± 2.5^B	20.5 ± 1.3^B	24.6 ± 0.2^B	8.2 ± 0.2^A	6.3 ± 0.2^A	7.2 ± 0.1^A	151.5 ± 2.8^B	55.9	13.5	16.2	4.2	10.2
PC2	70.8 ± 2.9^A	20.2 ± 3.3^AB	26.5 ± 3.3^BC	ND	ND	ND	117.5 ± 5.5^C	60.3	17.2	22.6		
DC2	81.6 ± 1.1^B	19.7 ± 2.6^B	26.5 ± 1.8^C	5.7 ± 0.2^B	6.4 ± 0.2^A	6.9 ± 0.4^B	146.8 ± 3.3^D	55.6	13.4	18.1	4.4	8.6
Model 3												
IR	56.5 ± 5.0^A	40.2 ± 5.6^A	23.7 ± 3.7^A	ND	ND	ND	120.4 ± 8.4^A	46.9	33.4	19.7		
CR	80.7 ± 4.5^B	42.8 ± 1.8^A	31.5 ± 2.4^BC	ND	ND	0.8 ± 0.1^A	155.8 ± 5.4^B	51.8	27.5	20.2		0.5
TR1	76.8 ± 4.1^BC	42.7 ± 5.0^A	31.0 ± 1.7^BC	ND	ND	ND	150.5 ± 6.7^BC	51.0	28.4	20.6		
TR2	73.2 ± 1.9^C	40.9 ± 1.3^A	29.0 ± 0.9^B	ND	ND	0.4 ± 0.1^A	143.5 ± 2.5^C	51.0	28.5	20.2		0.3
TR3	78.1 ± 1.1^B	39.7 ± 1.4^A	30.9 ± 0.9^C	ND	ND	1.0 ± 0.8^A	149.7 ± 2.0^B	52.2	26.5	20.6		0.7
TR4	73.7 ± 0.7^C	42.4 ± 1.9^A	28.8 ± 1.1^B	ND	ND	ND	144.9 ± 2.3^C	50.9	29.3	19.9		

PC, proximal colon reactor; DC, distal colon reactor; IR, inoculum reactor; CR, control reactor; TR, test reactor
Data are means ± SD of three last days at the end of the stabilization period; samples were analyzed in duplicate. ND, not detected
Values with different letters are significantly different within one model (P < 0.05)

doi:10.1371/journal.pone.0142793.t002

The concentrations of SCFA tested in reactors were model and reactor (-proximal and distal colon conditions) dependent, while intermediate products lactate and formate remained undetected in fermentation effluents. Acetate was the main metabolite in reactor effluents of all models, followed by butyrate and propionate which were generally produced at similar levels within a reactor. Butyrate concentrations were around 10 mM higher than propionate in CR and TR reactors of model 3. The molar ratios of acetate, butyrate and propionate in IR of the three models were different. In IR of model 1, operated at pH 5.5, a higher acetate fraction was produced (87/6/6) compared to IR of model 2 (60/23/17) and 3 (47/33/20) which were operated at a higher pH of 5.7. Higher concentrations of acetate and propionate were measured in TC and DC reactors of model 1, and DC reactors of models 2 and 3 compared to reactors IR and PCs of the same models, operated with proximal colon conditions. Butyrate concentrations increased along the reactors of the 3-stage model 1, but remained unchanged between PC and DC reactors of models 2 and 3. The mean concentrations of acetate, butyrate and propionate in the 5 distal reactors of model 3, were 76.5 ± 6.5, 41.7 ± 6.0 and 30.2 ± 3.4 mM, respectively, with small (less than 8 mM) but significant differences among reactors for acetate and propionate. Valerate was only detected in DC reactors from model 2.

Branched-chain fatty acids (BCFA) were not detected in any PC reactors. Isovalerate and isobutyrate were present in effluents samples from DC reactors of model 1 and model 2, at higher concentration in the latter. For model 3 isovalerate was only measured at low concentrations (≤ 1.0 mM) close to the detection limit in some DC reactors.

Correlations between Microbiota Composition and Metabolite Production

Pyrosequencing data on the genus level and metabolite concentrations measured by HPLC were investigated to test significant correlations between phylogenetic groups and metabolic activity. For model 2, significant negative correlations were calculated between isobutyrate, isovalerate and valerate concentrations and the dominant genera *Ruminococcus* and *Bacteroides* (Fig 6). Butyrate was positively correlated with the dominant genus *Roseburia* and unclassified members of *Ruminococcaceae* and *Lachnospiraceae*. In contrast, a dominant unclassified member of *Enterococcaceae* was negatively correlated with butyrate and positively correlated with all other metabolites detected. Furthermore many genera detected at less than 1% (*Dialister*, *Anaerococcus* and unclassified genera of *Rikenellaceae* and *Mogibacteriaceae*) showed positive correlations with isobutyrate, isovalerate and valerate with the exception of*Oscillospira*, *Peptoniphilus* and an unclassified genus of *Peptostreptococcaceae* (with abundances above 1% but in distal colon reactors only). Correlations between metabolites (only acetate and propionate) and phylogenetic groups were also found for model 3 (S1 Fig).

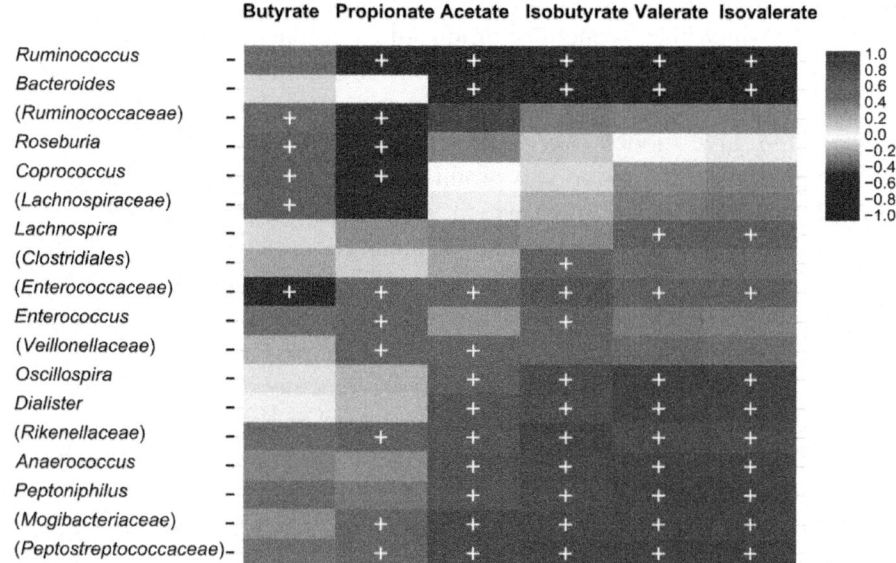

Figure 6. Correlations between genus-level phylogenetic groups and metabolites (SCFA, BCFA) of three last days at the end of stabilization period of model 2. The correlations, assessed by Spearman are indicated by either red (positive) and blue (negative), the significant correlations (q < 0.05) are indicated by '+'. Only genus related phylotypes > 0.1% and with at least one significant correlation with metabolites are depicted. Parentheses indicate an unknown genus belonging to a family or order.

DISCUSSION

Colonic fermentation models are useful tools to investigate factors that can influence the composition and metabolism of the gut microbiota, such as diet, antibiotic treatment, and bacterial infections *in vitro* and independent of the host [9, 10]. An important aspect for *in vitro* studies is the rational design of models and conditions, considering host target, model characteristics and limits, and the recognition that models are not perfect representation of reality. Therefore differences are often observed between fermentation samples and donor's feces. A major discriminatory factor of *in vitro* models is the technique used for fecal inoculation. For most models, a fresh liquid fecal suspension is inoculated whereas this has been shown to lead to limited stability (washout of less competitive or slow-growing bacteria), cell density and difficulty to reproduce both the planktonic (lumen) and sessile (food particule and mucus associated) microbiota of the colon.

In the present study, we report the first-time investigation of continuous fermentation models with fecal microbiota obtained from different healthy volunteers aged between 71 and 78 years using in-depth characterization methods of the microbial diversity. We immobilized the fecal microbiota and inoculated the fecal biocatalysts in the inoculum reactor of the tested models with different designs. A dense and diverse microbiota could be established in PolyFermS models, with reproducible microbial composition and metabolic activity for downstream test and control reactors within a model.

During collection and immobilization of fecal microbiota from each elderly donor special attention was paid to keep anaerobic conditions from donor to reactor, in order to reproduce both the planktonic and sessile forms of bacteria in the colon, as previously suggested [10]. Gel beads can provide a protective microenvironment for the bacteria and allow the growth of complex and stable gut ecosystems at high cell densities of up to ca. log 11 cells per mL effluent as observed in the present study with elder gut microbiota, preventing the loss of slow growing bacteria. As expected from the lack of water reabsorption total bacteria numbers in reactor effluents of models were up to 1.0 $\log_{10}$ lower compared to the corresponding donor's fecal sample. All bacterial groups tested in the fecal inoculum with qPCR were present in the corresponding models. The main differences in the bacterial composition and metabolic activity amongst models can be assigned to the different fecal inoculum used. In particular, *Roseburia*spp. was not detected in the feces and effluents of model 1 and this may explain the high acetate and low butyrate concentrations in this model (Table 2) since *Roseburia* is a main contributor for the conversion of acetate into butyrate [39, 40].

The pH of IR of model 1 was set to 5.5 in order to replicate the pH set in previous elder gut fermentation models [15, 17]. However, this pH is in the low range for the human proximal colon *in vivo*, [29, 41] and for models 2 and 3, the pH in IR and PCs was set to 5.7. This pH elevation induced an increase in total metabolites by approximately 20%, in agreement with previous observations of pH effect made in the PolyFermS model with child microbiota [13]. In contrast, butyrate concentrations did not increase in model 1 (pH 5.5) relative to model 2 and 3 (pH 5.7), as would have been expected from the stimulation of butyrate production at the lower pH, as observed in the previous study [13]. This is likely due to the lack of *Roseburia* spp. and the low *F. prausnitzii* numbers in model 1 which are the main butyrate producers in the human gut microbiota [42]. In models 2 and 3, the microbiota composition tested with qPCR was very similar between IR or PC reactors and distal reactors, while some limited changes were measured with pyrosequencing. In contrast, most targeted populations by qPCR significantly increased from PC (IR) to DC of model 1, suggesting that the low pH of 5.5 limited the growth of the targeted groups.

With qPCR we detected high *Enterobacteriaceae* copy numbers in reactor effluents compared to feces for all three models. This was observed in previous gut fermentation models [14, 21] and may be due to competitive advantage of these fast growing and robust bacteria that allows them to occupy niches during the immobilization process and the succeeding batch fermentation. The low levels of SCFA in the beginning of batch fermentation may further explain the increase in *Enterobacteriaceae* in reactors, as SCFA have inhibitory effects against*Enterobacteriaceae*, such as *Escherichia coli* [43].

The microbial composition of models 2 and 3 and corresponding fecal inocula may be considered more representative of the elderly population than the fecal inoculum of model 1, which did not harbor *Roseburia* spp., although the genus *Roseburia* was assigned at approximately 3% in fecal samples from elderly Irish subjects [8].

454 pyrosequencing was performed using the V5-V6 hypervariable region that was previously used to profile gut microbiota [44–46]. Sun *et al.* [47] recently reported that intragenomic heterogeneity for the V6 region may introduce overestimation of prokaryotes diversity. However in our study pyrosequencing data were used to compare of composition of donor and reactor samples within a model which should not be affected by this possible bias. In general, similar microbial profiles between effluent samples of model 2 and 3 and corresponding fecal donors were obtained. However, in both models the ratio of Firmicutes:Bacteroidetes was decreased when compared to the fecal sample, likely due to host-related factors including water and metabolite absorption and intestinal cells and host interaction, both of which are lacking in the fermentation models [13, 48]. Changes in microbiota composition and diversity may also reflect a possible loss of bacteria from donor to reactor, even though great care was taken to protect viability (strict anaerobiosis, mild conditions, short time), and the use of fecal microbiota to inoculate the inoculum reactor run in proximal colon conditions. Furthermore, the strictly controlled environmental factors, such as pH, transit time and medium composition in the *in vitro* models do no fully represent the specific donor conditions, thereby further contributing to*in vitro* and *in vivo* variations [10]. Despite the increase in Bacteroidetes in models 2 and 3, the Firmicutes:Bacteroidetes ratios of the fecal donor and the models were all in the range of previously reported data recorded during a large-scale *in vivo* study with elderly Irish people [8]. Indeed large inter-individual variations were observed in this study; however, the Bacteroidetes:Firmicutes ratio was

shown to be higher in the elder, relative to the adult population. The majority of the reads was assigned to Firmicutes and Bacteroidetes while only low levels of Proteobacteria and Actinobacteria were detected in concordance with *in vivo* findings [8, 49]. Many of the predominant genera (abundance > 1%) including *Bacteroides,Faecalibacterium, Roseburia* and *Ruminococcus* were also found above 1% in elderly Irish subjects [8]. No in-depth characterization of the microbiota in donor and effluent samples was reported in previous investigations of fermentation models of the elderly microbiota, in which only traditional plating methods [15, 18] or FISH [19] were used.

SCFA are mainly produced from carbohydrate fermentation and to a lesser extent via degradation of proteins and amino acids; the effects of SCFA on the host are well documented [1]. It was previously found that the major SCFA found in stools of healthy volunteers between 14–74 years of age were: acetate, propionate and butyrate at an approx. ratio of 3:1:1 [50]. In our study, similar ratios were found in distal colon reactors of models 2 (4:1.5:1) and 3 (3:1.5:2), whereas in model 1 the acetate fraction was considerably higher (7:1:1), likely due to the low pH of 5.5 in IR and the fecal microbiota used in this model as discussed above. *In vivo* investigations are, however, hampered by the continuous absorption of metabolic products, which results in less than 5% of total production excreted in the feces [51] along with the difficulty associated with obtaining samples from different regions of the colon. Therefore, metabolite concentrations and ratios in feces are not indicative for the colonic microbiota activity. In contrast *in vitro* modeling allows accurate measuring of the metabolic activity of the gut microbiota for the tested model conditions. Stable SCFA concentrations were obtained throughout the fermentation in the untreated IR's of the PolyFermS models demonstrating maintenance of gut microbial activity over the entire fermentation of 55 and 80 days for model 2 and 3, respectively (Fig 5).

BCFA are products of protein and amino acid fermentation but the formation of BCFA and associated species is not well studied [52]. In the colon of elders, an increase in proteolytic activity and a decrease in concentrations of SCFA were reported [53, 54]. Metabolites of amino acid fermentation can have toxic effects on the colonic lumen and were associated with several gut disorders [52]. In the tested models, BCFA were solely detected in significant levels within the distal reactors of model 1 and 2. This observation is consistent with the understanding that the distal colon is the major site for proteolysis whereas carbohydrate fermentation is the main energy yielding process in the proximal colon, resulting in a lower pH in this section [43, 55]. In model 2, genera with abundances of less than 1% were positively correlated with isobutyrate, valerate and isovalerate, suggesting that the dominant bacteria were mainly responsible for saccharolytic fermentation while proteins were degraded by the subdominant populations. Very low or no BCFA were detected in CR and TR reactors of model 3 which was set to mimic fermentation of transverse-distal colon sections within one reactor. This may be explained by the microbiota composition of model 3 that was different from model 1 and 2. The microbiota-dependent production of BCFA suggests the importance of using individual microbiota for inoculating intestinal fermentation models instead of pooling microbiota from different donors, as done in many studies for inoculation of gut fermentation models.

A major feature of the PolyFermS models over the three-stage model (model 1) is that several treatments can be investigated simultaneously and compared to a control inoculated with the same microbiota, thus generating reproducible and accurate data rather than when treatments are applied during consecutive periods. In both PolyFermS models microbial diversity and metabolic activity was very similar between control and test reactors. Model 2 built with two sets of proximal and distal colon reactors can be used for a broad range of studies, in proximal and distal colon conditions, such as the effect of an altered diet and administration of antibiotics on the gut microbiota in old age. Furthermore, the model is applicable for the *in vitro* investigation of the elderly microbiota in combination with health-related questions such as the manipulation of the gut microbiota using pro- and prebiotics [56]. PolyFermS model 3 built with multiple parallel distal colon reactors can be especially useful to study the effect of factors related to age on microbial metabolism in the lower colon, such as promotion of putrefaction due to low fiber intake. The PolyFermS intestinal platform has potential to be scaled down and adapted with multi-reactors to enhance screening efficiency.

To conclude, in the present study we showed the stability and reproducibility of PolyFermS continuous colonic fermentation models inoculated with immobilized elderly microbiota. Immobilization requires only small amounts of high quality fecal material to prime a gut model that can be stably operated over several months for testing parallel treatments in consecutive blocks [56]. The PolyFermS platform should be suitable for a range of *in vitro* gut microbiota investigations, from classical microbe interaction studies to complex ecological studies of the elderly gut microbiome investigated by in-depth analysis of the microbial diversity.

SUPPORTING INFORMATION

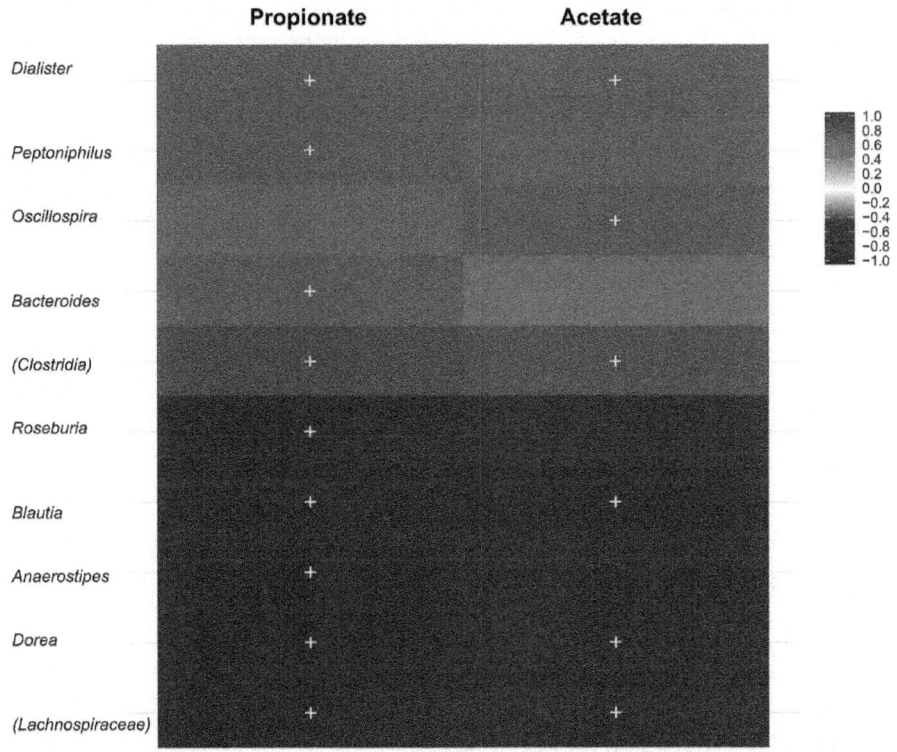

S1 Fig. Correlations between genus-level phylogenetic groups and acetate and propionate of three last days at the end of the stabilization period of model 3.

The correlations, assessed by Spearman are indicated by either red (positive) and blue (negative), the significant correlations (q < 0.05) are indicated by '+'. Only genus related phylotypes > 0.1% and with at least one significant correlation with metabolites are depicted.

S1 Table. Primers used for enumeration of bacterial groups by qPCR

Name	Sequence-5'-3'	Target-gene/purpose	Reference
Eub338F	ACTCCTACGGGAGGCAGCAG	Total-16S-rRNA-genes	(1)
Eub518R	ATTACCGCGGCTGCTGG		
Bac303F	GAAGGTCCCCCACATTG	Bacteroides-spp.	(2)
Bfr-Femrev	CGCKACTTGGCTGGTTCAG		
RrecF	GCGGTRCGGCAAGTCTGA	Roseburia-spp./E.-rectale	(2)
Rrec630mR	CCTCCGACACTCTAGTMCGAC		
Clep866mF	TTAACACAATAAGTWATCCACCTGG	Clostridium-cluster-IV	(2)
Clep1240mR	ACCTTCCTCCGTTTTGTCAAC		
F_Lacto-05	AGCAGTAGGGGAATCTTCCA	Lactobacillus/Pediococcus/Leuconostoc-spp.	(3)
R_Lacto-04	CGCCACTGGTGTTCYTCCATATA		
Fprau223F	GATGGCCTCGCGTCCGATTAG	Faecalibacterium-prausnitzii	(4)
Fprau420R	CCGAAGACCTTCTTCCTCC		
xfp-fw	ATCTTCGGACCBGAYGAGAC	Bifidobacterium-phosphoketolase	(5)
xfp-rv	CGATVACGTGVACGAAGGAC		
Eco1457F	CATTGACGTTACCCGCAGAAGAAGC	Enterobacteriaceae	(4)
Eco1652R	CTCTACGAGACTCAAGCTTGC		
Met915F	AGGAATTGGCGGGGGAGCAC	Methanobacteriales	(6)
1'100AR	TGGGTCTCGCTCGTTG		

ACKNOWLEDGMENTS

We thank Liselotte Wallquist, Sophie Poeker, Lukas Meile, Simon Galenda and Markus Reichlin for technical assistance.

AUTHOR CONTRIBUTIONS

Conceived and designed the experiments: SF CC MCH CL. Performed the experiments: SF MCH CC. Analyzed the data: SF CC MCH MD CL. Contributed reagents/materials/analysis tools: CL CF. Wrote the paper: SF CC MD CL.

REFERENCES

1. Chassard C, Lacroix C. Carbohydrates and the human gut microbiota. Curr Opin Clin Nutr Metab Care. 2013;16(4):453–60. Epub 2013/05/31. doi: 10.1097/MCO.0b013e3283619e63 pmid:23719143.

2. Robles Alonso V, Guarner F. Linking the gut microbiota to human health. Brit J Nutr. 2013;109 Suppl 2:S21–6. Epub 2013/02/13. doi: 10.1017/ S0007114512005235 S0007114512005235 [pii]. pmid:23360877.

3. Duncan SH, Flint HJ. Probiotics and prebiotics and health in ageing populations. Maturitas. 2013;75(1):44–50. Epub 2013/03/16. S0378-5122(13)00049-2 [pii] doi: 10.1016/j.maturitas.2013.02.004 pmid:23489554.

4.   Power SE, O'Toole PW, Stanton C, Ross RP, Fitzgerald GF. Intestinal microbiota, diet and health. Brit J Nutr. 2014;111(3):387–402. Epub 2013/08/13. doi: 10.1017/S0007114513002560 S0007114513002560 [pii]. pmid:23931069.

5.   O'Toole PW, Claesson MJ. Gut microbiota: Changes throughout the lifespan from infancy to elderly. Int Dairy J. 2010;20(4):281–91. doi: 10.1016/j.idairyj.2009.11.010 pmid:ISI:000275924800012.

6.   Biagi E, Candela M, Turroni S, Garagnani P, Franceschi C, Brigidi P. Ageing and gut microbes: Perspectives for health maintenance and longevity. Pharmacol Res. 2013;69(1):11–20. Epub 2012/10/20. doi: 10.1016/j.phrs.2012.10.005 S1043-6618(12)00192-2 [pii]. pmid:23079287.

7.   Ottman N, Smidt H, de Vos WM, Belzer C. The function of our microbiota: who is out there and what do they do? Front Cell Infect Mi. 2012;2:104. Unsp 104 doi: 10.3389/Fcimb.2012.00104 pmid:WOS:000324630500002; PubMed Central PMCID: PMC3417542.

8.   Claesson MJ, Cusack S, O'Sullivan O, Greene-Diniz R, de Weerd H, Flannery E, et al. Composition, variability, and temporal stability of the intestinal microbiota of the elderly. Proc Natl Acad Sci U S A. 2011;108 Suppl 1:4586–91. Epub 2010/06/24. doi: 10.1073/pnas.1000097107 1000097107 [pii]. pmid:20571116; PubMed Central PMCID: PMC3063589.

9.   Lacroix C, de Wouters T, Chassard C. Integrated multi-scale strategies to investigate nutritional compounds and their effect on the gut microbiota. Curr Opin Biotechnol. 2015;32C:149–55. doi: 10.1016/j.copbio.2014.12.009 pmid:25562815.

10. Payne AN, Zihler A, Chassard C, Lacroix C. Advances and perspectives in in vitrohuman gut fermentation modeling. Trends Biotechnol. 2012;30(1):17–25. Epub 2011/07/19. doi: 10.1016/j.tibtech.2011.06.011 S0167-7799(11)00115-6 [pii]. pmid:21764163.

11. Cinquin C, Le Blay G, Fliss I, Lacroix C. Immobilization of infant fecal microbiota and utilization in an in vitro colonic fermentation model. Microb Ecol. 2004;48(1):128–38. Epub 2004/04/16. doi: 10.1007/s00248-003-2022-7 pmid:15085302.

12. Cinquin C, Le Blay G, Fliss I, Lacroix C. New three-stage in vitro model for infant colonic fermentation with immobilized fecal microbiota. FEMS Microbiol Ecol. 2006;57(2):324–36. Epub 2006/07/27. FEM117 [pii] doi: 10.1111/j.1574-6941.2006.00117.x pmid:16867149.

13. Zihler Berner A, Fuentes S, Dostal A, Payne AN, Vazquez Gutierrez P, Chassard C, et al. Novel Polyfermentor intestinal model (PolyFermS) for controlled ecological studies: validation and effect of pH. PLoS One. 2013;8(10):e77772. Epub 2013/11/10. doi: 10.1371/journal.pone.0077772 PONE-D-13-16596 [pii]. pmid:24204958; PubMed Central PMCID: PMC3813750.

14. Tanner SA, Zihler Berner A, Rigozzi E, Grattepanche F, Chassard C, Lacroix C. *In vitro*continuous fermentation model (PolyFermS) of the swine proximal colon for simultaneous testing on the same gut microbiota. PLoS One. 2014;9(4):e94123. doi: 10.1371/journal.pone.0094123 pmid:24709947; PubMed Central PMCID: PMC3978012.

15. Baines SD, Freeman J, Wilcox MH. Effects of piperacillin/tazobactam on *Clostridium difficile* growth and toxin production in a human gut model. J Antimicrob Chemother. 2005;55(6):974–82. Epub 2005/04/30. dki120 [pii] doi: 10.1093/jac/dki120 pmid:15860551.

16. Baines SD, O›Connor R, Saxton K, Freeman J, Wilcox MH. Activity of vancomycin against epidemic *Clostridium difficile* strains in a human gut model. J Antimicrob Chemother. 2009;63(3):520–5. Epub 2008/12/30. dkn502 [pii] doi: 10.1093/jac/dkn502 pmid:19112083.

17. Freeman J, Baines SD, Saxton K, Wilcox MH. Effect of metronidazole on growth and toxin production by epidemic *Clostridium difficile* PCR ribotypes 001 and 027 in a human gut model. J Antimicrob Chemother. 2007;60(1):83–91. Epub 2007/05/08. dkm113 [pii] doi: 10.1093/jac/dkm113 pmid:17483547.

18. Crowther GS, Chilton CH, Todhunter SL, Nicholson S, Freeman J, Baines SD, et al. Development and validation of a chemostat gut model to study both planktonic and biofilm modes of growth of *Clostridium difficile* and human microbiota. PLoS One. 2014;9(2):e88396. Epub 2014/02/12. doi: 10.1371/journal.pone.0088396 PONE-D-13-46790 [pii]. pmid:24516647.

19. Likotrafiti E, Tuohy KM, Gibson GR, Rastall RA. An *in vitro* study of the effect of probiotics, prebiotics and synbiotics on the elderly faecal microbiota. Anaerobe. 2014;27:50–5. doi: 10.1016/j.anaerobe.2014.03.009 pmid:24685554.

20. Payne AN, Chassard C, Banz Y, Lacroix C. The composition and metabolic activity of child gut microbiota demonstrate differential adaptation to varied nutrient loads in an *in vitro* model of colonic fermentation. FEMS Microbiol Ecol. 2012;80(3):608–23. Epub 2012/02/14. doi: 10.1111/j.1574-6941.2012.01330.x pmid:22324938.

21. Zihler A, Gagnon M, Chassard C, Hegland A, Stevens MJ, Braegger CP, et al. Unexpected consequences of administering bacteriocinogenic probiotic strains for*Salmonella* populations, revealed by an *in vitro* colonic model of the child gut. Microbiology. 2010;156(Pt 11):3342–53. Epub 2010/08/07. mic.0.042036–0 [pii] doi: 10.1099/mic.0.042036–0 pmid:20688827.

22. Macfarlane GT, Macfarlane S, Gibson GR. Validation of a three-stage compound continuous culture system for investigating the effect of retention time on the ecology and metabolism of bacteria in the human colon. Microb Ecol. 1998;35(2):180–7. Epub 1998/12/16. MECOJKF97-5 [pii]. pmid:9541554. doi: 10.1007/s002489900072

23. Michel C, Kravtchenko TP, David A, Gueneau S, Kozlowski F, Cherbut C. *In vitro*prebiotic effects of Acacia gums onto the human intestinal microbiota depends on both botanical origin and environmental pH. Anaerobe. 1998;4(6):257–66. doi: 10.1006/anae.1998.0178 pmid:16887651.

24. Nagengast FM, van der Werf SD, Lamers HL, Hectors MP, Buys WC, van Tongeren JM. Influence of age, intestinal transit time, and dietary composition on fecal bile acid profiles in healthy subjects. Digest Dis Sci. 1988;33(6):673–8. pmid:3371139. doi: 10.1007/bf01540429

25. Madsen JL. Effects of gender, age, and body mass index on gastrointestinal transit times. Digest Dis Sci. 1992;37(10):1548–53. Epub 1992/10/01. pmid:1396002. doi: 10.1007/bf01296501

26. Madsen JL, Graff J. Effects of ageing on gastrointestinal motor function. Age Ageing. 2004;33(2):154–9. doi: 10.1093/ageing/afh040 pmid:14960431.

27. Merkel IS, Locher J, Burgio K, Towers A, Wald A. Physiologic and psychologic characteristics of an elderly population with chronic constipation. Am J Gastroenterol. 1993;88(11):1854–9. Epub 1993/11/01. pmid:8237932.

28. Evans JM, Fleming KC, Talley NJ, Schleck CD, Zinsmeister AR, Melton LJ 3rd. Relation of colonic transit to functional bowel disease in older people: a population-based study. J Am Geriatr Soc. 1998;46(1):83–7. Epub 1998/01/22. pmid:9434670. doi: 10.1111/j.1532-5415.1998.tb01018.x

29. Lawley TD, Walker AW. Intestinal colonization resistance. Immunology. 2013;138(1):1–11. doi: 10.1111/j.1365-2567.2012.03616.x pmid:23240815; PubMed Central PMCID: PMC3533696.

30. Dostal A, Fehlbaum S, Chassard C, Zimmermann MB, Lacroix C. Low iron availability in continuous *in vitro* colonic fermentations induces strong dysbiosis of the child gut microbial consortium and a decrease in main metabolites. FEMS Microbiol Ecol. 2013;83(1):161–75. Epub 2012/08/01. doi: 10.1111/j.1574-6941.2012.01461.x pmid:22845175; PubMed Central PMCID: PMC3511601.

31. Cleusix V, Lacroix C, Vollenweider S, Le Blay G. Glycerol induces reuterin production and decreases *Escherichia coli* population in an *in vitro* model of colonic fermentation with immobilized human feces. FEMS Microbiol Ecol. 2008;63(1):56–64. Epub 2007/11/22. FEM412 [pii] doi: 10.1111/j.1574-6941.2007.00412.x pmid:18028400.

32. Andersson AF, Lindberg M, Jakobsson H, Backhed F, Nyren P, Engstrand L. Comparative analysis of human gut microbiota by barcoded pyrosequencing. PLoS One. 2008;3(7):e2836. Epub 2008/07/31. doi: 10.1371/journal.pone.0002836 pmid:18665274; PubMed Central PMCID: PMC2475661.

33. Caporaso JG, Kuczynski J, Stombaugh J, Bittinger K, Bushman FD, Costello EK, et al. QIIME allows analysis of high-throughput community sequencing data. Nat Methods. 2010;7(5):335–6. doi: 10.1038/nmeth.f.303 pmid:20383131; PubMed Central PMCID: PMC3156573.

34. Li W, Godzik A. Cd-hit: a fast program for clustering and comparing large sets of protein or nucleotide sequences. Bioinformatics. 2006;22(13):1658–9. doi: 10.1093/bioinformatics/btl158 pmid:16731699.

35. Haas BJ, Gevers D, Earl AM, Feldgarden M, Ward DV, Giannoukos G, et al. Chimeric 16S rRNA sequence formation and detection in Sanger and 454-pyrosequenced PCR amplicons. Genome Res. 2011;21(3):494–504. doi: 10.1101/gr.112730.110 pmid:21212162; PubMed Central PMCID: PMC3044863.

36. Lozupone C, Knight R. UniFrac: a new phylogenetic method for comparing microbial communities. Appl Environ Microb. 2005;71(12):8228–35. doi: 10.1128/AEM.71.12.8228–8235.2005 pmid:16332807; PubMed Central PMCID: PMC1317376.

37. Lozupone CA, Hamady M, Kelley ST, Knight R. Quantitative and qualitative beta diversity measures lead to different insights into factors that structure microbial communities. Appl Environ Microb. 2007;73(5):1576–85. doi: 10.1128/AEM.01996-06 pmid:17220268; PubMed Central PMCID: PMC1828774.

38. Lahti L, Salonen A, Kekkonen RA, Salojarvi J, Jalanka-Tuovinen J, Palva A, et al. Associations between the human intestinal

microbiota, *Lactobacillus rhamnosus* GG and serum lipids indicated by integrated analysis of high-throughput profiling data. PeerJ. 2013;1:e32. doi: 10.7717/peerj.32 pmid:23638368; PubMed Central PMCID: PMC3628737.

39. Duncan SH, Barcenilla A, Stewart CS, Pryde SE, Flint HJ. Acetate utilization and butyryl coenzyme A (CoA):acetate-CoA transferase in butyrate-producing bacteria from the human large intestine. Appl Environ Microb. 2002;68(10):5186–90. pmid:12324374; PubMed Central PMCID: PMC126392. doi: 10.1128/aem.68.10.5186-5190.2002

40. Belenguer A, Duncan SH, Calder AG, Holtrop G, Louis P, Lobley GE, et al. Two routes of metabolic cross-feeding between *Bifidobacterium adolescentis* and butyrate-producing anaerobes from the human gut. Appl Environ Microb. 2006;72(5):3593–9. doi: 10.1128/AEM.72.5.3593–3599.2006 pmid:16672507; PubMed Central PMCID: PMC1472403.

41. Nugent SG, Kumar D, Rampton DS, Evans DF. Intestinal luminal pH in inflammatory bowel disease: possible determinants and implications for therapy with aminosalicylates and other drugs. Gut. 2001;48(4):571–7. pmid:11247905; PubMed Central PMCID: PMC1728243. doi: 10.1136/gut.48.4.571

42. Walker AW, Duncan SH, McWilliam Leitch EC, Child MW, Flint HJ. pH and peptide supply can radically alter bacterial populations and short-chain fatty acid ratios within microbial communities from the human colon. Appl Environ Microb. 2005;71(7):3692–700. Epub 2005/07/08. 71/7/3692 [pii] doi: 10.1128/AEM.71.7.3692–3700.2005 pmid:16000778; PubMed Central PMCID: PMC1169066.

43. Duncan SH, Louis P, Thomson JM, Flint HJ. The role of pH in determining the species composition of the human colonic microbiota. Environ Microbiol. 2009;11(8):2112–22. Epub 2009/04/29. EMI1931 [pii] doi: 10.1111/j.1462-2920.2009.01931.x pmid:19397676.

44. Zhang J, Guo Z, Xue Z, Sun Z, Zhang M, Wang L, et al. A phylo-functional core of gut microbiota in healthy young Chinese cohorts across lifestyles, geography and ethnicities. ISME J. 2015;9(9):1979–90. doi: 10.1038/ismej.2015.11 pmid:25647347.

45. Montassier E, Batard E, Massart S, Gastinne T, Carton T, Caillon J, et al. 16S rRNA gene pyrosequencing reveals shift in patient faecal microbiota during high-dose chemotherapy as conditioning regimen for bone marrow transplantation. Microb Ecol. 2014;67(3):690–9. doi: 10.1007/s00248-013-0355-4 pmid:24402367.

46. De Filippo C, Cavalieri D, Di Paola M, Ramazzotti M, Poullet JB, Massart S, et al. Impact of diet in shaping gut microbiota revealed by a comparative study in children from Europe and rural Africa. Proc Natl Acad Sci U S A. 2010;107(33):14691–6. doi: 10.1073/pnas.1005963107 pmid:20679230; PubMed Central PMCID: PMC2930426.

47. Sun DL, Jiang X, Wu QL, Zhou NY. Intragenomic heterogeneity of 16S rRNA genes causes overestimation of prokaryotic diversity. Appl Environ Microb. 2013;79(19):5962–9. doi: 10.1128/AEM.01282-13 pmid:23872556; PubMed Central PMCID: PMC3811346.

48. Van den Abbeele P, Grootaert C, Marzorati M, Possemiers S, Verstraete W, Gerard P, et al. Microbial community development in a dynamic gut model is reproducible, colon region specific, and selective for Bacteroidetes and *Clostridium* cluster IX. Appl Environ Microb. 2010;76(15):5237–46. doi: 10.1128/AEM.00759-10 pmid:20562281; PubMed Central PMCID: PMC2916472.

49. Biagi E, Nylund L, Candela M, Ostan R, Bucci L, Pini E, et al. Through ageing, and beyond: gut microbiota and inflammatory status in seniors and centenarians. PLoS One. 2010;5(5):e10667. Epub 2010/05/26. doi: 10.1371/journal.pone.0010667 pmid:20498852; PubMed Central PMCID: PMC2871786.

50. Schwiertz A, Taras D, Schafer K, Beijer S, Bos NA, Donus C, et al. Microbiota and SCFA in lean and overweight healthy subjects. Obesity. 2010;18(1):190–5. doi: 10.1038/oby.2009.167 pmid:19498350.

51. den Besten G, van Eunen K, Groen AK, Venema K, Reijngoud DJ, Bakker BM. The role of short-chain fatty acids in the interplay between diet, gut microbiota, and host energy metabolism. J Lipid Res. 2013;54(9):2325–40. doi: 10.1194/jlr.R036012 pmid:23821742; PubMed Central PMCID: PMC3735932.

52. Nyangale EP, Mottram DS, Gibson GR. Gut microbial activity, implications for health and disease: the potential role of metabolite analysis. J Proteome Res. 2012;11(12):5573–85. Epub 2012/11/03. doi: 10.1021/pr300637d pmid:23116228.

53. Woodmansey EJ. Intestinal bacteria and ageing. J Appl Microbiol. 2007;102(5):1178–86. Epub 2007/04/24. JAM3400 [pii] doi: 10.1111/j.1365-2672.2007.03400.x pmid:17448153.

54. Tiihonen K, Tynkkynen S, Ouwehand A, Ahlroos T, Rautonen N. The effect of ageing with and without non-steroidal anti-inflammatory drugs on gastrointestinal microbiology and immunology. Brit J Nutr.

2008;100(1):130–7.Epub2008/02/19.doi:10.1017/S000711450888871X S000711450888871X [pii]. pmid:18279548.

55. Cummings JH, Macfarlane GT. The control and consequences of bacterial fermentation in the human colon. J Appl Bacteriol. 1991;70(6):443–59. pmid:1938669. doi: 10.1111/j.1365-2672.1991.tb02739.x

56. Tanner SA, Chassard C, Zihler Berner A, Lacroix C. Synergistic effects of*Bifidobacterium thermophilum* RBL67 and selected prebiotics on inhibition of*Salmonella* colonization in the swine proximal colon PolyFermS model. Gut Pathog. 2014;6(1):44. doi: 10.1186/s13099-014-0044-y pmid:25364390; PubMed Central PMCID: PMC4215022.

57. Guo X, Xia X, Tang R, Zhou J, Zhao H, Wang K. Development of a real-time PCR method for Firmicutes and Bacteroidetes in faeces and its application to quantify intestinal population of obese and lean pigs. Lett Appl Microbiol. 2008;47(5):367-73.

58. Ramirez-Farias C, Slezak K, Fuller Z, Duncan A, Holtrop G, Louis P. Effect of inulin on the human gut microbiota: stimulation of Bifidobacterium adolescentis and Faecalibacterium prausnitzii. Brit J Nutr. 2009;101(4):541-50.

59. Furet JP, Firmesse O, Gourmelon M, Bridonneau C, Tap J, Mondot S, et al. Comparative assessment of human and farm animal faecal microbiota using real-time quantitative PCR. FEMS Microbiol Ecol. 2009;68(3):351-62.

60. Bartosch S, Fite A, Macfarlane GT, McMurdo ME. Characterization of bacterial communities in feces from healthy elderly volunteers and hospitalized elderly patients by using real-time PCR and effects of antibiotic treatment on the fecal microbiota. Appl Environ Microb. 2004;70(6):3575-81.

61. Cleusix V, Lacroix C, Dasen G, Leo M, Le Blay G. Comparative study of a new quantitative real-time PCR targeting the xylulose-5-phosphate/fructose-6-phosphate phosphoketolase bifidobacterial gene (xfp) in faecal samples with two fluorescence in situ hybridization methods. J Appl Microbiol. 2010;108(1):181-93.

62. Tymensen LD, McAllister TA. Community structure analysis of methanogens associated with rumen protozoa reveals bias in universal archaeal primers. Appl Environ Microb. 2012;78(11):4051-6.

Chapter 10

A ROLE FOR PROGRAMMED CELL DEATH IN THE MICROBIAL LOOP

Mó nica V. Orellana[1,2]., Wyming L. Pang1[3]., Pierre M. Durand[4,5], Kenia Whitehead[1,6], Nitin S. Baliga[1,7]

[1] Institute for Systems Biology, Seattle, Washington, United States of America

[2] Polar Science Center, Applied Physics Laboratory, University of Washington, Seattle, Washington, United States of America

[3] Genomatica, Inc., San Diego, California, United States of America

[4] Department of Molecular Medicine, University of the Witwatersrand and National Health Laboratory Service, Parktown, South Africa

[5] Department of Ecology and Evolutionary Biology, University of Arizona, Tucson, Arizona, United States of America

[6] Integral Consulting Inc., Seattle, Washington, United States of America

[7] Department of Microbiology, University of Washington, Seattle, Washington, United States of America

ABSTRACT

The microbial loop is the conventional model by which nutrients and minerals are recycled in aquatic eco-systems. Biochemical pathways in different organisms become metabolically inter-connected such that nutrients are utilized, processed, released and re-utilized by others. The result is that unrelated individuals end up impacting each others' fitness directly through their metabolic activities. This study focused on the impact of programmed cell death (PCD) on a population's growth as well as its role in the exchange of carbon between two naturally co-occurring halophilic organisms. Flow cytometric, biochemical, ^{14}C radioisotope tracing assays, and global transcriptomic analyses show that organic algal photosynthate released by Dunalliela salina cells undergoing PCD complements the nutritional needs of other non-PCD D. salina cells. This occurs in vitro in a carbon limited environment and enhances the growth of the population. In addition, a co-occurring heterotroph Halobacterium salinarum re-mineralizes the carbon providing elemental nutrients for the mixoheterotrophic chlorophyte. The significance of this is uncertain and the archaeon can also subsist entirely on the lysate of apoptotic algae. PCD is now well established in unicellular organisms; however its ecological relevance has been difficult to decipher. In this study we found that PCD in D. salina causes the release of organic nutrients such as glycerol, which can be used by others in the population as well as a co-occurring halophilic archaeon. H. salinarum also re-mineralizes the dissolved material promoting algal growth. PCD in D. salina was the mechanism for the flow of dissolved photosynthate between unrelated organisms. Ironically, programmed death plays a central role in an organism›s own population growth and in the exchange of nutrients in the microbial loop.

INTRODUCTION

The microbial loop model is fundamental to our understanding of biogeochemical cycling of nutrients and minerals in aquatic eco-systems [1], [2]. Metabolic processes are distinctly coupled in microbial communities: photosynthetic primary producers (bacteria and/or unicellular algae) release C- and N-based dissolved organic matter (DOM) comprising various organic compounds and amino acids that are readily assimilated and re-mineralized by heterotrophic bacteria/archaea and protozoa [3], [4]. Phytoplankton and bacteria are the main sources of DOM [5], [6], [7], [8], [9]. These biopolymers are produced by several mechanisms including direct release [10], mortality by viral lysis [11], [12], [13], [14], [15], regulated exocytosis of metabolites and polymergels [16], grazing [17], [18] and apoptosis [19]. Apoptosis, the commonest phenotype of programmed cell death (PCD), is well documented in chlorophytes[20], [21] allowing cellular materials to become dissolved in the environment.

Much is known about C and N cycling in the aquatic microbial loop; however, in hypersaline environments the interactions are largely unexplored. The physiological complexity of the microbial population in these environments offers the potential for a staggering number of interactions and biogeochemical interdependencies [22]. To address this gap in knowledge we focused on one such environment, the Great Salt Lake (GSL) in Utah, USA. Photosynthetic eukaryotes and halophilic archaea (hereon «haloarchaea" in the GSL metabolize vast amounts of C (145 gC m-^{2} year^{-1}) [23]. Important players in this environment are Dunalliela salina a unicellular photosynthetic chlorophyte, highly adapted to large changes in salinity, pH and temperature and Halobacterium salinarum, a halophilic archaeon that also thrives in environments with a range of salinities ($116.88–292.2$ PSU, temperatures

(30° C–50° C), light intensities, oxygen tension and nutrients [24]. For both organisms glycerol is a major component of the C cycle. In response to salinity stress, D. salina enhances CO_2 assimilation and channels the carbon and energy resources towards synthesis of glycerol (reaching internal concentrations as high as 7 M [25]), which it uses as an osmoprotectant [22]. When stressed, up to 17% of this glycerol can be found extracellularly and concentrations can reach as high as 30 μM after the demise of Dunaliella blooms [26]. H. salinarum is also well adapted to hypersaline conditions [27], [28], [29], [30]. The archeon responds well to a range of environmental stresses [31], [32] including low oxygen tensions [33], [34], [35] and fluctuating nutrients [30], [36]. Due to their adaptations to these environments, Halobacterium spp. can reach high abundances. H. salinarum and D. salina co-habit and are the most important organisms involved in hypersaline biogeochemistry [24]. Little is known about their physiological interactions, which became the immediate focus of this study.

While examining the flow of C (using glycerol as a proxy of DOM [26] between H. salinarum andD. salina a second question emerged. It was observed that PCD occurred in D. salina and its role in the nutrient exchanges between the two organisms was examined further. Algal release of DOM by a process that results in cell death has been reported and discussed primarily as a response to control cell growth under stressful and nutrient limiting conditions [19], [37], [38],[39], [40], [41], [42] (although not under carbon limiting conditions as described in this manuscript). However, the implications of these arguments are seldom examined in any detail and may rightly be criticized as naïve group selectionist thinking [43]. There are many possible evolutionary explanations (both adaptive and non-adaptive) for PCD in the unicellular world[42]. The fitness effects of unicellular PCD have also been examined in at least two model organisms: Chlamydomonas reinhardti [44], a relative chlorophyte of Dunalliela, and the yeast Saccharomyces cerevisiae [37] and been considered in phytoplankton ecology [45]. Here, we examined the role of PCD on a chlorophyte›s population growth and the exchange of nutrients between two co-occurring organisms.

Surprisingly, we found that the growth rate of a D. salina culture can increase even though a significant proportion of its own population 55%+/−15 undergoes PCD at night after daytime growth. The benefit of releasing photosynthetically fixed C by PCD is achieved by increased growth measured as C-assimilation due to a positive metabolic feedback loop from itself when in pure culture, as well as from a co-habiting heterotrophic archaeon. Furthermore, the heterotrophic haloarchae Halobacterium salinarum can subsist entirely on the lysate of apoptotic algae. These data indicate that not only can PCD in D.

salina benefit others of its own kind, but carbon released in PCD materials is recycled between the chlorophyte population and co-occurring unrelated haloarchaea.

MATERIALS AND METHODS

Ecological Sampling

Water was sampled at the South (Lat: 40°43′N Long: 112°13′W) and North (Lat: 41°26′N Long: 112°40′W) Arms of the Great Salt Lake (GSL; UT, USA) with autoclaved bottles and transported to the laboratory on ice. These studies do not involve protected species and no specific permits were required for the described field studies in the public areas of the GSL.Dunaliella spp. and haloarchaeal cells were isolated by enrichment in Minimal Media (MM1) liquid media (Dunaliella) at salinities ranging from 36, 75, 100, 150, 200, and 250 PSU. MM2-agar containing plates at the same salinities as for Dunaliella were used for haloarchaeal cells (see Supplement below).

Species Identification by DNA Sequencing

To survey the diversity of organisms that naturally co-exist in hypersaline environments we enriched photoautotrophs from the South-Arm of the GSL by inoculating water samples into MM1 media containing only salts at different concentrations (75, 100, 150, 200, 250 PSU) (Fig. S1 A). Exponentially growing Dunaliella spp. cells were harvested by centrifugation (2300 rcf×4 min); DNA extraction and purification were done using DNeasy Plant Mini Kit (Qiagen), and quantified spectrophotometrically. Species-specific internal transcribed spacers (ITS-1, and ITS-2) of the RNA genes, were amplified using the primers described in [46]. Secondly, to identify co-existing microbial species, aliquots of GSL water were plated onto nutrient rich solid agar medium over a similar salinity range as for the autotrophs. Bright pink, red and orange colonies grew (Fig. S1 B) that have characteristic colony morphologies of haloarchaea due to their high cellular content of bacterioruberins, carotenoids, and rhodopsins. Isolated colonies of haloarchaeal cells (Fig. S1 B) were then grown to exponential phase and harvested by centrifugation. 16 s rRNA sequences were amplified from genomic DNA extracts according to the PCR protocol described by [47]. All amplicons were sequenced using an ABI 3730xl DNA Analyzer.

Culture Conditions

For single and co-culture experiments, axenic batch cultures of Dunaliella salina (Culture Collection of Algae and Protozoa, UK, CCAP 19/18) were grown in artificial sea water (ASW[48] containing salts reaching a total of 200 PSU and enriched with nutrients as in f/2 media[49] called MM1. Cultures were grown under a 13 h:11 h light:dark photoperiod and at 150 µmol photons m^{-2} s^{-1} (verified by a Li-Cor 191SA (Li-Cor Inc.) at 30°C and 100 rpm shaking. Cells were entrained to this light regimen for 3 tandem culture transfers until the growth rate did not change (3 weeks) prior to the experiments. Growth rates were determined by the change in cell number and/or change in red autofluorescence (680 nm) over time. H. salinarum NRC-1 was maintained in CM media [50] at a salinity of 200 PSU and for co-culture experiments cells were grown in MM1 enriched with nutrients akin to DOM, including a mixture of amino acids (see supplement Table S1, for details) and acclimated to the same light:dark photoperiod as D. salina. The inoculation density of co-cultures was 10^5 ml^{-1} D. salina cells, and 10^8 ml^{-1} H. salinarum cells. All experiments were run in triplicate. D. salina cells were counted by flow cytometry, H. salinarum was measured as optical density at 600 nm (OD$_{600}$ of 1.0=8×10^8cells/ml).

Duirnally Synchronized Syntrophic Interaction between *D. salina* and *H. salinarum*

To investigate the physiological exchange and interplay between D. salina and H. salinarum, supernatant from a light/dark adapted D. salina culture was collected and seeded with H. salinarum. Culture growth was monitored at 37°C for 7 days using a BioscreenC high throughput microbial growth analysis instrument. Sterile MM1, MM1 with 400 mM glycerol and MM1 supplemented with amino acids (60 µM; see supplement Table S1 for a list of amino acids) commonly found in hypersaline ecosystems were used as controls.

Single Cell Analysis and Fluorescent Cell Stains

An Influx flow cytometer (Cytopeia) using a Coherent Innova 305C argon ion laser excitation source tuned at 488 nm and 200 mW was used to quantify cell abundance. Cell count was based on forward angle light scatter (FALS) and red fluorescence collected with a 610 nm long pass and a 700/50 nm band pass filter. Sampling was done every two hours with a calibrated robot. The cells were fixed in paraformaldehyde (0.1%) and kept at 4°C until counting. Yellow/green fluorescent 1µm microspheres (Polysciences, Warrington, PA, USA) were used to calibrate gain and object detection threshold settings and as an internal fluorescence standard for normalization and counting.

Intracellular glycerol-containing vesicles in D. salina cells were observed according to standard protocols by staining with the highly specific aminoacridine dye 1 µM Quinacrine [16], [51], [52]for 15 minutes and washed twice in 200 PSU saline. Cells were analyzed for green fluorescence of quinacrine (ex: 488 nm, em: 504–523 nm, 1 µM) using a 500 nm long pass (LP) and a 530/30 nn band pass (BP) filter and red autofluorescence of chlorophyll using a 610 nm LP and a 715/50 nm BP filters. Loss of quinacrine fluorescence from pre-labeled cells indicates vesicle secretion.

DNA in live cells was detected by staining with SYBR I (λex=497 nm/ λ_{em}=520 nm, 3 µM Invitrogen) [53] or DAPI (4',6-diamidino-2-phenylindole, dihydrochloride) (ex: 350 nm em: 450 nm, 3 µM) and analyzed with a BD FACS ARIAII flow sorter and a Delta Vision confocal microscope. SYTOX® blue (ex: 440 nm, em: 525 nm, 3 µM), an impermeable probe, was used to detect dead cells with compromised plasma membranes. It was used in combination with the Annexin V stain for PCD detection (see below). Dunaliella sub-populations were defined by a combination of their forward scattering characteristics and their red autofluorescence of chlorophyll a. Populations were also identified by their red autofluorescence of chlorophyll a in combination with the fluorescence of the labeled probes used (FITC-Annexin, SYTOX® blue) and subsequently sorted. After sorting with the BD FACS ARIAII, the cell›s membrane and nucleus integrity were evaluated and photographed with a Delta Vision confocal microscope. Data analysis was done using FlowJo analytical software (Tree Star).

Programmed Cell Death (PCD) Analyses

Three independent markers of PCD in chlorophytes were used [54], [55]: phosphatidylserine (PS) externalization, caspase activity and observation of the typical morphological changes associated with PCD in algae. Caspase activation and morphology are qualitative assessments; PS externalization allows qualitative and quantitative (cell enumeration) evaluation. The TUNEL assay (detection of DNA fragmentation flow cytometrically) was avoided for cell quantitation as it may detect non-specific double stranded DNA breaks, which occur during necrosis [56] and its specificity has been questioned by [45].

1. PS externalization was detected by fluorescein isothiocyanate (FITC)-labeled Annexin V adapted to the manufacturer›s suggested protocol for Dunaliella (BD Pharmingen). Annexin is a a Ca^{2+}-dependent phospholipid-binding protein that binds with high affinity for PS. PS is located on the cytoplasmic surface of the cell membrane, except in PCD when PS migrates to the outer surface of the plasma membrane.

FITC-Annexin was used in combination with SYTOX® blue (SB), an impermeable probe that only detects dead cells when the plasma membrane is disrupted. Experiments were done in triplicate. Cells were photographed in a Delta Vision microscope.

2. Cysteine protease activity of caspase-3, which is specifically associated with PCD inDunalliela spp [54], [55], was measured using a Caspase-3 Fluorometric Assay Kit (Assay Designs Catalog No 907-014) according to the manufacturer)s instructions. Caspase-3 exists as a proenzyme, becoming activated during the PCD pathway. This assay measures the conversion of a non-fluorogenic peptide Ac-DEVD-AMC substrate for caspase-3 to a fluorogenic product that emits light at 400 nm when excited at 360 nm. Caspase-3 activity was calibrated with a solution of 7-Amino-4-methyl coumarin at 5 μM in reaction buffer at 30°C and expressed as units of fluorescence relative to the equivalent fluorescence obtained from 5.56 units of fully active Caspase-3 when reacting with the Caspase-3 substrate according to the assay kit.

3. Cellular morphological changes associated with PCD including cell shrinkage, vacuolization, plasma membrane blebbing, nuclear condensation and ejection of the nucleus were examined using a Delta Vision confocal microscope. Microscopic detection of PS externalization was also performed.

Radioisotope Tracing

C-flux was determined with radioisotope tracing by addition of $NaH^{14}CO_3$ to an activity of $1\mu Ci^{14}C$ ml^{-1} to light:dark acclimated pure and co-cultures of D. salina and H. salinarum at an inoculating ratio of 10^5 D. salina cells: 10^8 H. salinarum cells. Total activity was determined using phenylethylamine that stabilizes radioactive counts [57]. D. salina and H. salinarum cells were collected by 2 μm and 0.22 μm filtration, respectively, and washed in 200 g/L saline. ^{14}C uptake was halted with 250 μL 6 M HCl, incubated at room temperature (RT) for 30 min and prepared for counting with the addition of Ecosint (National Diagnostics). Samples were counted (disintegrations per minute, DPM) using a Tri Carb 2810 TR (Perkin Elmer) scintillation counter and carbon incorporation calculated according to [58] with modifications.

D. salina Glycerol Production and Release

Glycerol, an important constituent of DOM in hypersaline environments [26], was measured as follows. For bulk glycerol release experiments, 2 ml samples from pure D. salina and D. salinaplus H. salinarum co-cultures were collected in 14 mL falcon tubes at each time point. A 500 μL aliquot was removed and fixed in paraformaldehyde to a final concentration of 0.1% for cell counting via a hemocytometer and flow cytometry. One-mL of the remaining sample was centrifuged (3500 rcf) for 15 min and the supernatant removed for extracellular glycerol measurements. The D. salina cell pellet was resuspended in 500 μL ddH$_2$O to lyze cells and release intracellular glycerol. Glycerol measurements were done using a free glycerol detection kit (Sigma F6428) as per the manufacturer's instructions.

Transcriptional Response of *H. salinarum* to *D. salina* Dissolved Organic Matter (DOM)

Microarrays were generated at the Institute for Systems Biology Microarray Facility and processed according to standard protocols [59]; [60]. Each microarray slide contained 70mer oligonucleotides for each of the 2400 genes of Halobacterium salinarum [61] spotted in quadruplicate at two spatially distinct locations. Labeling, hybridization and washing were performed as described previously [31]. Statistical significance of differential gene expression was determined using the maximum likelihood method [62]. Data reported in this paper has been deposited in the Gene Expression Omnibus (GEO) database record GSE45752 (http://www.ncbi.nlm.nih.gov/geo/query/acc.cgi?acc=GSE45752).

Carbonate System.

Total CO$_2$ and Alkalinity were calculated according to Dickson et al. [63](http://andrew.ucsd.edu/co2qc/index.html), pH was measured spectrophotometrically [64], and the salinity and temperature of the cultures (200 PSU, 30°C), and ambient pCO$_2$ measurements (pCO$_2$=400) were used for calculations.

Nutrients.

Phosphate, nitrate, nitrite, and ammonia were measured at the University of Washington Marine Chemistry Laboratory (http://oceanweb.ocean.washington.edu/services/techservices.html).

Growth Model

A growth model for both pure and co-cultures of D. salina and D. salina + H. salinarum was developed using the technical computing environment MATLAB (Mathworks, Inc.). The model was defined as a single birth-death ODE for the change in total cell density,

$$\dot{N} = \gamma \left(1 - \frac{N}{\kappa M} \right) NI - \delta(1 - I)$$

(1)

with light irradiance,

$$I = \begin{cases} 0 & , \quad \text{dark} \\ 1 & , \quad \text{light} \end{cases}$$

(2)

a "memory" variable,

$$\dot{M} = \begin{cases} 0 & , \quad I = 1 \\ (N - M) - \delta M(1 - I) & , \quad I = 0 \end{cases}$$

(3)

and parameters γ, δ, and κ representing the maximum cell growth rate, cell death rate, and a saturation scaling factor, respectively.

Cellular growth, defined by the first term of Eq. (1), only allows N to increase during the light phase (I=1) with a saturation level defined by κM (a Droop growth model) [65]. Importantly, Eq. (3) is such that $\dot{M} = 0$ for light phases, providing a constant saturation threshold therein. Conversely, M resets to the value of N during dark phases. Thus, M allows the saturation level to increase with each diurnal cycle, a characteristic observed in experimental data. Cell death, defined by the second term of Eq. (1), only occurs during the dark phase (I=0), and is modeled as simple exponential decay based on the assumption that the cells have little to no tolerance to darkness.

Light irradiance (I, Eqn. (2)) was simulated as a square wave over a range of 0 (dark) and 1 (light) with a duty cycle equivalent to the nominal experimental photocycle of 13 hr:11 hr light:dark (45.8%).

The model was fit to experimental data by iteratively adjusting the values of γ, δ and κ until the sum of squared errors between the predicted trajectory for N (numerical solution of Eqns. (1)–(3)) and experimental data was minimized. Model fits were done individually on each experimental replicate within each culture set to generate statistics for each parameter.

RESULTS

Microbial Organisms (Chlorophytes and Bacteria) in the Great Salt Lake

The enriched algal population (**Fig. S1A**)was identified as predominantly belonging toDunaliella spp, single-celled wall-less photosynthetic chlorophytes, specifically D. salina, D. pseudosalina, D. viridis, D. parva, and a few other species (**Fig. 1B; Fig. S2, A & B**). The microbial populations were all confirmed as halobacteria, with several direct matches to H. salinarum (**Fig.**

S2 C), a halophilic photoheterotrophic archaeon that thrives over a wide range of salinity (135–300 PSU). These results confirm that Dunaliella spp co-exist with diverse halobacteria in the South Arm of the GSL [24]. Dunaliella spp density fluctuated over several months between 10^3–10^8/L and the density of haloarchaea fluctuated between 10^6 and 10^9/ml. Thus we choose D. salina and H. salinarum as a model system to study their interaction in laboratory experiments.

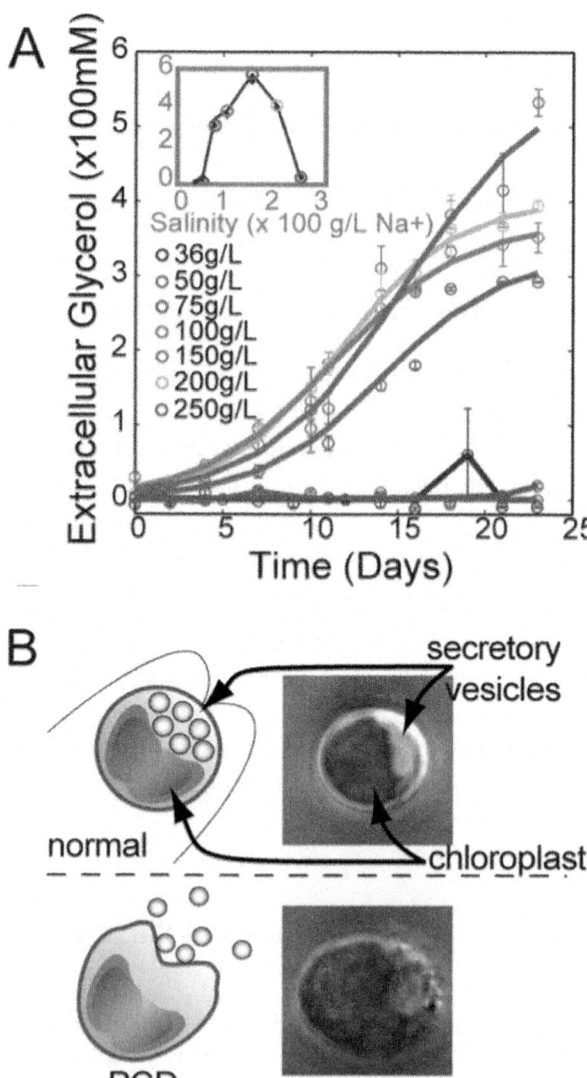

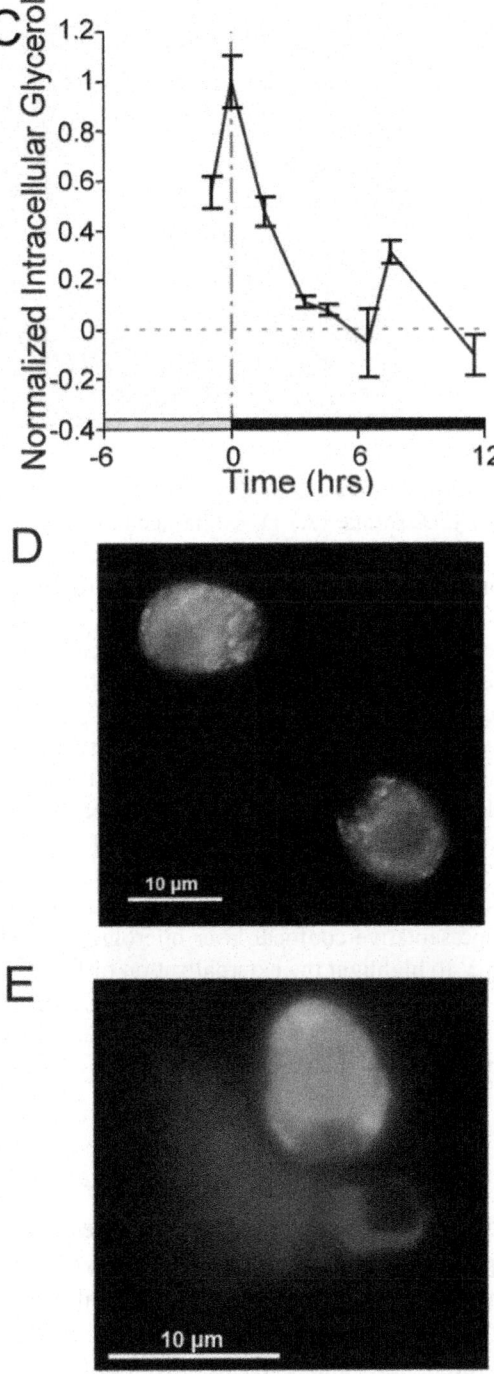

F

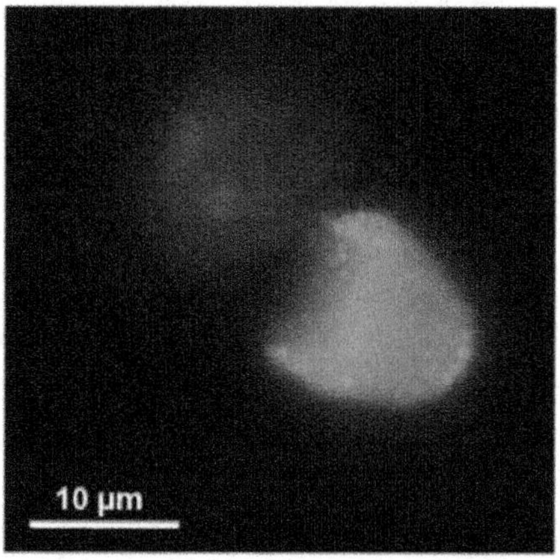

Figure 1. Cell death upon exposure to darkness or H.salinarum **cells triggers the release of glycerol by** D. salina. (A) D. salina accumulates and utilizes glycerol as an osmoprotectant in hypersaline growth conditions. Accumulation of glycerol in D. salina cultures is correlated to increasing salinity in the growth medium peaking at 150 PSU (inset). (B) D. salinareleases glycerol by cell death. Illustrated and merged phase contrast/fluorescence photomicrographs of a D. salina cell undergoing cell death. D. salina stores glycerol and other byproducts of photosynthesis inside secretory vesicles that are localized to the apical flagellar pole (top). The green color of the vesicles is due to quinacrine staining of glycerol and the red fluorescence corresponds to chloroplasts. The image show dramatic disruption of the cell membrane and complete loss of internal glycerol in a cell that has undergone cell death (bottom). (C) A shift of live light acclimated cultures (100–150 μmol-photon m^{-2}sec^{-1}) to complete darkness (0 μmol-photon m^{-2} sec^{-1}) results in release of glycerol by D. salina. The intracellular glycerol is measured by flow cytometry analysis of quinacrine stained vesicles D. salina cells. (D) Representative confocal laser micrograph of D. salina cells stained with FITC-Annexin V to highlight the externalization of PS (green fluorescence), and red corresponds to red chlorophyll autofluorescence. (E) Representative confocal laser micrographs of dead D. salina stained with FITC-Annexin V and SYTOX® blue highlighting PS completely externalized and the ejection of the nucleus indicating cell death.

Mechanism of DOM Production by *D. salina*

The mechanism for release of photosynthetically sequestered C by D. salina, the predominant primary producer in halophilic environments [24], [26], was investigated. In response to routine increases in salinity, D. salina enhances CO_2 assimilation to produce and store glycerol for use as an osmoprotectant [66]. This was corroborated here and glycerol production by D. salinapeaked at its preferred salinity range between 150–200 PSU (**Fig. 1A**). Microscopy and quinicrine fluorescence revealed that glycerol stores are localized in vesicles **(Fig. 1B)** in D. salina. It is known that moieties for export are stored in vesicles [16], [67] and in natural settings, up to 17% of this glycerol reserve is released along with other uncharacterized metabolites into the environment [26].

To characterize the timing of glycerol release as a proxy of DOM release with respect to the diurnal cycle, intracellular glycerol was measured during the transition from day to night. There was a precipitous drop in intracellular glycerol upon transfer of light acclimated($\geq$150 μmol photon m^{-2} s^{-1}) culture to complete darkness (0 μmol photon m^{-2} s^{-1}) [40] **(Fig. 1 B & C)**. A high percentage of cells releasing glycerol and other uncharacterized C and N metabolites died during this process and exhibited membrane damage. The percentage of FITC-annexin V in combination with SYTOX® blue labeled cells **(Fig. 1 D & E)** detected the externalization of PS, confirming that 65+/−15% of cells were undergoing PCD. The proportion of apoptotic cells increased after the onset of darkness with a mean value of 59%. Previous observations also revealed that D. salina cells undergo PCD when placed in complete darkness and that this process is blocked by inhibitors of caspases, which are the effectors of apoptosis[40]. The cells stained with FITC-Annexin V showed the range of PCD-related morphological changes, from membrane blebbing to ejection of the cell nucleus followed by complete dissolution of the cells into the media **(Fig. 1 E & F)**. In addition, a one-to-one correlation between numbers of cells undergoing PCD and amount of extracellular glycerol was observed (see below). These findings indicate that glycerol and other uncharacterized PCD materials are released by active cell death in D. salina in hypersaline environments. The stimulus for PCD and subsequent glycerol and DOM release was the onset of darkness.

Duirnally Synchronized Syntrophic Interaction between *D. salina* and *H. salinarum*

Neither the MM1 media nor glycerol supplementation alone supported H. salinarum growth. The PCD supernatant from the D. salina culture containing DOM (glycerol, plus unknown C and N containing nutrients not measured), on the other hand, was nearly as supportive of haloarchael growth as MM1 supplemented with amino acids (Fig. 2). Thus, DOM released by diurnally synchronized cultures of D. salina fully complements nutritional requirements of H. salinarum.

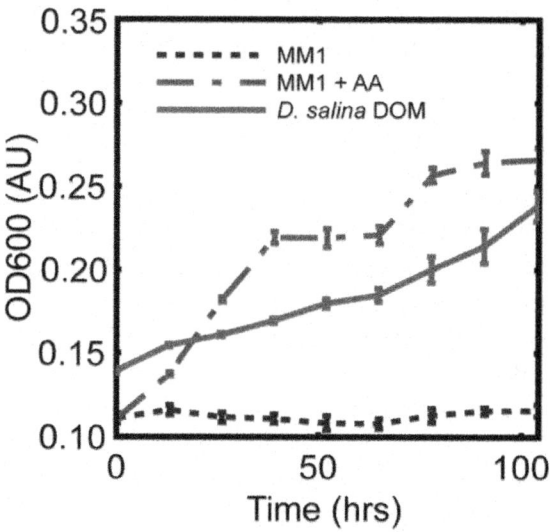

Figure 2. Dissolved organic material (DOM or photosynthate) released by D.salina **fully complements nutritional requirements of** H. salinarum. Supernatant of D. salina culture in artificial seawater (MM1) supported H. salinarum growth at a level that was comparable to its growth in MM1 supplemented with amino acids at naturally occurring concentrations.

Next, the hypothesis that D. salina benefits from its association with H. salinarum was explored. The intracellular glycerol concentration in both pure and co-cultures cultures, showed a slight drop in concentration during the night, and ranged between 5–2 M over the three day experiment (Fig. 3A). While the extracellular glycerol and possibly other nutrients in the DOM accumulated in both cultures over time, the rate of accumulation was significantly slower in co-cultures with H. salinarum relative to pure cultures of D. salina (Fig. 3B). This slower accumulation of glycerol and nutrients in co-cultures was partly explained by radioisotope incorporation and tracing, which revealed that a) D. salina incorporated its own released DOC (dissolved organic carbon) at night when in pure culture (Fig. 3C) and that b) H. salinarum incorporated ^{14}C-labeled nutrients from the algal DOM and released metabolized products. However, H. salinarum did not incorporate NaH^{14}CO$_3$ by itself (Fig. 3D). Subsequently at nightD. salina re-assimilated DOM, specifically dissolved carbon, in the presence of haloarchaea as shown by at least two-fold higher productivity (C/cell/h) relative to pure culture (Fig. 3C). This is consistent with the heterotrophic capabilities of Dunaliella spp. [68] and known abilities of other marine microalgae to nocturnally assimilate organic compounds suggesting that this might be a general behavior and not unique to hypersaline algae [69].

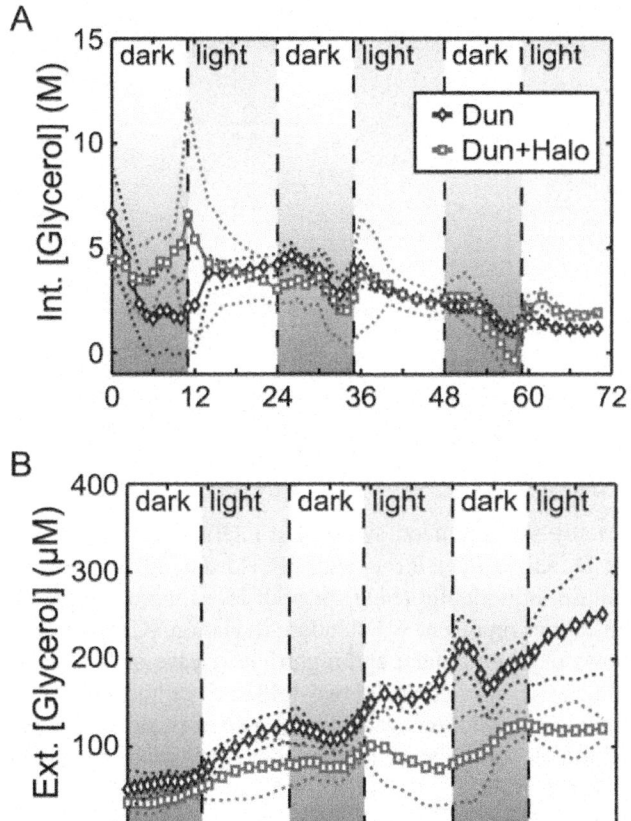

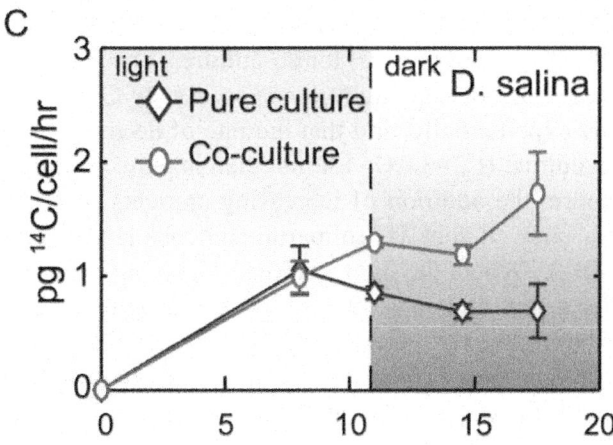

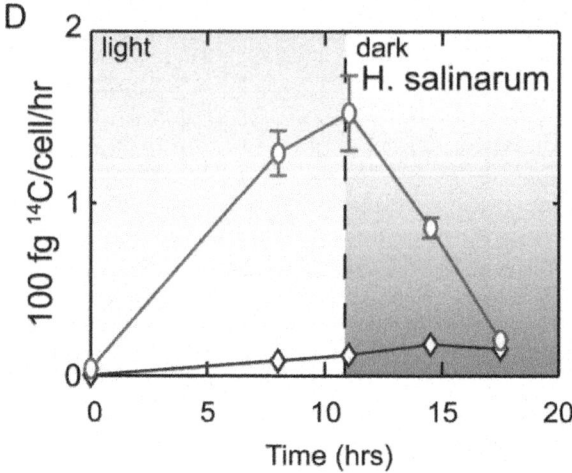

Figure 3. Diurnally synchronized syntrophic interaction with H.salinarumincreases productivity of D. salina. (A) Intra- and (B) extra-cellular glycerol concentrations in D. salina culture individually (blue) or with H. salinarum (red) over several day: night cycles, dash lines represent +/− standard deviation. (C) Radiolabel incorporation and tracing shows daytime uptake and nighttime release of ^{14}C by D. salina. Uptake of ^{14}C by D. salina at night is enhanced two-fold in co-cultures relative to pure cultures indicating nighttime assimilation of ^{14}C in presence of H. salinarum. (D) Simultaneous tracing of C within H. salinarum cells demonstrates uptake and processing of ^{14}C in sync with the diurnal cycle.

   To understand the implications of darkness induced cell death on algal population dynamics we constructed a growth model from experimental measurements of total cell counts (see methods) (**Fig. 4A**). The fitted parameters are shown in **Table 1**. The R-squared values (adjusted for degrees of freedom) for model fits to experimental data were 0.794+/−0.04 (s.e.m.) for pure and 0.702+/−0.09 (s.e.m.) for co-cultures. Our previous experimental data shows cell death during nighttime (**Fig. 4A & C**). Our model predicts that this occurs exponentially, and that the rate of decay is significantly faster (p<0.05) in co-culture ($t_{1/2}$=5.9+/−1.4 hrs) than in pure culture ($t_{1/2}$=10.1+/−2.8 hrs). Furthermore, the addition of increasing amounts of H. salinarum to D. salina cultures showed that H. salinarum independently induces algal cell death (**Fig. 4B**). When daylight returns, surviving cells resume growth, saturating after a few divisions. Notably, the light induced burst growth rate is similar for both pure (t_{doub}=1.1+/−0.6 hrs) and co-cultures (t_{doub}=1.3+/−0.6 hrs). This process iterates over sequential day/night cycle resulting in daytime regeneration and nighttime drops in the algal population. Overall, both pure and co-cultures maintain a net positive growth rate, with the algal population doubling approximately every 13 hrs and 20 hrs in pure and co-culture, respectively, consistent with previously reported growth rates for D. salina [70].

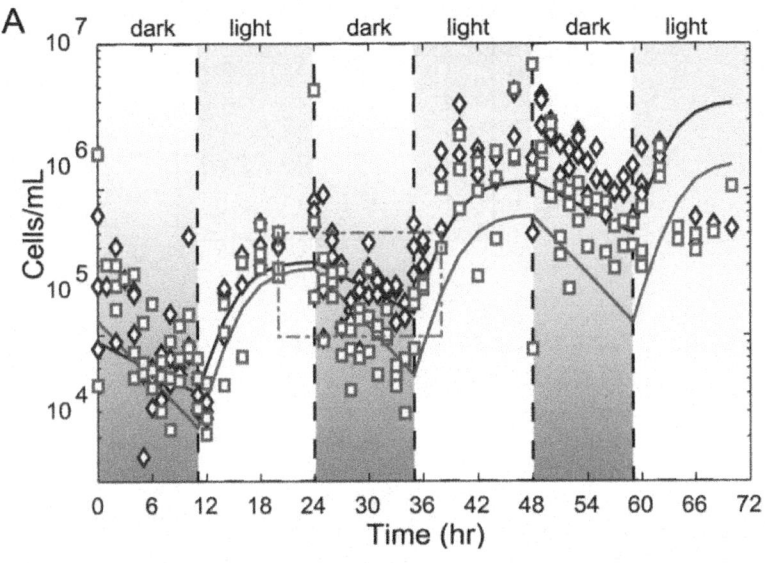

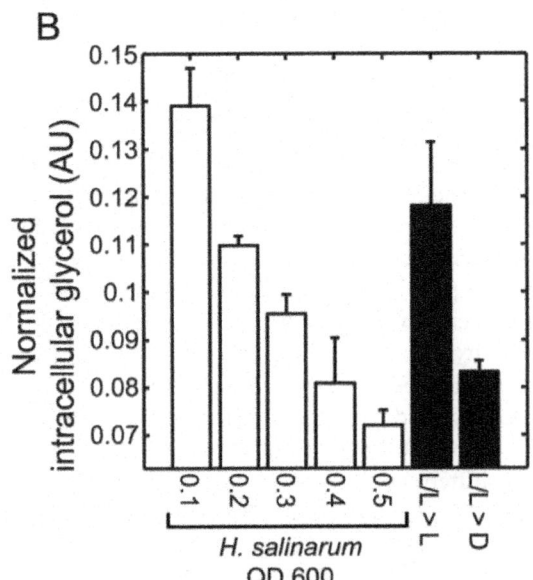

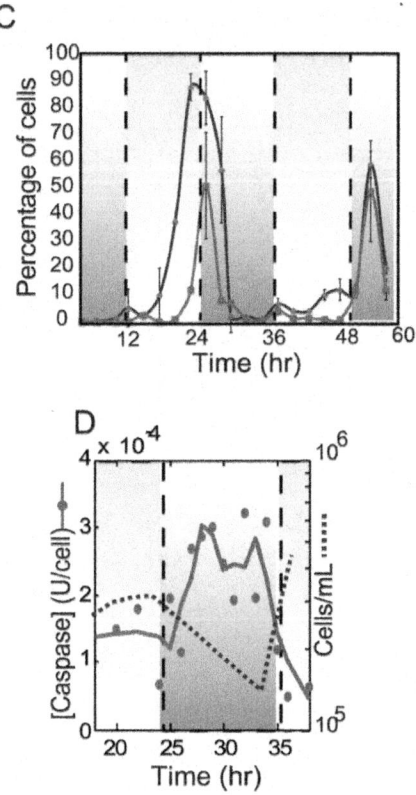

Figure 4. Cell death is triggered at nighttime as part of the diurnal synchronized program of D.salina. (A) Cell numbers for D. salina in pure and co-cultures with H. salinarum over several diurnal cycles. Live cell concentration measured using flow cytometry are indicated with blue (pure culture) and red (co-culture) points while lines are fitted model simulations. Green boxed region indicates time frame reported in Fig. 3D over which caspase-3 activity was assayed. (B) H. salinarum induces cell death in D. salina under continuous light regime. Intracellular glycerol within D. salina was stained with quinacrine and quantified with flow cytometry. Decrease of intracellular glycerol proportionally with higher cell density of H. salinarum. Unstimulated (pure D. salina culture and dark shifted samples are shown as controls. (L/L>L, cultures grown on a 24 h constant light regime maintained in the light during the measurements, LL>D, cultures grown in constant light shifted to dark conditions (0 μmoles m²s⁻¹). (C) The decrease in D. salina cell number in the model (**Fig. 4A** (blue line) due to cell death is supported by the time course of annexin V labeled cells (blue line) indicating percentage of cells exhibiting externalization of PS and SYTOX® blue stained cells indicating the percent dead cells (red line). (D) The decrease in cell number in the model (blue dotted line) due to cell death is also supported by higher levels of caspase-3 during nighttime. Red line is Savitsky-Golay smoothed (span of 5) average of two replicate measurements for each time point.

Table 1. Initial and fitted parameter values for model of growth for pure D. salina and D. salina + H. salinarum **co-cultures**

	Description	Units	Pure Culture			Co-Culture		
			Initial	Fitted	±CI$_{95\%}$	Initial	Fitted	±CI$_{95\%}$
N_0	Initial cell density	Cells/mL	9000	–	–	12500	–	–
M_0	Initial cell density saturation threshold	Cells/mL	9000	–	–	12500	–	–
γ	Burst growth rate	hr^{-1}	0.6262	0.678	0.406	0.4940	0.551	0.280
δ	Death rate	hr^{-1}	0.0957	0.069	0.019	0.1141	0.119	0.027
κ	Saturation scaling factor	–	12.5813	8.10	2.73	12.5813	9.93	10.13

"±CI$_{95\%}$" are parameter 95% confidence intervals such that the lower and upper bound of estimated values are X-CI$_{95\%}$ and X+CI$_{95\%}$, respectively.
doi:10.1371/journal.pone.0062595.t001

Furthermore, the D. salina cells externalized PS (a marker of PCD) a few hours before the actual death, the cells start dying just before the night period starts, with a peak immediately after the lights change into the dark period. These data indicate that death occurs at night, while cell numbers double during the day (**Fig. 4C**). This was also confirmed by measurements of caspase-3 activity, which were higher during times of cell death and low during the growth phase (**Fig. 4D**), and the morphological changes (**Fig. 1D & E**). Importantly, the intracellularly measured glycerol (~4 M, assuming a volume of a secretory granule=1 μm [71]) content in D. salina combined with the number of cells observed to undergo nocturnal cell death (~3×10^5 mL^{-1}) accounted for the majority of the extracellularly measured glycerol (~70 μM).

We also analyzed the transcriptional response of H. salinarum when challenged with the DOM released by D. salina (**Fig. 5**). At least 50 genes were differentially expressed and upregulated, of which nearly half are not annotated and are of unknown function [72], [73]. Those that had some putative functional assignment were mostly up regulated and included genes for siderophore biosynthesis, proteases, transport, metabolism, and cell division. Differential expression of a subset of these functional categories is consistent with known mechanisms for stimulation of algal productivity, nutrient uptake, and growth [74]. Similarly, up regulation of DMSO fermentation enzymes [75] and cytochrome oxidase suggests putative energy production mechanisms in haloarchaea engaged in this interaction, although DMSO production by D. salina is yet to be demonstrated. The upregulation of hemolysin a pore forming enzyme damaging to chlorophyte membranes may have accelerated D. salina's cell death. Furthermore the induction of a putative protein transporter, hydrolases and several proteases might be important for the hydrolysis of macromolecular components in the algal photosynthate. Notably, proteases and other hydrolytic enzymes are known to be abundant in the marine and freshwater DOM [2], 76 In terms of the physical architecture of this interaction, we also observed that haloarchaeal cells were often physically associated with the algal cells both in co-cultures of laboratory strains and in the natural environment (**Fig. S3 A and B**), analogous to cellular interactions observed in the phycosphere [77], [78].

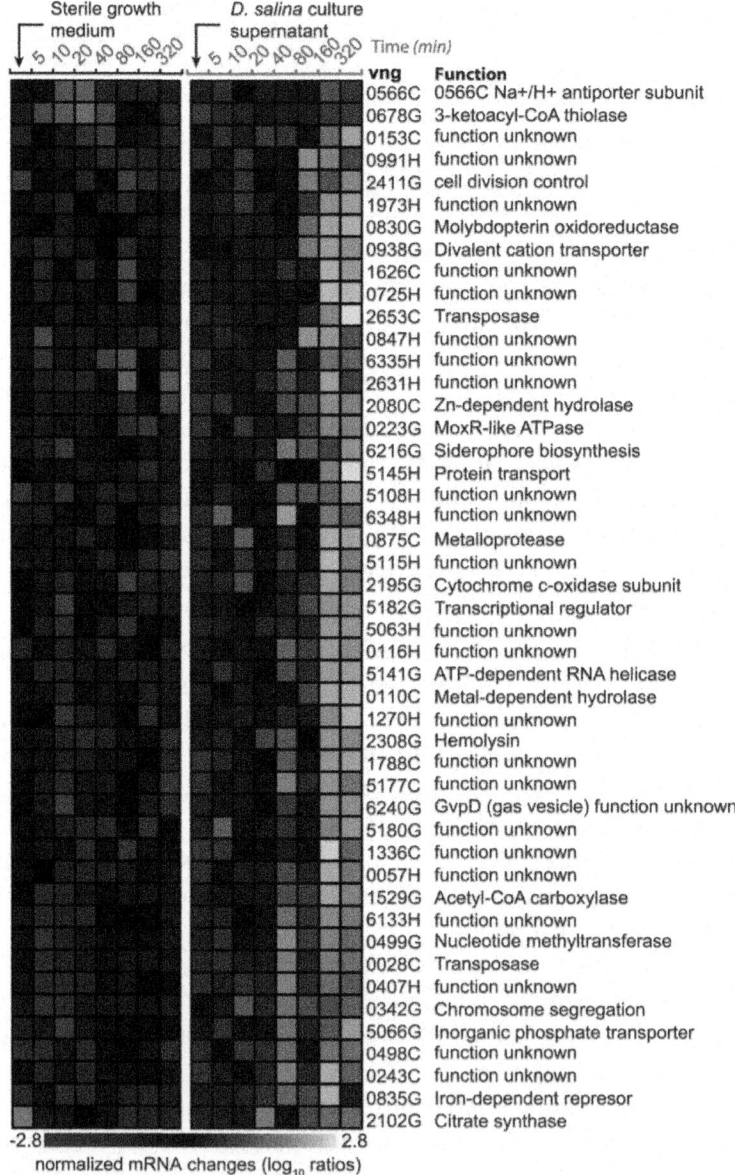

Figure 5. Mechanisms of communication and interactions in the syntrophic interaction. Transcriptional response of H. salinarum NRC-1 to D. salina conditioned artificial seawater amended with nutrients (MM1).

DISCUSSION

Interactions between members of different species play a fundamentally important role in all ecosystems and are driven by natural selection [79]. Interactions between bacteria and phytoplankton are well studied; bacteria exist as free living organisms in the phycosphere [78]attached to the surface of algal cells [77], intracellularly as algal symbionts [80] or attached to aggregates of cells and phytoplankton exopolymeric products [81] or microgels [82]. In contrast, archaeal-phytoplankton interactions, including their role in biogeochemical cycles, are poorly understood. Given the abundance of archaea on our planet (~20% of total biomass on earth[83]), eukaryotic-archaeal syntrophic interactions and their metabolic mechanisms are central to understanding the cycling of C, N, and P in aquatic systems.

To explore this gap in knowledge, the cycling of C (in the form of glycerol) between D. salinaand H. salinarum in a hypersaline environment, pure cultures and co-cultures of these species were used as a model. We found that darkness-induced PCD in D. salina, results in the active release of cell constituents such as glycerol and other unknown C and N, and P nutrients that can be used by others. At first glance, this seemingly altruistic behavior might appear perplexing. However, at high salinities (200 PSU) and high temperatures (30°C) CO_2, the main source of carbon for photosynthesis, is limiting in the laboratory cultures and the GSL [84]. $CO_2$is very insoluble at high salinities (200 PSU) and high temperatures (30°C)[85], reaching a total concentration equal to 6.4 µM and an alkalinity equal to 76.8 µM in the experimental conditions (see methods [63]). This is similar to the titratable alkalinity in the GSL of 78 µM [23]. Furthermore, in the oceans even at lower salinities TCO_2 is known to fluctuate from 10–15 µM, already posing a carbon limitation on phytoplankton cells [86]. Although D. salina possesses a carbon concentrating mechanism (CCM) [87], microbial mats from brine pools and hypersaline lakes have a diffusion-limited assimilatory pathway in which the isotopic fractionation by ribulose1,5- bisphosphate carboxylase is completely suppressed indicating CO_2 limitation [85]. Furthermore CCMs can be modulated by temperature [88]. Thus, under the laboratory conditions designed to mimic the conditions experienced at GSL when other nutrients are present (NO_3: 160 µM), carbon appears to play the role of a limiting nutrient [14], [19], [89],[90]. These data show that nighttime induction of cell death in a fraction of the algal population is a regulated process with no long term detriment to the net size of the algal population, on the contrary, PCD provides a carbon source to others that enhances population growth (**Fig. 4A**) in a carbon limited environment. This is an apparent evolutionary adaptation to live in such a hypersaline, carbon deplete environment.

Interestingly, D. salina photosynthate also complements the nutritional needs of H. salinarum, which independently induces algal cell death (**Fig. 4B**). The archaeon is able to metabolize and remineralize DOM produced by PCD and carbon metabolites thereby providing elemental nutrients for the mixoheterotrophic chlorophyte (**Fig. 3C**). These data describe previously unknown physiological interactions in the hypersaline microbial loop and serendipitously, provided new molecular insights into the sociobiology of programmed cell death. Surprisingly, PCD in D. salina played an unexpected role in nutrient supplementation of its own population in a carbon limited environment (**Fig. 4C**). H. salinarum was able to exploit this process and use the DOM released by PCD to supplement its own population with carbon and nitrogen needs. Diurnally synchronized programmed death of D. salina was the mechanism for releasing sequestered C in the form of glycerol and other DOMs (amino acids and other unknown N and C and P containing compounds). This facilitated nutrient exchange with others in its own population to maintain positive growth (**Fig. 3C, Fig. 4A**) as well as a heterotrophic archaeon and possibly other organisms in the field not discovered in this study (**Fig. 6**). Our flow cytometry, biochemical and morphological data indicate that the death phenotype bears the hallmarks of apoptosis-like PCD [91]. In addition, detection of PS externalization permitted the quantification of PCD cells indicating that a high percentage of cells die via PCD before the onset of darkness suggesting an anticipation to photosynthesis decline and the cell's need for a heterotrophic carbon source (**Fig. 4C**). PS externalization is followed by other PCD markers like the loss of membrane integrity and nuclear condensation and eventually death as determined by SYTOX® blue staining (**Fig. 1 D&E**). Furthermore we also measured caspase 3 activity, which has previously shown to be associated with PCD in chlorophytes (**Fig. 4D**) [54], [55]. A mean of 57%+/−17% of the algal cells died each night and recovered at an equivalent rate with 2–3 cell divisions the subsequent morning (**Fig. 4A**). The equilibrium between nighttime population loss and daytime regeneration is a strong indicator that these are linked events.

The DOM released by algal programmed death complements the nutritional needs of co-inhabiting H. salinarum, which in turn (indirectly) supplements the requirements for further algal growth (**Figs 2 & 3C**). While the precise nutritional dependencies of D. salina and H. salinarum have yet to be characterized, algal dependency on bacteria and archaea generated metabolites such as vitamin B is well known [92], [93], as is the reciprocal archaeal and bacterial need for algal metabolites [94], [95]. It is significant in this regard that haloarchaea can further stimulate cell death in a density dependent manner (**Fig. 4 B**), to simultaneously regulate DOM production and acquire its nutritional necessities, controlling D. salina population dynamics by decelerating the algal population growth [96] (**Figs 4A & B**) and to control perhaps its own density suggesting a predator –prey dynamic maintaining ratio dependent densities which predicts proportional increases of both populations. While prey dependent models predict that only prey benefit from increase prey production [97]. In this case, D. salina population keeps on growing (**Fig. 4A**).

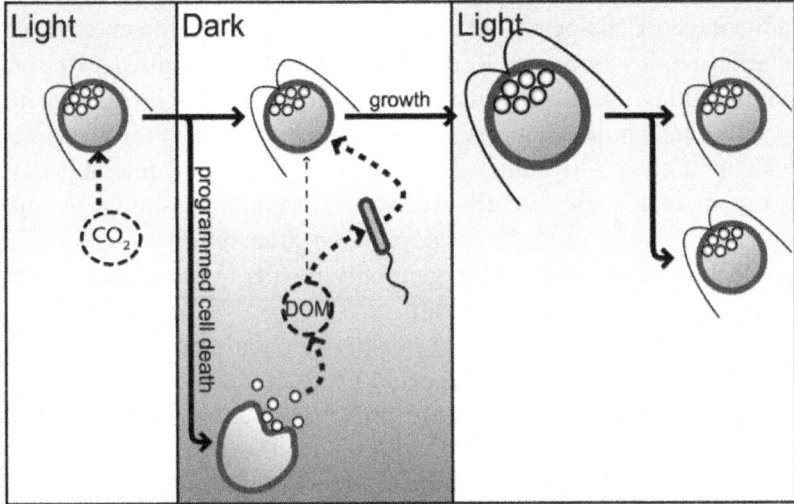

Figure 6. Diurnally synchronized cell death drives C-flux in an algal-archaeal syntrophic interaction. At night a stochastic process determines the fate of each algal cell resulting in up to 74% of cells undergoing death to release DOM (byproducts of photosynthetic C assimilation) into the surrounding media. The DOM are further metabolized and remineralized by archaea into a form that is readily consumed by algae. With onset of the subsequent day cycle, the algal population rapidly regenerates with up to 3 doublings with a cell division rate of 1.4 hrs. This entire process iterates over the next diurnal cycle.

Thus, the increased rate of cell death in the co-cultures was perhaps due to an additive effect of cell death stimulation, by both changes in illumination (darkness induced PCD) and presence of haloarchaea producing hemolysin a pore-forming toxin to further lyse D. salina and increase the number of dead cells (**Fig. 5**). Although it was known previously that photoautotrophic algae and photoheterotrophic haloarchaea [60] independently synchronize their physiologies with the light/dark cycle, our results demonstrate how such entrainment might enable their physiological cooperation. Furthermore, since bacteria/archaea can only take up molecules <650 Da [98], macromolecules can only be utilized if they are first hydrolysed outside the cell. This explains the upregulation of proteases and hydrolases by haloarchaea and the uptake of DOM during the day after it is released at night by D. salina.

A key evolutionary interest here is the role of PCD in this system. Programmed death in D. salina provides nutrients for its own population in a carbon limited environment. The haloarchaea exploit the process in a density dependent manner (**Fig. 4B**) and in turn (indirectly) supplements the requirements for further algal growth (**Fig. 3C**). Evolutionary theory predicts that such interactions can easily become unstable as "cheaters" (mutants that take advantage of "helpers") take over the population. However, an alternate explanation comes from the Red Queen Hypothesis applied to populations and communities [99]. When natural selection acts at a level other than the individual organism interactions that increase the overall fitness of interacting species can achieve a dynamic equilibrium. Specifically, interaction with H. salinarum increases the overall fitness of D. salina as shown by improved primary productivity of the algal population after remineralization of DOM (**Fig. 3C**). Increased productivity generally results in increased growth rates, which taken together with the death of 57%+/−17% of its members might also help to maintain a young and healthy population of D. salina. In other words, while H. salinarum is supported by the photosynthetic activity of D. salina, it in turn becomes a helper to prevent the exhaustion of resources by improving algal productivity and, ultimately, the carrying capacity of the environment. Such behaviors are frequent when food is scarce and species are in competition [74], [100]. In hypersaline environments where carbon is highly insoluble and limiting, nutrients are rapidly exhausted after spring blooms [101]. However, if the chlorophyte-chlorophyte and chlorophyte-haloarchaeal interactions are stable year round [24], then such a strategy is evolutionarily advantageous for both species in this carbon limited environment. In these interactions we can then predict that the physical clumping of the algal and algal-haloarchaeal cells is likely to be critical for regulating this process and maintaining the structure and equilibrium of these interactions. Based on the fitness measurements (algal or haloarchaeal growth), PCD benefits both species. Since PCD is clearly not beneficial to the actor, this study supports the argument that the phenomenon is an altruistic adaptation at a level of organization other than the individual cell. Our data show that PCD may be beneficial for population growth in nutrient-limited environments; however there may be other reasons not explored here. For example cellular aging or population regulation as well as non-adaptive pleiotropy or epistasis may make as yet undiscovered contributions.

On a grander scale, by mediating such interplay with archaea and other organisms, diurnally synchronized cell death of algae is potentially an important determinant of microbial diversity and dynamics, succession of species, and biogeochmical cycles of C, N and P in natural aquatic ecosystems [95], [102]. As such, it also offers a new dimension to the structure of interspecies interactions in the microbial loop [1], [2] by adding to the diversity of mechanisms by which microbial interactions can occur. By characterizing the microbial loop in this manner we can now explore new strategies for predicting and managing the flux of C, N and other key elements of life with artificial (e.g. caspase inhibitors) or natural (e.g. archaeal mutants) modulators of algal cell death. Such strategies could also have interesting biotechnology applications to the controlled harvest of algal products. Ironically, programmed death takes center stage in the complexity of a living system that is the microbial loop.

SUPPORTING INFORMATION

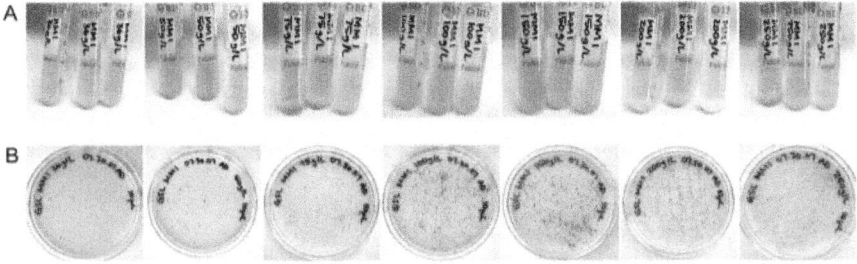

Figure S1. Enrichment of halophilic photoautorophs and heterotrophs from Great Salt Lake. (A) Tubes containing Great Salt Lake (GSL) field samples enriched for photoautotrophs in MM1 with salinities from 36 to 250 g/L (left to right)–characteristically defined by their green pigmentation. (B) Rich medium agar plates in salinities from 36 to 250 g/L (left to right) used to enrich bacterial/archaeal species from GSL water samples.

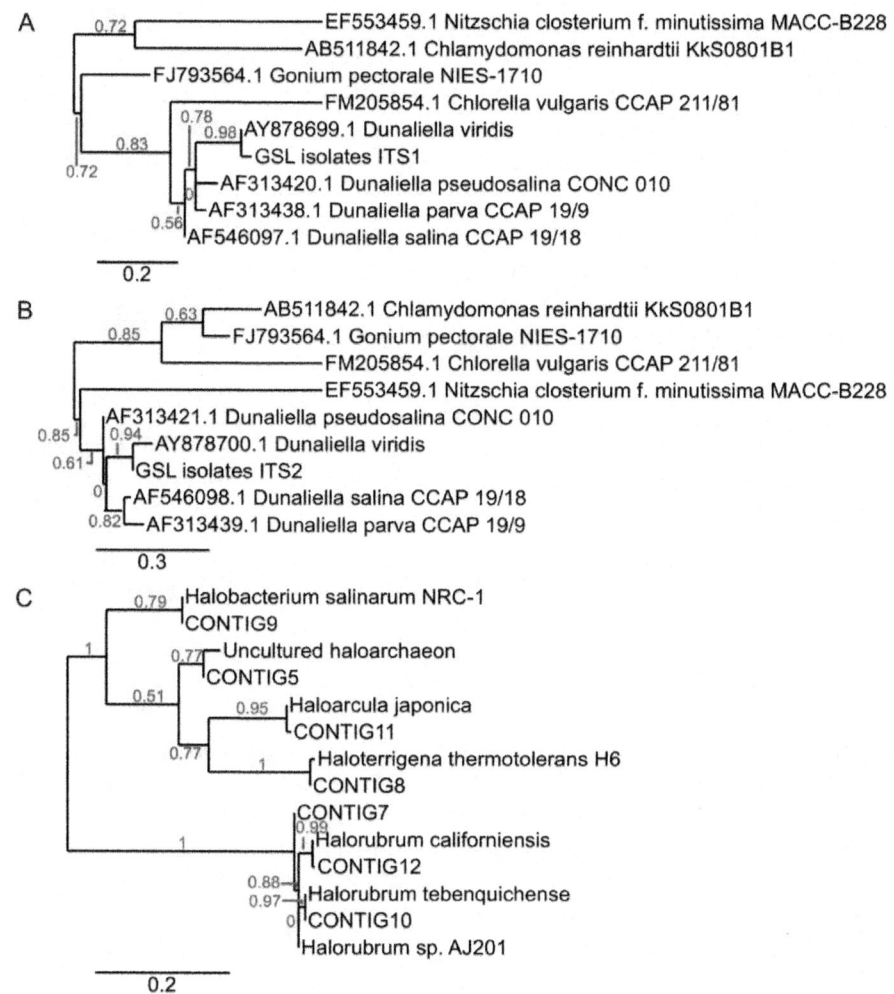

Figure S2

Figure S2. Halophilic biodiversity of the Great Salt Lake, Utah, USA. Phylogenetic dendrograms of algal 18S rRNA regions ITS1 (A) and ITS2 (B) comparing GSL isolated photoautotrophs toDunaliella spp. and other phytoplankton. Similarly, 16S rRNA genotyping of microbial heterotrophs isolated from the GSL (C) confirms the presence of Halobacterium sp. (an exact match for current lab strains) along with other halophilic archaea.

Figure S3

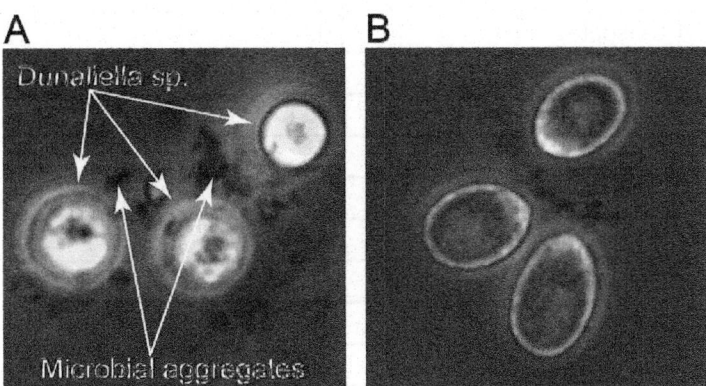

Figure S3. Mechanisms of communication and interactions in the syntrophic interaction. (A) Algal-bacterial/archaeal aggregates observed in (GSL) field samples enriched for photoautotrophs in 200 g/L salinity MM2. (B) Dunaliella salina and Halobacterium salinarum aggregates observed in 4-day old 13 hr:11 hr light:dark adapted cultures.

Table S1. MM1 Composition (All concentrations in mM unless otherwise specified).

Salts

Sodium chloride (NaCl)	3.22 M
Magnesium sulfate hepta-hydrate (MgSO$_4$*7H$_2$O)	40
3-morpholinopropane-1-sulfonic acid (MOPS)	40
Potassium chloride (KCl)	27
Sodium phosphate mono-basic (NaH$_2$PO$_4$)	167 μM

Vitamins

Folic acid	11.33 μM
Thiamine HCl	14.82 μM
Biotin	2.05 μM

MM1 Amino Acid Supplementation for H. salinarum growth

All concentrations in mM unless otherwise specified.

L-alanine	5.71
L-arginine HCl	1.9
L-asparagine	0.96
L-aspartate	1.88
L-glutamate	10.64
L-glutamine	5
L-glycine	1
L-histidine HCl	0.239
L-isoleucine	3.35
L-leucine	6.1
L-lysine HCl	1.04
L-methionine	0.603
L-phenylalanine	0.303
L-proline	0.4
L-serine	2.902
L-theorine	4.2
L-tryptophan	0.098
L-tyrosine	0.618
L-valine	2.135
L-cysteine	45

ACKNOWLEDGMENTS

We thank Aimee Desaki, Amardeep Kaur, Pavana, Anur, Manjula Bharadwaj, Joseph Horsman, Elijah Christensen, Amanda Pease, David Rodriguez, Zac Simon, and Noel Blake for experimental and technical support; Bruz Marzolf and Min Pan for microarray processing; Bonnie Baxter for performing a field survey and sample collection at the GSL; Karlyn Beer, Danielle Miller, and Serdar Turkarslan for sampling support during diurnal co-culture experiments. We thank Mary Jane Perry for her gracious gift of a Photosynthesis/Irradiance incubator and [14]C. We also thank Sacha Coesel, Danny Ionescu, J. Jeffrey Morris and an anonymous reviewer for their helpful comments.

AUTHOR CONTRIBUTIONS

Conceived and designed the experiments: MVO WLP PMD KW NSB. Performed the experiments: MVO WLP KW. Analyzed the data: MVO WLP

KW. Contributed reagents/materials/analysis tools: MVO WLP PMD NSB. Wrote the paper: MVO WLP PMD NSB.

REFERENCES

1. Azam F (1998) Oceanography: Microbial control of oceanic carbon flux: the plot thickens. Science 280: 694–696. doi: 10.1126/science.280.5364.694

2. Azam F, Malfatti F (2007) Microbial structuring of marine ecosystems. Nat Rev Microbiol 5: 782–791. doi: 10.1038/nrmicro1747

3. Pomeroy LR (1974) Oceans food web, a changing paradigm. Bioscience 24: 499–504. doi: 10.2307/1296885

4. Azam F (1983) The ecological role of water-column microbes in the sea. Mar Ecol Prog Ser 10: 257–263. doi: 10.3354/meps010257

5. Aluwihare LI, Repeta DJ, Chen RF (1997) A major biopolymeric component to dissolved organic carbon in surface sea water. Nature 387: 166–169. doi: 10.1038/387166a0

6. Jiao N, Herndl GJ, Hansell DA, Benner R, Kattner G, et al. (2010) Microbial production of recalcitrant dissolved organic matter: long-term carbon storage in the global ocean. Nat Rev Micro 8: 593–599. doi: 10.1038/nrmicro2386

7. Kaiser K, Benner R (2008) Major bacterial contribution to the ocean reservoir of detrital organic carbon and nitrogen. Limnol Oceanogr 53: 99–112. doi: 10.4319/lo.2008.53.1.0099

8. McCarthy MD, Hedges JI, Benner R (1998) Major bacterial contribution to marine dissolved organic nitrogen. Science 281: 231–234. doi: 10.1126/science.281.5374.231

9. Aluwihare LI, Repeta DJ, Pantoja S, Johnson CG (2005) Two chemically distinct pools of organic nitrogen accumulate in the ocean. Science 308: 1007–1010. doi: 10.1126/science.1108925

10. Decho AW (1990) Microbial exopolymer secretions in ocean environments: their role(s) in food webs and marine processes. Oceanogr Mar Biol Annu Rev 28: 73–153.

11. Fuhrman JA (1999) Marine viruses and their biogeochemical and ecological effects. Nature 399: 541–548. doi: 10.1038/21119

12. Proctor LM, Fuhrman JA (1990) Viral mortality of marine bacteria and cyanobacteria. Nature 343: 60–62. doi: 10.1038/343060a0

13. Suttle CA (2007) Marine viruses–major players in the global ecosystem. Nature Rev Microbiol 5: 801–812. doi: 10.1038/nrmicro1750

14. Vardi A, Van Mooy BAS, Fredricks HF, Popendorf KJ, Ossolinski JE, et al. (2009) Viral glycosphingolipids induce lytic infection and cell death in marine phytoplankton. Science 326: 861–865. doi: 10.1126/science.1177322

15. Vardi A, Haramaty L, Van Mooy BAS, Fredricks HF, Kimmance SA, et al. (2012) Host–virus dynamics and subcellular controls of cell fate in a natural coccolithophore population. Proc Nat Acad Sci U S A 109: 19327–19332. doi: 10.1073/pnas.1208895109

16. Chin W-C, Orellana MV, Quesada I, Verdugo P (2004) Secretion in unicellular marine phytoplankton: demonstration of regulated exocytosis in Phaeocystis globosa. Plant Cell Physiol 45: 535–542. doi: 10.1093/pcp/pch062

17. Nagata T, Kirchman D (1997) Role of submicron particles and colloids in microbial food webs and biogeochemical cycles within marine environments. Adv Microb Ecol 15: 81–103. doi: 10.1007/978-1-4757-9074-0_3

18. Strom SL, Ronald Benner, Ziegler S, Dagg MJ (1997) Planktonic grazers are a potentially important source of marine dissolved organic carbon. Limnol Oceangr 42: 1364–1374. doi: 10.4319/lo.1997.42.6.1364

19. Berman-Frank I, Bidle K, Haramaty L, Falkowski P (2004) The demise of the marine cyanobacterium, Trichodesmium spp., via an autocatalyzed cell death pathway. Limnol Oceanogr 49: 997–1005. doi: 10.4319/lo.2004.49.4.0997

20. Bidle KD, Falkowski PG (2004) Cell death in planktonic, photosynthetic microorganisms. Nat Rev Micro 2: 643–655. doi: 10.1038/nrmicro956

21. Berges JA, Falkowski P (1998) Physiological stress and cell death in marine phytoplankton: Induction of proteases in response to nitrogen or light limitation. Limnol Oceanogr 43: 129–135. doi: 10.4319/lo.1998.43.1.0129

22. Ben-Amotz A, Polle JEW, Subba Rao DV (2009) The alga Dunaliella: biodiversity, physiology, genomics and biotechnology; Ben-Amotz A, Polle JEW, Subba Rao DV, editors: Science Publishers.

23. Stephens DW, Gillespie DM (1976) Phytoplankton production in the Great Salt Lake, Utah, and a laboratory study of algal response to enrichment. Limnol Oceanogr 21: 74–87. doi: 10.4319/lo.1976.21.1.0074

24. Bardavid RE, Khristo P, Oren A (2008) Interrelationships between Dunaliella and halophilic prokaryotes in saltern crystallizer ponds. Extremophiles 12: 5–14. doi: 10.1007/s00792-006-0053-y

25. Liska AJ, Shevchenko A, Pick U, Katz A (2004) Enhanced photosynthesis and redox energy production contribute to salinity tolerance in Dunaliella as revealed by homology-based proteomics. Plant Physiol 136: 2806–2817. doi: 10.1104/pp.104.039438

26. Oren A (1993) Availability, uptake and turnover of glycerol in hypersaline environments. FEMS Microb Ecol 12: 15–23. doi: 10.1016/0168-6496(93)90020-8

27. Bogomolni RA, Spudich JL (1982) Identification of a third rhodopsin-like pigment in phototactic Halobacterium halobium. Proc Nat Acad of Sci U S A 79: 6250–6254. doi: 10.1073/pnas.79.20.6250

28. Spudich EN, Takahashi T, Spudich JL (1989) Sensory rhodopsins I and II modulate a methylation/demethylation system in Halobacterium halobium phototaxis. Proc Nat Acad of Sci U S A 86: 7746–7750. doi: 10.1073/pnas.86.20.7746

29. Spudich JL (1993) Color sensing in the Archaea: a eukaryotic-like receptor coupled to a prokaryotic transducer. J Bacteriol 175: 7755–7761.

30. Storch K-F, Rudolph J, Oesterhelt D (1999) Car: a cytoplasmic sensor responsible for arginine chemotaxis in the archaeon Halobacterium salinarum. EMBO J 18: 1146–1158. doi: 10.1093/emboj/18.5.1146

31. Baliga NS, Bjork SJ, Bonneau R, Pan M, Iloanusi C, et al. (2004) Systems level insights into the stress response to UV radiation in the halophilic ArchaeonHalobacterium NRC-1. Genome Res 14: 1025–1035. doi: 10.1101/gr.1993504

32. Kottemann M, Kish A, Iloanusi C, Bjork S, DiRuggiero J (2005) Physiological responses of the halophilic archaeon Halobacterium sp. strain NRC-1 to desiccation and gamma irradiation. Extremophiles 9: 219–227. doi: 10.1007/s00792-005-0437-4

33. Brooun A, Bell J, Freitas T, Larsen RW, Alam M (1998) An Archaeal aerotaxis transducer combines subunit I core structures of eukaryotic cytochrome c oxidase and eubacterial methyl-accepting chemotaxis proteins. J Bacteriol 180: 1642–1646.

34. DasSarma S, Arora P, Lin F, Molinari E, Yin LR (1994) Wild-type gas vesicle formation requires at least ten genes in the gvp gene cluster of Halobacterium halobium plasmid pNRC-100. J Bacteriol 176: 7646–7652.

35. Halladay JT, Jones JG, Lin F, MacDonald AB, DasSarma S (1993) The rightward gas vesicle operon in Halobacterium plasmid pNRC100: identification of the gvpA and gvpC gene products by use of antibody

probes and genetic analysis of the region downstream of gvpC. J Bacteriol 175: 684–692.

36. Kokoeva MV, Oesterhelt D (2000) BasT, a membrane-bound transducer protein for amino acid detection in Halobacterium salinarum. Mol Microbiol 35: 647–656. doi: 10.1046/j.1365-2958.2000.01735.x

37. Fabrizio P, Battistella L, Vardavas R, Gattazzo C, Liou L-L, et al. (2004) Superoxide is a mediator of an altruistic aging program in Saccharomyces cerevisiae. J Cell Biol 166: 1055–1067. doi: 10.1083/jcb.200404002

38. Bidle KD, Bender SJ (2008) Iron starvation and culture age activate metacaspases and programmed cell death in the marine diatom Thalassiosira pseudonana. Eukaryot Cell 223: 223–230. doi: 10.1128/ec.00296-07

39. Franklin DJ, Brussaard CPD, Berges JA (2006) What is the role and nature of programmed cell death in phytoplankton ecology? Eur J Phycol 41: 1–44. doi: 10.1080/09670260500505433

40. Segovia M, Haramaty L, Berges JA, Falkowski PG (2003) Cell Death in the unicellular Chlorophyte Dunaliella tertiolecta. A hypothesis on the evolution of apoptosis in higher plants and metazoans. Plant Physiol 132: 99–105. doi: 10.1104/pp.102.017129

41. Vardi A, Eisenstadt D, Murik O, Berman-Frank I, Zohary T, et al. (2007) Synchronization of cell death in a dinoflagellate population is mediated by an excreted thiol protease. Environ Microbiol 360–369. doi: 10.1111/j.1462-2920.2006.01146.x

42. Nedelcu AM, Driscoll WW, Durand PM, Herron MD, Rashidi A (2011) On the paradigm of altruistic suicide in the unicellular world. Evolution 65: 3–20. doi: 10.1111/j.1558-5646.2010.01103.x

43. Williams GC (1966) Adaptation and Natural Selection: A Critique of Some Current Evolutionary Thought. Princeton, N.J.: Princeton University Press.

44. Durand PM, Rashidi A, Michod RE (2011) How an organism dies affects the fitness of its neighbors. The American Naturalist 177: 224–232. doi: 10.1086/657686

45. Franklin DJ, Brussaard CPD, Berges JA (2006) What is the role and nature of programmed cell death in phytoplankton ecology? Eur J Phycol 41: 1–44. doi: 10.1080/09670260500505433

46. González MA, Coleman AW, Gómez PI, Montoya R (2001) Phylogenetic relationship among various strains of Dunaliella (Chorophyceae) based on nuclear ITS rDNA sequences. J Phycol 37: 604–611. doi: 10.1046/j.1529-8817.2001.037004604.x

47. Ng WV, Kennedy SP, Mahairas GG, Berquist B, Pan M, et al. (2000) Genome sequence of Halobacterium species NRC-1. Proc Natl Acad Sci U S A 97: 12176–12181. doi: 10.1073/pnas.190337797

48. Sverdrup Hu, Johnson M, Fleming R (1957) The Oceans: Prentice-Hall, Inc.

49. Chanley MH, Smith WL (1975) Culture of Marine Invertebrate Animals: Proceedings-1st Conference on Culture of Marine Invertebrate Animals Greenport: Springer.

50. Dassarma S, Fleischmann EM (1995) Archaea a Laboratory Manual: Halophiles: Cold Spring Harbor Laboratory Press.

51. Breckenridge LJ, Almers W (1987) Final steps in exocytosis observed in a cell with giant secretory granules. Nature 84: 1945–1949. doi: 10.1073/pnas.84.7.1945

52. Pralong WF, Bartley C, Wollheim CB (1990) Single islet beta-cell stimulation by nutrients: relationship between pyridine nucleotides, cytosolic Ca^{2+} and secretion. EMBO J 9: 53–60.

53. Marie D, Partensky F, Jacquet S, Vaulot D (1997) Enumeration and cell cycle analysis of natural populations of marine picoplankton by flow cytometry using the nucleic acid stain SYBR Green I. Appl Environ Microbiol 63: 186–186.

54. Jiménez C, Capasso JM, Edelstein CL, Rivard CJ, Lucia S, et al. (2009) Different ways to die: cell death modes of the unicellular chlorophyte Dunaliella viridis exposed to various environmental stresses are mediated by the caspase-like activity DEVDase. J Exp Bot 60: 815–828. doi: 10.1093/jxb/ern330

55. Moharikar S, D›Souza JS, Kulkarni AB, Rao BJ (2006) Apoptotic-like cell death pathway is induced in unicellular chlorophyte Chlamydomonas reinhardtii(Chlorophyceae) cells following UV irradiation detection and funtional analysis. J Phycol 42: 423–433. doi: 10.1111/j.1529-8817.2006.00207.x

56. Engelbrecht D, Durand PM, Coetzer TL (2012) On programmed cell death inPlasmodium falciparum: Status quo. J Trop Med 2012: 646534–646534. doi: 10.1155/2012/646534

57. Iverson RL, Bittaker HF, Myers VB (1976) Loss of radiocarbon in direct use of Aquasol for liquid scintillation counting of solutions containing ^{14}C-$NaHCO_3$. Limnol Oceanogr 21: 756–758. doi: 10.4319/lo.1976.21.5.0756

58. Strickland JDH, Parsons TR (1972) A practical handbook of seawater analysis. Ottawa Fisheries Research Board of Canada.

59. Kaur A, Pan M, Meislin M, Facciotti MT, El-Gewely R, et al. (2006) A systems view of haloarchaeal strategies to withstand stress from transition metals. Genome Res16: 841–854. doi: 10.1101/gr.5189606

60. Whitehead K, Pan M, Masumura K-i, Bonneau R, Baliga NS (2009) Diurnally entrained anticipatory behavior in Archaea. PLoS ONE 4. doi: 10.1371/journal.pone.0005485

61. Bonneau R, Facciotti MT, Reiss DJ, Schmid AK, Pan M, et al. (2007) A predictive model for transcriptional control of physiology in a free living cell. Cell 131: 1354–1365. doi: 10.1016/j.cell.2007.10.053

62. Ideker T, Thorsson V, Siegel AF, Hood LE (2004) Testing for differentially-expressed genes by maximum-likelihood analysis of microarray data. J Comput Biol 7: 805–817. doi: 10.1089/10665270050514945

63. Dickson AG, Sabine CL, Christian JR (2007) Guide to best practices for ocean CO$_2$measurements. 191 p. PICES Special Publication.

64. Zhang H, Byrne RH (1996) Spectrophotometric pH measurements of surface seawater at in-situ conditions: absorbance and protonation behavior of thymol blue. Mar Chem 52: 17–25. doi: 10.1016/0304-4203(95)00076-3

65. Droop MR (1973) Some thoughts on nutrient limitation in algae. J of Phycol 9: 264–272. doi: 10.1111/j.1529-8817.1973.tb04092.x

66. Ben-Amotz A, Avron M (1973) The role of glycerol in the osmotic regulation of the halophilic alga Dunaliella parva. Plant Physiol 51: 875–878. doi: 10.1104/pp.51.5.875

67. Orellana MV, Matrai PA, Janer M, Rauschenberg C (2010) DMSP storage in Phaeocystis secretory vesicles. J Phycol 47: 112–117. doi: 10.1111/j.1529-8817.2010.00936.x

68. Rao SDV (2009) Cultivation, growth media, division rates and applications of Dunaliellaspecies.; Ben-Amotz AJEWPaDVSR, editor. India: Science Pubublishers. 555 p.

69. Mary I, Garczarek L, Tarran GA, Kolowrat C, Terry MJ, et al. (2008) Diel rhythmicity in amino acid uptake by Prochlorococcus. Environ Microbiol 10: 1927–2190. doi: 10.1111/j.1462-2920.2008.01633.x

70. Tafreshi A, Shariati M (2006) Pilot culture of three strains of Dunaliella salina for β-carotene production in open ponds in the central region of Iran. World J Microbiol Biotechnol 22: 1003–1006. doi: 10.1007/s11274-006-9145-1

71. Fernandez JM, Villalón M, Verdugo P (1991) Reversible condensation of mast cell secretory products in vitro. Biophys J 59: 1022–1027. doi: 10.1016/s0006-3495(91)82317-7

72. Bonneau R, Baliga NS, Deutsch EW, Shannon P, Hood L (2004) Comprehensive de novo structure prediction in a systems-biology context for the archaea Halobacteriumsp. NRC-1. Genome Biology 5: R52. doi: 10.1186/gb-2004-5-8-r52

73. Van PT, Schmid AK, King NL, Kaur A, Pan M, et al. (2008) Halobacterium salinarumNRC-1 Peptide Atlas: toward strategies for targeted proteomics and improved proteome coverage. J Proteome Res 7: 3755–3764. doi: 10.1021/pr800031f

74. Falkowski PG, Raven JA (2007) Aquatic Photosynthesis.: Princeton University Press.

75. Muller JA, DasSarma S (2005) Genomic Analysis of anaerobic respiration in the archaeon Halobacterium sp. Strain NRC-1: Dimethyl sulfoxide and trimethylamine N-Oxide as terminal electron acceptors. J Bacteriol 187: 1659–1667. doi: 10.1128/jb.187.5.1659-1667.2005

76. Kirchman DL (2008) Microbial Ecology of the Oceans. Hoboken, New Jersey: John Wiley & Sons, Inc., Publication. 593 p.

77. Bidle KD, Azam F (1999) Accelerated dissolution of diatom silica by marine bacterial assemblages. Nature 397: 508–512. doi: 10.1038/17351

78. Blackburn N, Fenchel T, Mitchell J (1998) Microscale nutrient patches in planktonic habitats shown by chemotactic bacteria. Science 282: 2254–2256. doi: 10.1126/science.282.5397.2254

79. Leigh Jr EG (2010) The evolution of mutualism. J Evol Biol23: 2507–2528. doi: 10.1111/j.1420-9101.2010.02114.x

80. Lewis J, Kennaway G, Franca S, Alverca E (2001) Bacterium-dinoflagellate interactions: investigative microscopy of Alexandrium spp. (Gonyaulacales, Dinophyceae). Phycologia 40: 280–285. doi: 10.2216/i0031-8884-40-3-280.1

81. Azam F, Long RA (2001) Oceanography–sea snow microcosms. Nature 414: 495–498. doi: 10.1038/35107174

82. Verdugo P (2012) Marine microgels. Ann Rev Mar Sci 4: 9.1–9.25. doi: 10.1146/annurev-marine-120709-142759

83. DeLong EF, Pace NR (2001) Environmental diversity of bacteria and archaea. Syst Biol 50: 470–478.

84. Stephens DW, Gillespie DM (1976) Phytoplankton production in the Great Salt Lake, Utah, and a laboratory study of algal response to enrichment. Limnol and Oceanog 21: 74–87. doi: 10.4319/lo.1976.21.1.0074

85. Schidlowski M, Matzigkeit U, Krumbein WE (1984) Superheavy organic carbon from hypersaline microbial mats assimilatory pathway

and geochemical implications. Naturwissenschaften 71: 303–308. doi: 10.1007/bf00396613

86. Riebesell U, Wolf-Gladrow D, Smetacek V (1993) Carbon dioxide limitation of marine phytoplankton growth rates. Nature 361: 249–251. doi: 10.1038/361249a0

87. Kaplan A, Reinhold L (1999) CO_2 concentrating mechanisms in photosynthetic microorganisms. Annu Rev Plant Physiol Plant Mol Biol 50: 539–570. doi: 10.1146/annurev.arplant.50.1.539

88. Giordano M, Beardall J, Raven JA (2005) CO_2 concentrating mechanisms in algae: mechanism, environmental modulation, and evolution. Annu Rev Plant Biol 56: 99–131. doi: 10.1146/annurev.arplant.56.032604.144052

89. Segovia M, Haramaty L, Berges JA, Falkowski PG (2003) Cell death in the unicellular chlorophyte Dunaliella tertiolecta: a hypothesis on the evolution of apoptosis in higher plants and metazoans. Plant Physiol 132: 99–105. doi: 10.1104/pp.102.017129

90. Bidle KD, Bender SJ (2008) Iron starvation and culture age activate metacaspases and programmed cell death in the marine diatom Thalassiosira pseudonana. Eukaryot Cell 223: 223–230. doi: 10.1128/ec.00296-07

91. Kroemer G, Galluzzi L, Vandenabeele P, Abrams J, Alnemri ES, et al. (2008) Classification of cell death: recommendations of the Nomenclature Committee on Cell Death 2009. Cell Death Differ 16: 3–11. doi: 10.1038/cdd.2008.150

92. Croft MT, Lawrence AD, Raux-Deery E, Warren MJ, Smith AG (2005) Algae acquire vitamin B12 through a symbiotic relationship with bacteria. Nature 438: 90–93. doi: 10.1038/nature04056

93. Droop MR (2007) Vitamins, phytoplankton and bacteria: symbiosis or scavenging. J Plankton Res 29: 107–113. doi: 10.1093/plankt/fbm009

94. Lau WWY, Keil RG, Armbrust EV (2007) Succession and diel transcriptional response of the glycolate-utilizing component of the bacterial community during a spring phytoplankton bloom. Appl Environ Microbiol 73: 2440–2450. doi: 10.1128/aem.01965-06

95. Reimann L, Steward GF, Azam F (2000) Dynamics of bacterial community composition and activity during a mesocosm diatom bloom. Appl Environ Microbiol 66: 578–587. doi: 10.1128/aem.66.2.578-587.2000

96. Pace ML, Cole JJ (1996) Regulation of bacteria by resources and predation tested in whole-lake experiments. Limnol Oceanogr 41: 1448–1460. doi: 10.4319/lo.1996.41.7.1448

97. Rosenzweig ML (1977) Aspects of biological exploitation. Quarterly Rev Biol 52: 371–380. doi: 10.1086/410124

98. Payne JW (1980) Transport and utilization of peptides by bacteria. In: Paine JW, editor. Microorganisms and Nitrogen Sources.New York: John Wiley and Sons. pp. 212–256.

99. Liow LH, Van Valen L, Stenseth NC (2011) Red Queen: from populations to taxa and communities. Trends in ecology & evolution 26: 349–358. doi: 10.1016/j.tree.2011.03.016

100. Wilkerson CR (1987) Interoceanic differences in size and nutrition of coral reef sponge populations. Science 236: 16541657. doi: 10.1126/science.236.4809.1654

101. Post FJ (1981) Microbiology of the Great Salt Lake north arm. Hydrobiologia 81–82: 59–69. doi: 10.1007/bf00048706

102. Kirchman DL, Malmstrom RR, Cottrell MT (2005) Control of bacterial growth by temperature and organic matter in the Western Arctic. Deep-Sea Res II 52: 3386–3395. doi: 10.1016/j.dsr2.2005.09.005

Chapter 11

FUNCTIONALLY STABLE AND PHYLOGENETICALLY DIVERSE MICROBIAL ENRICHMENTS FROM MICROBIAL FUEL CELLS DURING WASTEWATER TREATMENT

Shun'ichi Ishii[1,2,4], Shino Suzuki[1] , Trina M. Norden-Krichmar[1] , Kenneth H. Nealson[1,3], Yuji Sekiguchi[2] , Yuri A. Gorby[3] , Orianna Bretschger[1]

[1] J. Craig Venter Institute, San Diego, California, United States of America

[2] Biomedical Research Institute, National Institute of Advanced Industrial Science and Technology (AIST), Tsukuba, Ibaraki, Japan

[3] University of Southern California, Los Angeles, California, United States of America

[4] Japan Society for the Promotion of Science (JSPS), Chiyoda-ku, Tokyo, Japan

ABSTRACT

Microbial fuel cells (MFCs) are devices that exploit microorganisms as biocatalysts to recover energy from organic matter in the form of electricity. One of the goals of MFC research is to develop the technology for cost-effective wastewater treatment. However, before practical MFC applications are implemented it is important to gain fundamental knowledge about long-term system performance, reproducibility, and the formation and maintenance of functionally-stable microbial communities. Here we report findings from a MFC operated for over 300 days using only primary clarifier effluent collected from a municipal wastewater treatment plant as the microbial resource and substrate. The system was operated in a repeat-batch mode, where the reactor solution was replaced once every two weeks with new primary effluent that consisted of different microbial and chemical compositions with every batch exchange. The turbidity of the primary clarifier effluent solution notably decreased, and 97% of biological oxygen demand (BOD) was removed after an 8–13 day residence time for each batch cycle. On average, the limiting current density was 1000 mA/m^2, the maximum power density was 13 mW/m^2, and coulombic efficiency was 25%. Interestingly, the electrochemical performance and BOD removal rates were very reproducible throughout MFC operation regardless of the sample variability associated with each wastewater exchange. While MFC performance was very reproducible, the phylogenetic analyses of anode-associated electricity-generating biofilms showed that the microbial populations temporally fluctuated and maintained a high biodiversity throughout the year-long experiment. These results suggest that MFC communities are both self-selecting and self-optimizing, thereby able to develop and maintain functional stability regardless of fluctuations in carbon source(s) and regular introduction of microbial competitors. These results contribute significantly toward the practical application of MFC systems for long-term wastewater treatment as well as demonstrating MFC technology as a useful device to enrich for functionally stable microbial populations.

INTRODUCTION

Presently, wastewater treatment is an energy intensive and expensive process. In the USA over 126 billion liters of domestic wastewater are treated daily at an annual cost of over $25 billion [1]. Therefore decreasing total energy consumption during wastewater treatment is an important goal that can be accomplished through several strategies including: 1) implementing energy efficient equipment and practices; 2) recovering energy during treatment processes; and 3) optimizing treatment methods to minimize overall disposal costs of wastewater effluents and biosolids.

Here we address the use of microbial fuel cells (MFCs) for the degradation of carbon sources in primary clarifier effluents from a conventional wastewater treatment plant. MFC treatment may be utilized to replace or supplement conventional secondary treatment systems and minimize the overall costs associated with aeration, secondary clarification, and secondary sludge treatment. MFC technology exploits biological fermentation and respiratory mechanisms to directly recover energy as electricity during the degradation of organic matter contained in wastewater and/or sludge [2], [3], [4]. Relative to conventional primary and secondary treatment processes, MFC systems also have the benefit of reducing overall operational costs because aeration is not needed [5]. In addition, lower overall sludge volumes can be realized because the growth of secondary biomass is limited under anaerobic MFC conditions [6].

A MFC reactor physically separates the oxidation and reduction reactions [7]. The biological oxidation of organic matter proceeds in the anode chamber of a MFC under anaerobic conditions. Reducing equivalents (electrons) liberated during the oxidation processes are biologically transferred to a conductive anode electrode where they flow as electrical current across the

MFC circuit to the neighboring cathode electrode (Fig. 1A). Protons resulting from the oxidation processes travel by diffusion to the cathode chamber where the terminal reduction reaction consumes the electrons, protons, and a given oxidant. The cathode reduction reaction is typically catalyzed by a noble metal substrate, but biocathodes have also been explored [8]. Both the biological and engineered MFC components influence the total performance (e.g., power density, coulombic efficiency and organic-loading rate) of a MFC system [7], and have been a subject for improvement [6].

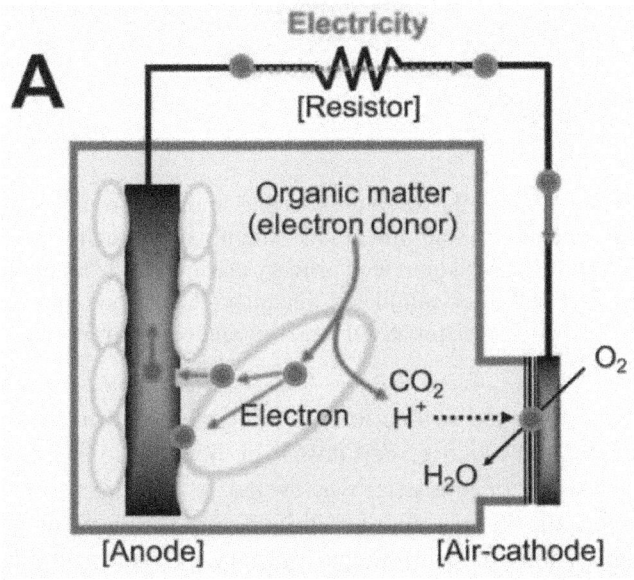

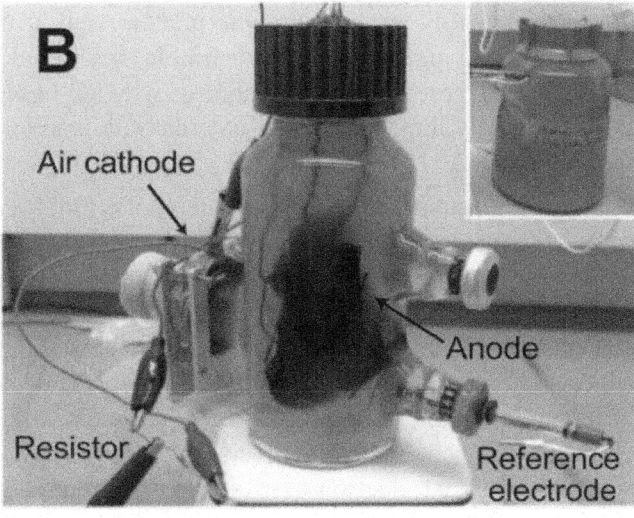

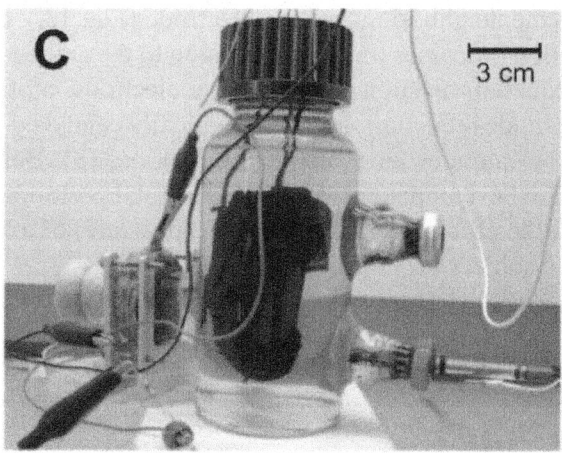

Figure 1. Microbial fuel cell (MFC) used in this study. Schematic diagram of microbial electricity generation in an air-cathode microbial fuel cell (A). The air-cathode microbial fuel cell filled with untreated primary clarifier effluent (shown in inset) (B). The microbial fuel cell after completed treatment (C). Anode and air-cathode were connected with a 750 Ω resistor. A Ag/AgCl reference electrode was used for linear sweep voltammetry.

Most of the reported evaluations of MFCs have utilized a single carbon source (e.g. acetate, glucose, cellulose), or homogenous effluents, such as synthetic wastewater or industrial wastewater (e.g. brewery waste) to explore the degradation of organic matter and energy recovery as electricity [2]. However, the treatment of municipal primary clarifier effluents is a greater challenge for MFC technology because the clarifier effluents generally contain low concentrations of highly variable organic matters, and at the same time contains highly diverse and temporally variable microbial populations. Consequently, the long-term functional stability of MFC-based wastewater treatment may be severely impacted by the inherent compositional variability of wastewater.

Sustained microbial catalytic activity and metabolic function are the most important factors to consider for long-term wastewater treatment using MFCs. Effective utilization of biological resources for flexible self-organization and self-regulation is needed to treat heterogeneous mixtures of organic and inorganic substrates contained in municipal wastewater. In particular, the microbial community in a MFC must play two critical roles: 1) rapid and complete degradation of various organic compounds through microbial metabolism; and 2) efficient recovery of energy through microbial energy transduction via extracellular electron transport to the anode. Controlling and optimizing these two biological roles is critical to enhance and stabilize MFC performance. To these ends, it is necessary to elucidate microbial function(s) from the enzyme level to the microbial community level.

Here we present the first year-long evaluation of microbial diversity and functionality in a MFC that used municipal primary clarifier effluent as the sole substrate and inoculum source. Microbial functional stability was evaluated by quantifying organic matter degradation and electricity production. Analyses of microbial population dynamics associated with refined statistics were concurrently performed so that we are able to present the first data set that contributes to understanding the population dynamics of a given electricity-generating microbial consortium during long-term MFC operation. Finally, through combined system performance and phylogenetic analyses, we are able to better understand how variable carbon sources and competing microbial communities affect the performance and biodiversity of MFC anode-associated microbial communities.

RESULTS

Long-Term MFC Operation

A single-chamber, air-cathode MFC [9] was used to explore the electrochemical performance and microbial functionality of a wastewater-utilizing MFC system (Fig. 1A–C). The bottle-type MFC was directly inoculated with primary clarifier effluent collected from the North City Water Reclamation Plant (San Diego, USA) (Fig. 1B). Current production was observed after a 13-day lag time and stabilized after two consecutive primary clarifier effluent exchanges (Fig. 2). Subsequently, primary clarifier effluent was collected bimonthly and used for repeat-batch operation with complete replacement of the anolyte after decreasing current generation was observed. After a month of operation under an applied external resistor of 750 Ω, the electric current stabilized at approximately 0.4 mA with 300 mV of the cell voltage. Some differences in current maxima were apparent when the solution was replaced with newly collected wastewater or stored wastewater (at 4°C for up to two weeks) likely due to the lower concentration of BOD contained in the stored samples (Fig. 2).

During the long-term MFC operation, current generation was also impacted by biofilm formation on the cathode surface. After three months of operation, a dense biofilm was visibly apparent at the cathode surface and led to a reduction of current generation due to an overall increase in internal resistance (Fig. 2). To restore previously observed current generation and internal resistance measurements, the cathode was mechanically treated to remove the attached biofilm. This treatment occurred twice during the 300 day operational period, the first time on day 88 and again at day 192. Subsequently, stable current production was observed to be approximately 0.25 mA, and the repeat-batch cycles were 8–15 days depending on the condition of primary clarifier effluents.

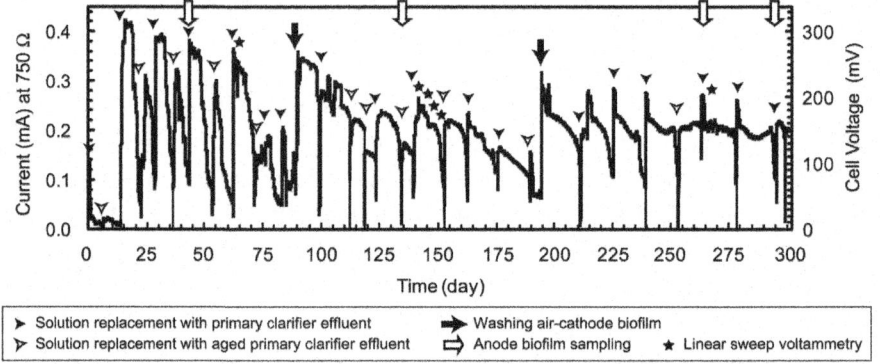

Figure 2. Long-term electricity generation in an air-cathode MFC operating with a 750 Ω external resistance. Filled arrow, removal of cathode biofilm; Open arrow, anode biofilm sampling for DNA extraction; Filled arrowhead, solution replacement with primary clarifier effluents collected on the same day; Open arrowhead, solution replacement with primary clarifier effluents stored at 4°C for 5–10 days; Solid star, linear sweep voltammetry to analyze anode performance.

Organic Matter Degradation

The purpose of repeat-batch cycling was to observe the reproducible organic matter degradation as measured by a decrease in turbidity, removal of chemical oxygen demand (COD), biological oxygen demand (BOD), and inorganic compounds. Although turbidity of the original sample was high, the anolyte solution was almost completely clear after the 8–13 day residence time (Fig. 1C). Organic matter as determined by COD also consistently decreased with current generation (Fig. S1). Our results show an average of 86±2% of COD was removed in a typical cycle. Organic matter as determined by BOD also significantly decreased with 97% of BOD being removed in a single batch (Table 1). These results suggest that the depletion of biodegradable organic chemicals coincided with the decrease in current generation. Electron recovery was calculated as coulombic efficiency, which was found to be stable at 26±1% over the repeat-batch cycles. That is, approximately 75% of electrons associated with the COD degradation of each sample were consumed by competing reactions such as oxygen respiration, anaerobic respiration with soluble electron acceptors such as sulfate or nitrate reduction, and/or biosynthesis. During long-term operation we did not visibly observe any liquid displacement or significant volume changes with gas production in the reactor suggesting that fermentation, methanogenesis and/or hydrogenesis could not occur at high rates.

Table 1. Chemical composition of primary clarifier effluent and MFC effluent

	Primary clarifier effluent[b]	MFC effluent[c]	Removal ratio (%)
COD (mg/L)[a]	263.3	36.8	86
BOD (mg/L)[a]	181.4	6.2	97
TSS (mg/L)[a]	116.0	22.3	81
Turbidity (NTU)	88.7	41.2	54
Nitrate-N (mg/L)	0.03	0.03	0
Nitrite-N (mg/L)	0.15	<0.01	>93
Ammonia-N (mg/L)	32.4	26.2	19
Sulfate (mg/L)	262.1	218.1	17
Heavy metal conc. (µg/L)			
Fe	1688	1027	39
Sr	1120	1010	10
Al	717	120	83
Mn	125	129	−3
Cu	82	23	72
Zn	79	33	58
Mo	12	3	75
Cr	6	1	80

[a]COD = chemical oxygen demand, BOD = biological oxygen demand, and TSS = total suspended solid.
[b]Primary clarifier effluent was collected on day 225 and was directly added to the MFC.
[c]The MFC effluent was collected on day 239.
doi:10.1371/journal.pone.0030495.t001

Other components of the primary clarifier effluent sample and treated MFC anolyte are compared in Table 1. These comparisons show that potential anaerobic electron acceptors such as sulfate and nitrite were slightly reduced during the MFC batch cycle. These data suggest that alternative electron acceptors existed in the primary clarifier wastewater but were not used as the preferential terminal electron acceptors during anaerobic respiration. Oxygen is known to permeate the air-breathing cathodes used in these experiments. However, thick biofilms were observed at the cathode surface (data not shown) and were likely responsible for the removal of oxygen at the cathode surface and maintenance of anaerobic conditions at the anode. The concentrations of several inorganic heavy metals (except for manganese and strontium) were significantly decreased in the process; however, the oxidation states of these metals were not quantified so it is unknown if the metals served as electron acceptors or if abiotic adsorption contributed to these observed values. The conductivity (1.7 mS/cm) and pH (7.57) of the MFC anolyte were unchanged from their values seen in the raw primary clarifier effluent, and were stable throughout each cycle.

High pressure liquid chromatography (HPLC) analysis of the raw primary clarifier effluent and treated MFC anolytes revealed that various organic chemicals were nearly completely degraded in the MFC process (Fig. S2). While there was a clear decrease in number and amounts of organics, many of the peaks could not be matched to known standards of volatile fatty acids due to the complex nature of the raw sample. The publically available, North City Water Reclamation Plant (NCWRP) annual report indicates that approximately 30 mg/l of n-hexane extractable material (organic solvents) and 8 mg/l of methylene blue active substances (MBAS) (surfactants) were present in the primary clarifier effluent [10]. It is possible that many of the unidentified peaks in the HPLC chromatographs may be related to organic solvents and/or surfactants.

Electricity Production

Performance of MFC systems are commonly evaluated in terms of power and current densities[11]. Electricity production was monitored as current versus time, but additional electrochemical measurements were performed to thoroughly characterize the MFC system. On day 32, anodic, cathodic and whole electrochemical cell polarization curves were determined by using a graded series of external resistors (Fig. 3). These curves showed an open circuit cell voltage of 550 mV, open circuit anode potential of -270 mV vs Standard Hydrogen Electrode (SHE), and open circuit cathode potential of $+280$ mV vs SHE. The maximum power density per projected anode surface area was 12.4 mW/m^2 (with 750 Ω of external resistance), while the maximum power output per reactor volume was 0.3 W/m^3. The limiting current density was 84 mA/m^{2}with 10 Ω of external resistance (Fig. 3A). The anodic and cathodic polarization curves apparently revealed that reactor performance was cathode limited (Fig. 3B), suggesting that the cell polarization curve did not reflect the available microbial biocatalytic activity at the anode[9], [12].

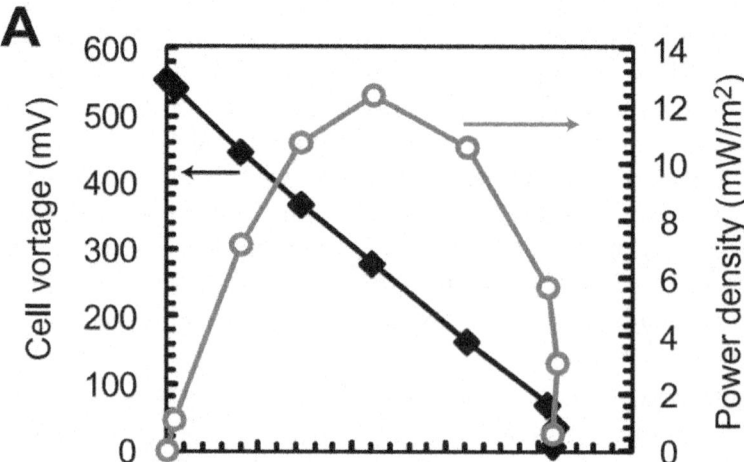

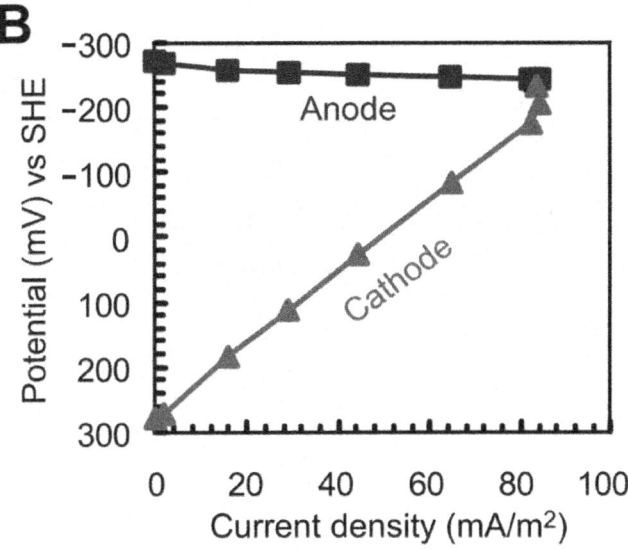

Figure 3. Power curve and polarization curves in the primary clarifier effluent-fed MFC. Power curve (A) and anodic and cathodic polarization (B) measured from the MFC after stable current production was observed on day 32. Anode potential (solid square), cathode potential (solid triangle), and cell voltages (solid diamond) were measured at various external resistances and plotted versus current density normalized to the projected anode surface area. Reactor power performance is represented by power density per anode surface area (open circle).

In order to better analyze the microbial current generating properties, the anode polarization curves were determined by linear sweep voltammetry (LSV) (Fig. 4). Using a potentiostat, the anode potential was varied from the open circuit anode potential to +300 mV vs SHE at a scan rate of 0.5 mV/sec, allowing measurement of anodic activity independent of cathode limitations[12]. The anode polarization curves on day 64 (Fig. 4A) revealed a limiting current density of approximately 600 mA/m². After another 3 months of operation, the limiting current density improved to 1,000 mA/m² (day 141), indicating that electricity generating microorganisms were further optimized during the repeat-batch operation. However, the limiting current density did not increase again after another 100 days of operation (measured on day 264), suggesting that electricity generating performance had stabilized between day 141 and day 264. This current density of 1,000 mA/m² represents the limiting anodic biocatalytic activity and may be a function of low ionic conductivity in the anolyte (1.7 mS/cm), increasing the ohmic losses in the system; and/or low substrate concentration at the biofilm surface (the COD was 260 mg/L) as a result of diffusion limitations.

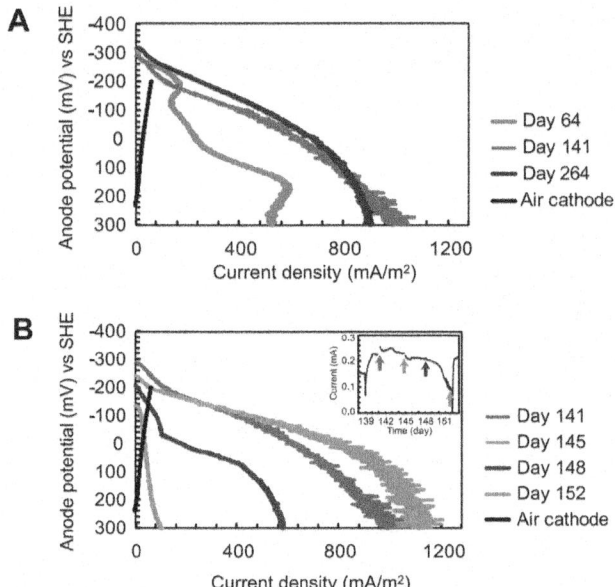

Figure 4. Anode polarization curves in the primary clarifier effluent-fed MFC determined by linear sweep voltammetry. The anode polarization curves during the enrichment process of electricity generating mixed community (A). The anode polarization curves during a single batch operation from day 139 to day 151, current generations are shown in the inset (B). Cathode polarization curve (black line) was determined on day 108.

Correlation between Anode Biomass and Electricity Production

Each repeat-batch cycle began with new and different suspensions of chemical compounds and microbes. The anode polarization curves were also measured during the representative single batch-feed from day 139 to day 152 (Fig. 4B). The polarization curves revealed similar trends through the initial 6 days, while the curves after day 9 showed an abrupt decrease in maximum current density and a more electropositive open circuit anode potential. These changes affirm the drop in current production as a result of the reduction of available organic matter.

The relationship between each new microbial cell suspension and the observed patterns of electricity production and organic matter degradation was also examined. The suspended cell concentrations decreased during the residence time associated with each batch cycle. During a representative cycle, it was found that the suspended cell counts decreased from 7.5×10^7 cells/ml at day 139 to 1.2×10^7 cells/ml at day 152, suggesting that planktonic microbial cells were not contributing to organic matter degradation or electricity generation in the MFC. This is also visibly evident in the turbidity changes shown in Fig. 1B and 1C. Scanning electron microscope (SEM) observations of the anode surface at day 152 clearly revealed a dense biofilm constructed on the surface (Fig. S3). The biomass density of the biofilm was determined as total protein concentration per anode surface area, resulting in 1.25 mg-protein/cm^2 on day 152. In order to analyze normalized biocatalytic activity of the anode biofilm, the per-biomass electron-donating rate was calculated to be 52.4 μmol-electron g-protein^{-1} min^{-1}.

These data, when taken together, suggest that microbial growth primarily occurred at the electrode surfaces and that the anode consortium was primarily responsible for organic matter degradation and current production in the MFC.

Phylogenetic Composition

To analyze the microbial community composition dynamics of the anode biofilm, we constructed 16 S rRNA gene clone libraries of anode biofilm samples collected at day 44 (W1), day 133 (W2), day 263 (W3), and day 294 (W4). Clone libraries prepared from raw primary clarifier effluent samples collected at day 14 (PC1), day 152 (PC2), and day 304 (PC4) were also analyzed because these samples served as carbon and inoculum sources for each repeat-batch cycle.

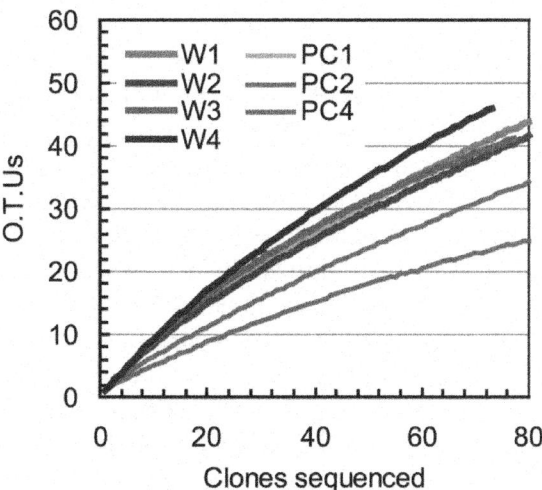

Figure 5. Rarefaction curves for the different phylotypes obtained from 16 S rRNA gene clone libraries. The raw primary clarifier effluent samples were collected at day 14 (PC1), day 152 (PC2), and day 304 (PC4). The anode biofilm samples were collected on day 44 (W1), day 133 (W2), day 263 (W3), and day 294 (W4).

The rarefaction-curve analysis for the MFC anode and primary clarifier samples showed that the anode biofilm community at any given time was more diverse than the communities of primary clarifier effluents (Fig. 5). This result indicates that counts of phylotypes in the anode biofilm were not reduced throughout the year long enrichment process. The Shannon's diversity index, Simpson diversity index, and Chao-1 richness also suggest this trend (Table 2).

Table 2. Diversity statistics and recovery of bacterial phyla from MFC anode biofilms and primary clarifier effluents

Source	Anode biofilm				Primary clarifier effluent		
Library ID	W1	W2	W3	W4	PC1	PC2	PC4
Sampling day	44	133	263	295	14	152	304
Shannon's Index	3.5	3.4	3.5	3.6	3.3	2.8	2.0
Simpson Diversity Index (1 - D)	0.96	0.94	0.96	0.97	0.94	0.80	0.66
Number of clones sequenced	88	84	77	73	72	146	91
Number of O.T.U. (99% cutoff)	47	43	41	46	40	54	27
Chao1 Richness	228±65	203±59	77±15	119±27	96±22	180±42	45±10
Sørensen's similarity coefficients							
W1		0.18	0.11	0.11	0.09	0.06	0.11
W2			0.21	0.14	0.12	0.10	0.14
W3				0.32	0.03	0.00	0.00
W4					0.00	0.02	0.03
PC1						0.28	0.21
PC2							0.20

doi:10.1371/journal.pone.0030495.t002

Results from the seven different 16 S rRNA clone library analyses are shown in Fig. 6 and summarize the phylogenetic affiliations to taxa at the phylum or class level. The analyses indicate that all clones were affiliated with the domain *Bacteria*. The relatively high abundant phyla in the 16 S rRNA clone libraries of both the anode biofilm and primary clarifier effluents were *Proteobacteria*, *Firmicutes*, and *Bacteroidetes*. Those phylotypes are considered as relatively abundant species in the microbial communities. Of the *Proteobacteria*, members of the class *Deltaproteobacteria* were abundant in the anode biofilm samples and significantly increased in abundance throughout the long-term MFC operation. In contrast, those of the*Betaproteobacteria, Gammaproteobacteria*, and *Epsilonproteobacteria* decreased in abundance within the anode biofilm samples even though these members were abundant in every characterized primary clarifier effluent sample. These results indicate that deltaproteobacterial species are important for electricity generation in the anode biofilm. Furthermore, the frequency of the phylum *Bacteroidetes* in the anode biofilm was slightly higher than that of primary clarifier effluents, suggesting that *Bacteroidetes* could also be important for efficient biofilm function.

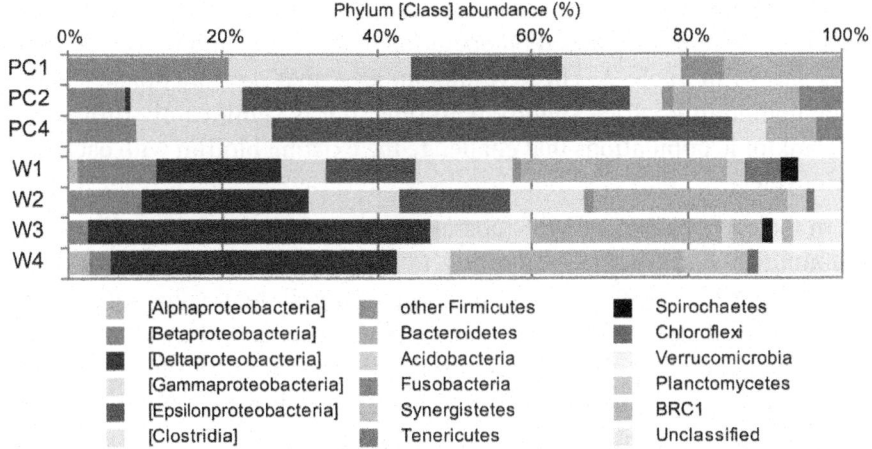

Figure 6. Phylum level taxonomic distribution of 16 S rRNA community profile. The taxonomic profiles were analyzed for the original primary clarifier effluents (PC1, PC2, PC4) and the MFC anode biofilms (W1–W4). Phylum *Proteobacteria* and *Firmicutes* are divided into class level taxonomies.

In order to further compare the microbial communities, a multidimensional scale plot (MDS) was created based on genus level taxonomy (Fig. 7). The plot suggests that the microbial communities associated with primary clarifier effluents were more diverse than those of the anode biofilm, and there was a significant difference between those two groups. Sørensen›s similarity coefficient was also used to statistically compare the similarity of these two types of microbial communities.

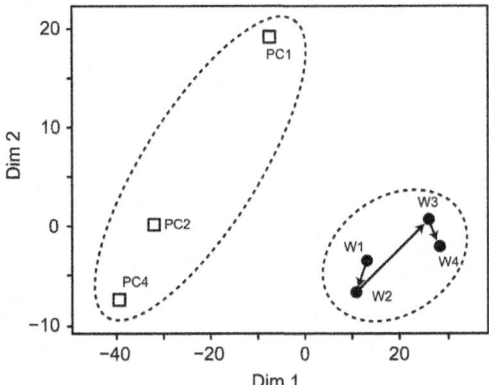

Figure 7. Multidimensional scale plot comparing bacterial communities based on genus level taxonomy. Open square, the bacterial communities of original primary clarifier effluents (PC1, PC2, PC4); Filled circle, the bacterial communities of anode biofilm enriched in the primary clarifier effluent-fed MFC (W1–W4).

The summary of Sørensen›s similarity coefficients among the seven microbial communities based on operational taxonomy unit (OTU) at 99% cut off is shown in Table 2. These results indicate that the functionally enriched anode community varies slightly with time, but is significantly different than the planktonic populations that challenge the existing biofilm with each repeat-batch cycle.

In the early phases of MFC operation (W1 and W2), the anodic microbial communities had slight similarities to the microbial communities of the primary clarifier effluents. These results suggest that some of the initial bacteria introduced to the MFC reactor attached and then established the electricity-generating microbial community in the anode biofilm. Furthermore, the data indicate that the established biofilm was not susceptible to invasion by new microbial populations introduced with the primary clarifier repeat-batch anolyte replacements.

Electricity-Generating Bacteria

Table 3 shows the abundant phylotypes, containing more than 5 clones, from the anode electricity-generating biofilm samples W1–W4. Large portions (~50%) of the biofilm clones were categorized into the minor phylotypes, which are summarized in Table S1.

Table 3. Abundant phylotypes obtained from the enriched electricity-generating consortia[a]

Phylum-Class Phylotype	No. of clone in the library				% Database match (Accession No.)
	W1	W2	W3	W4	
Proteobacteria - Deltaproteobacteria					
W1_B02	11	-	-	-	98% *Geobacter sp.* Ply1 (EF527233)
W2_M05	-	13	-	-	99% *Geobacter lovleyi* SZ (CP001089)
W3_M01	-	1	6	3	96% *Geobacter lovleyi* SZ (CP001089)
W2_A03	-	1	7	8	99% *Desulfuromonas acetexigens* (U23140)
W3_A19	-	-	4	2	99% *Desulfobacter postgatei* DSM 2034 (AF418180)
W3_O19	-	-	3	3	99% *Desulfocapsa thiozymogenes* (X95181)
W3_O03	-	-	2	3	91% *Desulfobulbus rhabdoformis* Mic5c02 (AB546246)
Proteobacteria - Betaproteobacteria					
W1_F12	3	2	-	-	100% *Acidovorax sp.* PPs-5 (FJ605421)
Proteobacteria - Gammaproteobacteria					
W2_C01	-	5	1	-	100% *Pseudomonas sp.* SMT-9 partial (AM689953)
Proteobacteria - Epsilonproteobacteria					
W1_J08	9	10	-	-	100% *Arcobacter cryaerophilus* (U34387)
Bacteroidetes					
W2_I23	2	1	3	3	88% *Ruminofilibacter xylanolyticum* S1 (DQ141183)
W1_B10	4	3	1	1	87% *Cellulophaga tyrosinoxydans* EM41 (EU443205)
W1_J04	4	-	1	-	93% *Parabacteroides goldsteinii* JCM13446 (EU136697)
W2_C17	-	3	2	-	90% *Prolixibacter bellariivorans* F2 (AY918928)
W4_O10	-	-	4	3	90% *Cytophaga fermentans* (M58766)
W3_A07	-	-	4	2	88% *Cytophaga fermentans* (M58766)
Firmicutes - Clostridia					
W1_H18	3	3	1	-	97% *Fusibacter sp.* SA1 (AF491333)
Synergistetes					
W3_A23	-	-	3	4	90% *Aminomonas paucivorans* (AF072581)
Others[b]	52	42	35	41	
Total clone	88	84	77	73	

[a]Abundant phylotype was defined as a phylotype containing more than 5 clones within four clone libraries.
[b]For a list of other phylotypes (minor phylotypes), refer to the supplementary materials (Table S1).
doi:10.1371/journal.pone.0030495.t003

Deltaproteobacteria .

As mentioned above, *Deltaproteobacteria* are potentially important bacteria for extracellular electron transfer in the anode biofilm, however, the dominant phylotypes within this class changed over time. During the early stage of MFC operation (day 44, W1 clone library) the *Deltaproteobacteria* were primarily comprised of a phylotype W1_B02, which is closely related to *Geobacter* sp. Ply1. However, during the following stage (day 133, W2 clone library) the dominant phylotype was W2_M05, which is more related to *Geobacter lovleyi* SZ. Over 200 days of operation, the matured biofilm samples represented in W3 (day 263) and W4 (day 294) clone libraries showed that the dominate phylotypes at the early stages had diminished, and another five abundant phylotypes were apparent in the class *Deltaproteobacteria*. The phylotype W2_A03, which appeared first in the W2 clone library, subsequently became dominant as the biofilm continued to mature, was found to be closely related to *Desulfuromonas acetexigens*.

Epsilonproteobacteria .

A phylotype W1_J08, which is closely related to *Arcobacter cryaerophilus*, was highly abundant in only the early stages of anode microbial community development (Table 3). This phylotype was also abundantly present in the primary clarifier effluent (Table S2), suggesting that the bacterium was simply introduced from the wastewater but did not thrive at the anode surface during prolonged electricity generating conditions.

Bacteroidetes .

The phylum *Bacteroidetes* was also abundant in the anode biofilm and various abundant phylotypes were observed throughout the long-term MFC operation (Table 3). Interestingly, the*Bacteroidetes* phylotypes that were found in the anode-associated consortia were not observed in the raw primary clarifier effluent (Table S2); furthermore, the phylotypes were not closely related to isolated *Bacteroidetes* strains (Table 3).

DISCUSSION

The present study successfully investigated the performance and phylogenetic diversity of an MFC community fed solely with raw primary clarifier effluent from a municipal wastewater treatment plant for over 300 days and 31 solution changes (Fig. 2). Despite the regular changes in organic matter composition [10] and repeated exposure to new microbial populations (Table S2), the reported MFC system was able to continuously generate a current of approximately 0.25 mA and degraded the chemical constituents associated with primary clarifier effluents with reproducibility (Fig. 2). These results suggest that the MFC can be a robust tool for wastewater treatment and for selecting functionally stable microbial communities from diverse and dynamic inoculum sources. The anodic biocatalytic activity that is one of the most important parameter for developing electrochemical fuel cells was also reproducible after day 141, yielding a current density of approximately 1,000 mA/m^2 (Fig. 4A), which if modified may lead to useful energy recovery at a treatment plant scale.

To consider the implementation of MFC technology in a wastewater treatment facility, it is important to demonstrate that removal rates and electricity generation are not affected by seasonal changes in wastewater effluents or the variability of chemical compounds and concentrations. This can only be proven by long-term MFC evaluations of the microbial community dynamics and resulting current production and COD removal rates. Here we present a comprehensive data set that contributes to a better understanding about how electrogenic microbes respond to real wastewater conditions. Using this knowledge, we can begin to make system design improvements to increase current densities, bioaugment or pre-seed anode associated communities for rapid MFC enrichment, and ultimately speed COD degradation rates. Although an 8–13 day treatment time is not acceptable for practical application to wastewater treatment, we believe the COD removal rates will significantly increase with design modifications to the system, such as lowering internal resistance and improved cathode catalysts, and using our enriched electricity-generating microbial communities as anode catalysts.

Several reports have recently appeared in the literature exploring MFC performance during treatment of primary clarifier effluents [13], [14], [15], [16]. Di Lorenzo et al. described the phylogenetic composition of an electricity-generating microbial community associated with a MFC anode by using denaturing gradient gel electrophoresis (DGGE) based on 16 S rRNA gene fragments. However, the microbial community was only analyzed as a function of different anode surfaces within the system and not as a function of time, wastewater inoculum sources, or MFC performance [13]. Therefore, there is much to understand about the microbial populations responsible for catalyzing rapid degradation of organic compounds and electricity production, and how microbial populations assemble and functionally maintain in MFCs. Here, we conducted electrochemical analyses and clone analyses based on 16 S rRNA gene sequences to achieve a more detailed understanding about the composition of the microbial population responsible for electricity production and the degradation of organic compounds, and how the population is correlated to biological and chemical heterogeneity of the influents throughout long-term operation. In addition, we periodically compared the anode biofilm to the introduced microbial community from the primary clarifier effluents to discern how the community diverged from the planktonic inoculum and substrate sources.

Three distinct microbial communities existed in the MFC system: anode biofilm (anaerobic metabolism), cathode biofilm (aerobic/ facultative metabolism), and suspended (planktonic) cells (facultative metabolism) [17], [18]. Each population may contribute to overall system operation in terms of organic degradation rates and electricity generation. During each batch cycle, the suspended cell density decreased by one-sixth, suggesting that planktonic cells were not significantly contributing to direct electron transfer or COD/BOD removal. Further, our repeat-batch approach decreased the accumulation of mediators (electron shuttles) in the anolyte so the presence of such compounds would need to be within the biofilm matrix, or in very close proximity to the electrode surface in order to contribute to the observed current densities.

Although it is known that the air-cathode we used allows the diffusion of oxygen into the MFC system [19], [20], a robust biofilm was found to thrive at this surface. It is likely that this biofilm was responsible for capturing oxygen, which was important to maintaining the anaerobic conditions required for efficient electricity generation at the anode. The success of this process can be noted through the presence of strict anaerobes that thrived at the anode surface, but were not present in high abundance in the primary clarifier effluents. Consequently, it can be inferred that the anode biofilm was mainly responsible for electricity production through direct extracellular electron transfer from biofilm constituents to the anode surface (Fig. 1A).

In a previously reported study using a cellulose-fed MFC, the diversity of the anode microbial community decreased with time while system performance remained stable [21]. However, the electricity-generating microbial community in the primary clarifier effluent-fed MFC did not show a decrease in diversity even though the system operation remained stable (Fig. 5 and Table 2). This result indicates that a complex microbial community was necessary to degrade diverse and variable organic substrates. Interestingly, the introduction of new microbes with each repeat-batch cycle did not appear to impact the electricity-generating biofilm. Throughout MFC operation the anode-associated population was found to change, but each generation was more closely related to its surface-attached predecessors and was very different from the microbial populations that were introduced with the new primary clarifier effluents.

Although the anode-associated microbial community was still diverse after the enrichment process, the taxonomic distribution of the electricity-generating microbial community clearly reached a stable population (Fig. 7). The early stage microbial community was slightly affected by the introduction of primary clarifier microbes including *Acidovorax* sp. and *Arcobacter* spp., which were present in relatively high abundance in the primary clarifier effluent (Table S1). Over time, the community shifted to a more "conventional" anode-respiring population (Table 3) including *Geobacter* and *Desulfuromonas* spp., indicating that a longer-term enrichment process facilitated the adaptation to the electrode-reducing conditions while oxidizing complex heterogeneous substrates in the primary clarifier effluent.

The enriched electricity-generating anode microbial community was mostly comprised of the class *Deltaproteobacteria* and the phylum *Bacteroidetes* (Fig. 6), which have been frequently observed as dominant taxa in both sediment MFCs [22], [23] and in MFC reactors fed with industrial wastewaters [24], artificial wastewater [17], or with defined chemicals such as acetate and glucose [9], [18], [25]. The class *Deltaproteobacteria* includes various dissimilatory solid metal reducing bacteria [26], some of which are also reported as electricity generators in MFC systems [27], [28], [29].

The *Geobacter* spp. are well-known electricity generating bacteria [27], [28], indicating that the phylotypes observed within the anode-associated community are likely playing an important role in electricity production. Recently, many *Geobacter* strains have been observed as part of electricity-generating microbial communities, especially in acetate-fed MFC anodes [18], [25],[30], suggesting that the *Geobacter* strains are oxidizing acetate during anode respiration. Various *Geobacter* strains including *Geobacter metallireducens* can also utilize a wide variety of electron donors including toluene and benzoate [31], suggesting that the abundant *Geobacter* strains in the described system may also be playing a role in the oxidation of more complex primary effluent substrates.

In the highly enriched electricity-generating anode microbial communities (W3 and W4), four other *Deltaproteobacteria* phylotypes closely related to genera *Desulfuromonas*, *Desulfobacter*, *Desulfocapsa*, and *Desulfobulbus* were also observed (Table 3). The most abundant phylotype W2_A03 in the mature biofilm was found to be closely related with *Desulfuromonas acetexigens*, which has been reported as a solid iron/electrode reducer [32]. The strain has also been observed in electrically active anode biofilms in sediment MFCs [22], [23], and may therefore be contributing to electricity generation in our system. Other prevalent phylotypes were closely associated to *Desulfocapsa* and *Desulfobulbus*, which have both been previously reported as potential electrode reducers [22], [29], [33].

The prevalence of different electricity generating phylotypes in our primary clarifier MFC implies that several species within the *Deltaproteobacteria* class were syntrophically cooperating to produce electricity from the wide varieties of chemical compounds in the primary clarifier effluents. Interestingly, the dominant *Deltaproteobacteria* species changed with time, but the electricity generating performance and chemical oxidation rates remained stable. This phenomenon clearly suggests that the anode-associated microbial population can be functionally maintained for the treatment of primary clarifier effluents and sustain energy recovery in the process.

We hypothesize that the phylotypes associated with *Deltaproteobacteria* species were primarily responsible for direct electricity production; however the other highly abundant anode-associated phylotypes were closely related to the phylum *Bacteroidetes* and observed in all stages of anode biofilm enrichment. While the phylum *Bacteroidetes* has mainly been described as a fermentor in the human gut [34], Shimoyama *et al.* recently demonstrated that*Bacteroidetes* was an abundant phylum correlated with electricity generation from artificial wastewater treatment in a continuous-flow cassette-electrode MFC [17]. However, our phylotypes were not closely related to the other previously reported *Bacteroidetes* strains including the clone CE38 abundantly observed in the cassette-electrode MFC. These results suggest that MFC enriched *Bacteroidetes* species may possess diverse functional traits and may thus represent an interesting phylum worthy of further study. To address the potential roles of these dominant phylotypes classified to *Bacteroidetes* within our MFC system, we will attempt to isolate these strains and analyze their genomic and functional characteristics in future work.

Other less dominant phylotypes observed in the anode-associated biofilm were very diverse and functionally unknown (Table S1). Some phylotypes could be contaminants from the primary clarifier effluent (Table S2), while others could be contributing to the degradation of various types of organic chemicals. The long-term survival of these less abundant phylotypes suggest that they have a functional role in organic compound degradation and perhaps even within extracellular electron transfer.

In summary, while MFC performance was very reproducible, the phylogenetic analyses of anode-associated electricity-generating biofilms demonstrated that the microbial populations fluctuated temporally while maintaining high biodiversity throughout the year-long experiment. These results suggest that MFC operation induces a self-optimizing process toward functional performance from diverse, heterogeneous microbial communities, and therefore can be used to reproducibly select for functional microbial communities regardless of carbon source. These results contribute significant knowledge toward the practical application of MFC systems for long-term wastewater treatment, while demonstrating the utility of MFCs for enrichment of functionally stable microbial populations capable of organic compound degradation and extracellular electron transfer.

Additional impact of the reported results is provided by the ability to observe the metabolic activities and energy transduction within a complex consortium. This has significant benefits to the field of microbial ecology and may yield insight into numerous geochemical cycles that involve biological transformations of extracellular material, such as iron and manganese oxides. Future studies will apply metagenomics and metatranscriptomics to the anode biofilm to describe gene expression profiles associated with carbon metabolism and extracellular electron transfer during the degradation of complex organic substrates. Gene expression data will further elucidate metabolic networks and energy transduction in complex consortia.

MATERIALS AND METHODS

MFC Configuration and Operation

A single-chamber, air-cathode MFC was used for municipal sewage wastewater treatment with power generation. The MFC was a bottle-type reactor (350 ml in capacity), with two joined anode electrodes made of carbon cloth (7 cm×3 cm, or 84 cm² total projected surface area per reactor; TMIL, Japan) [9]. The air-cathode was made with a 30 wt% wet-proofed carbon cloth (type B-1B, E-TEK) coated with platinum (0.5 mg/cm²), Nafion, and PTFE as described elsewhere [20]. The air-cathode was placed at the side port, providing a total projected cathode surface area (one side) of 4.9 cm².

After sterilization of the fully assembled MFC, the chamber was filled with municipal wastewater collected from the primary clarifier at the North City Water Reclamation Plant (San Diego, USA) without any pretreatment except the mechanical removal of grit, rags and scum. The sole inoculum source consisted of those microorganisms present in the primary clarifier effluents. The MFC was gently mixed with a magnetic stirrer, and incubated at room temperature (22°C±3°C) throughout the duration of testing.

The anode and cathode electrodes were connected with an external resistor of 750 Ω. Cell voltages across the resistor were recorded every 30 min using a voltage recorder (GL200A, Graphtec) and the corresponding electric current was calculated using Ohm's law (V=IR). When the electric current decreased due to depletion of the organic matter in the wastewater, the anode solution was fully discarded and the reactor was refilled with either the fresh wastewater collected that day or with aged wastewater stored at 4°C. This repeat-batch process occurred twice monthly with fresh wastewater, and weekly with aged wastewater, for 300 days. Each wastewater sample introduced to the reactor included the naturally occurring microorganisms and various chemical compounds, no filtration or additional pretreatment was conducted. On days 88 and 195, the biofilm formed on the cathode surface was mechanically removed to recover the cathode performance.

Polarization Analyses

To obtain polarization and power density curves, an Ag/AgCl reference electrode (+200 mV vs SHE, RE-5B, BASi) was placed in the side port of the MFC. The external resistance across the circuit was then changed stepwise from 3.3 kΩ to 10 Ω and the cell voltage, the anode potential, and the cathode potential were recorded after they had stabilized over a period of at least 7 min [9]. Current density per projected anode surface area was calculated from the voltage measured across the known resistor. Power density was calculated as the product of current density and the cell voltage.

In order to obtain anode polarization curves without cathodic reaction limitation, linear sweep voltammetry analyses were conducted using a potentiostat (Reference 600™, Gamry) [12]. The anode potential was swept from open circuit anode potential to +300 mV vs SHE at a scan rate of 0.5 mV/sec and the corresponding anodic current, resulting from the active biofilm, was recorded.

Chemical Analyses

Chemical oxygen demand (COD) was determined using a potassium chromide assay according to the manufacturer's instructions (Orion CODHP0, Thermo Scientific). Coulombic efficiency, CE (%), was calculated as $CE=C_p/C_{th}\times100$, where C_p (C) is the total charge passed during a single batch, and C_{th} (C) is the theoretical amount of charge allowable from a complete COD decrease (assuming that reducing one mole of oxygen requires the transfer of four electrons). Biological oxygen demand (BOD), total suspended solid (TSS), turbidity, nitrate-N, nitrite-N, ammonium-N, sulfate, and heavy metal concentrations were determined in accordance with US EPA and state of California requirements by CRG Marine Laboratories, Inc (Torrance CA, USA). Conductivity of the solution was determined by portable pH/ORP/DO/ionic meter (Orion 1215000, Thermo Scientific). Acetate and other volatile fatty acids were determined using a high-pressure liquid chromatography (HPLC) machine equipped with DI detector (Agilent 1200 series) and a packed

C18 column (Epic Polar, ES Industries). The eluant was 50 mM phosphoric acid (pH 1.87) at a flow rate of 1.0 ml/min.

In order to determine the total bacterial cell density on the anode surface, part of the anode (7 mm×7 mm) containing cells was removed from the MFC (n=3). Total protein was extracted from the electrodes as described elsewhere [9], [28]. Bacterial cell concentrations in the solution were also determined by direct cell counts. Cells were stained with 2 mg/L of 4,6-diamidino-2-phenylindole (DAPI) for 5 min then observed using an AX10 fluorescence microscope (Carl Zeiss).

Scanning Electron Microscopy (SEM)

A small portion of carbon cloth was collected from the anode, fixed with 1.25% gultaraldehyde, dehydrated using a graded series of ethanol solutions, and dried using a critical point drier (Autosamdri 815, Tousimis) [35]. The specimen was coated with Pt/Pd and imaged at 2 kV on a LEO 1540XB Field Emission SEM (Carl Zeiss SMT AG). The imaging was conducted at the Western Nanofabrication Facility, University of Western Ontario (Canada).

PCR Amplification, Cloning and Sequencing of 16 S rRNA Gene Fragments

Total DNA was extracted from the biofilm associated with the carbon cloth anode or from the suspended cells in the primary clarifier effluent. All DNA extractions were performed using the UltraClean® Soil DNA Isolation Kit (MO bio) according to manufacturer instructions, which employed physical cell disruption. PCR amplification of 16 S rRNA gene fragments was performed using Taq DNA polymerase (ExTaq, Takara) with universal primers U27f (5′-AGAGTTTGATCCTGGCTCAG-3′) and U1492r (5′-GGTTACCTTGTTACGACTT-3′) [36]. The amplification conditions were as follows: an initial step of 94°C for 3 min, 25 cycles consisting of 94°C for 30 sec, 55°C for 30 sec and 72°C for 90 sec, and a final elongation step at 72°C for 10 min. Amplified fragments were ligated into a pGEM-T vector (Promega) and cloned into*Escherichia coli* JM109 competent cells. PCR-amplified 16 S rRNA gene fragments were recovered by PCR using primers M13f and M13r (the primers targeted the pGEM-T vector sequences flanking the insertion), then sequenced by ABI 3730xl sequencers using primer U907r (5′-CCGYCAATTCMTTTRAGTTT-3′) [37]. The nucleotide sequences reported in this paper have been deposited in the GSDB, DDBJ, EMBL and NCBI nucleotide sequence databases under accession numbers HQ688300 to HQ688420 for the primary clarifier effluents, and HQ688421 to HQ688596 for the anode biofilm.

Phylogenetic Analyses

Sequences of partial 16 S rRNA genes determined in this study were aligned to each other using CLC genomics work bench version 3.6.5 (CLC bio), and assigned to phylotypes (classified as an operational taxonomic unit, >99% cut-off). Database searches for related 16 S rRNA gene sequences were conducted using the BLAST program [38]. Checks for chimeric sequences and a multidimensional scale (MDS) plot were conducted using JCVI 16 S/18 S small sub-unit analysis pipeline. A rarefaction analysis was conducted using the Analytic Rarefaction program [21]. Chao1 richness was calculated using web-based software (http://www2.biology.ualberta.ca./jbrzusto/rarefact.php). A Shannon's index, Simpson diversity index, and Sorensen similarities among the bacterial communities were calculated using Estimate S [39].

SUPPORTING INFORMATION

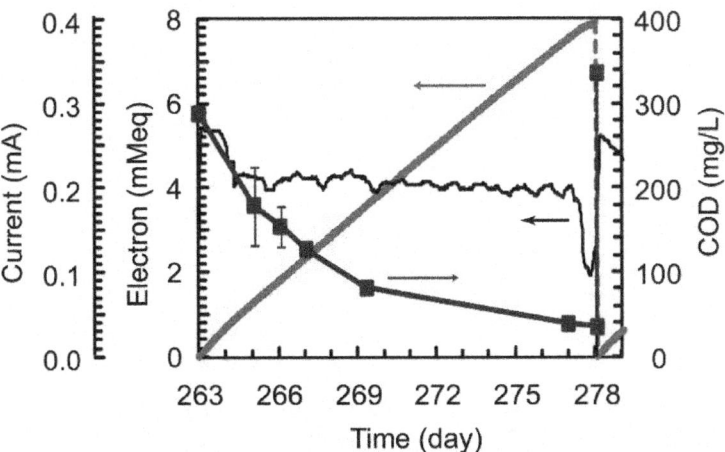

Figure S1. Typical batch cycle of current generation and COD concentrations in the primary clarifier effluent-fed MFC. Thin black line, electric current (mA); Thick red line, accumulated electron production expressed as 'mM equivalent (eq.)' calculated as the number of total electrons passing across the circuit from the available COD in solution; solid square blue line, total COD (mg/L) in solution.

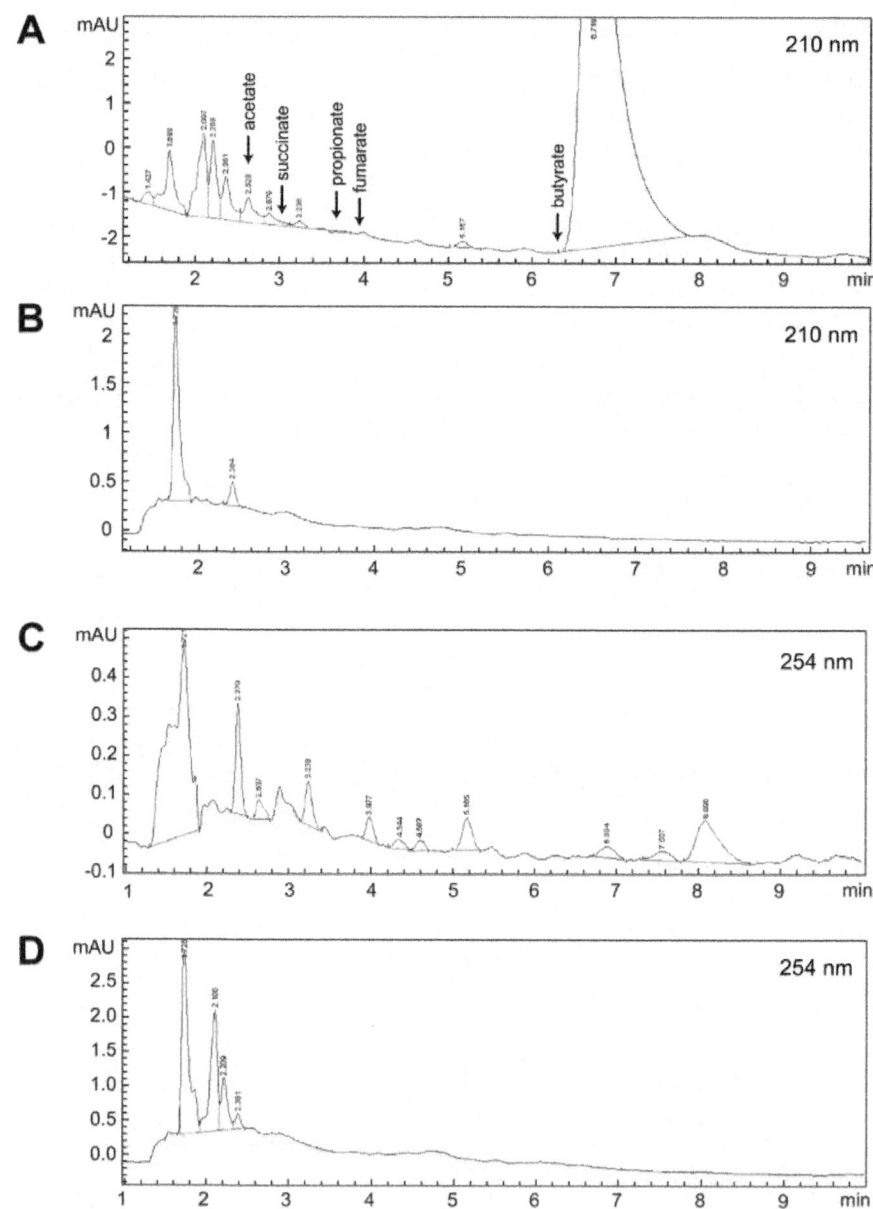

Figure S2. HPLC chromatographs of wastewater samples before and after MFC treatment. Organic compounds were detected at wavelengths of 210 nm (A, B) or 254 nm (C, D). The untreated primary clarifier effluent chromatographs indicate the presence of several different compounds (A, C), most of which were no longer present in the MFC treated samples (B, D).

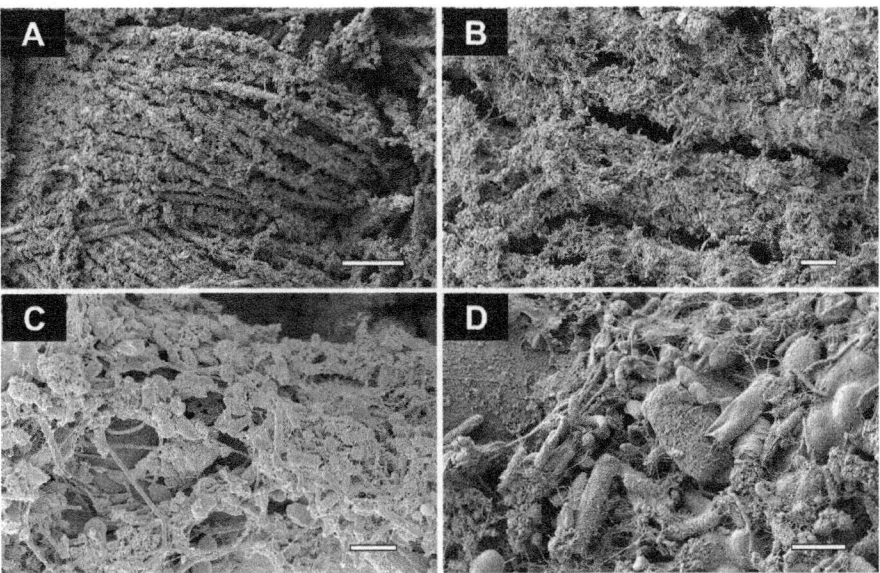

Figure S3. FE-SEM images for anode biofilms adhering onto carbon cloth anodes (day 152). Bar in panel A is 100 μm, bar in panel B is 10 μm, and bars in panel C and D are 2 μm.

Table S1. All phylotypes obtained from the enriched electricity-generating consortia

Phylum-Class Phylotype	W1	W2	W3	W4	% match	Best matched sequence	Accession No.	PC1	PC2	PC4
Proteobacteria - Deltaproteobacteria										
W1_B02	11	-	-	-	98	Geobacter sp. Ply1	EF527233	-	-	-
W2_M05	-	13	-	-	99	Geobacter lovleyi SZ	CP001089	-	-	-
W3_M01	-	1	6	3	96	Geobacter lovleyi SZ	CP001089	-	-	-
W2_A03	-	1	7	8	99	Desulfuromonas acetexigens	U23140	-	-	-
W3_A19	-	-	4	2	99	Desulfobacter postgatei DSM 2034	AF418180	-	-	-
W3_O19	-	-	3	3	99	Desulfocapsa thiozymogenes	X95181	-	-	-
W3_O03	-	-	2	3	91	Desulfobulbus rhabdoformis Mic5c02	AB546248	-	-	-
W4_M24	-	-	-	1	96	Geobacter hephaestius	AY737507	-	-	-
W2_A17	-	1	-	-	95	Geobacter sp. Ply4	EF527234	-	-	-
W3_M21	-	-	1	-	93	Desulfatirhabdium butyrativorans strain HB1	DQ146482	-	-	-
W4_O24	-	-	-	1	94	Desulfatirhabdium butyrativorans strain HB1	DQ146482	-	-	-
W3_C13	-	-	2	-	93	Algidimarina propionica strain AK-P	AY851291	-	-	-
W3_K17	-	-	1	-	96	Delta proteobacterium S2551	AF177428	-	-	-
W3_G11	-	-	1	2	93	Olavius crassitunicatus delta-proteobacterial	AJ620510	-	-	-
W1_L08	1	-	-	1	98	sulfate-reducing bacterium R-ButA1	AJ012596	-	-	-
W3_O07	-	-	2	-	98	sulfate-reducing bacterium R-ButA1	AJ012596	-	-	-
W1_B06	1	-	-	-	97	Desulforegula conservatrix	AF243334	-	-	-
W1_L06	1	-	-	-	97	Bacterium ROME95Asa	AY998140	-	-	-
W3_C15	-	-	3	-	98	Delta proteobacterium JS_SRB400Ace	AM774323	-	-	-
W3_O15	-	-	1	1	97	Desulfobulbus propionicus	AY548789	-	-	-
W2_A01	-	1	-	-	99	Desulfomicrobium sp. ADR26	AM419442	-	-	-
W4_I14	-	-	-	1	100	Desulforhabdus sp. DDT	EF442978	-	-	-
W3_C01	-	-	1	-	99	Uncultured delta proteobacterium clone TDNP_Wbc97_127_1_32	FJ517009	-	-	-
W4_C08	-	-	-	1	96	Syntrophobacter wolinii	X70906	-	-	-
W2_O15	-	1	-	-	91	Anaerotruncus sp. NML 070203	EU815226	-	-	-

Proteobacteria - Betaproteobacteria										
W1_F12	**3**	**2**	-	-	100	Acidovorax sp. PPs-5	FJ605421	**6**	**2**	**2**
W2_C11	-	1	-	-	97	Hydrogenophaga sp. GPTSA14	DQ854970	-	-	-
W2_G23	-	1	-	-	98	Beta proteobacterium HIBAF001	AB452981	-	-	-
W1_F02	1	-	-	-	97	Comamonas sp. BF-3	GQ245981	-	-	-
W2_O13	-	2	-	-	100	Comamonas sp. T108	FJ719342	-	-	-
W1_H02	1	-	-	-	99	Imtechium assamiensis strain BPTSA16	AY544767	-	-	-
W3_C21	-	-	2	-	99	Janthinobacterium sp. Lc50-4	GU244366	-	-	-
W1_L22	1	-	-	-	97	Bacterium N2441	AY928207	-	-	-
W2_C03	-	1	-	-	96	Candidatus Accumulibacter phosphatis clade IIA	CP001715	-	-	-
W1_F16	1	-	-	-	95	Azoarcus sp. DS 30	EF494194	-	-	-
W1_B04	1	-	-	1	96	Dechloromonas aromatica RCB	CP000089	-	1	-
W1_H10	1	-	-	-	99	Dechloromonas sp. JDS6	AY084087	1	-	-
W2_G03	-	1	-	-	99	Aminomonas aminovorus C2A1	AY027801	-	-	-
W4_E18	-	-	-	1	96	Beta proteobacterium pACH94	AY297809	-	-	-
Proteobacteria - Gammaproteobacteria										
W2_C01	-	**5**	**1**	-	100	Pseudomonas sp. SMT-9	AM689953	-	-	-
W1_L02	1	-	-	-	100	Pseudomonas sp. GN33-1	GU994886	-	-	-
W2_G21	-	3	-	-	100	Pseudomonas fragi strain CS11RH1	EU255303	-	-	-
W1_B14	1	-	-	-	99	Pseudomonas pseudoalcaligenes strain 23	EU780001	-	-	-
W1_F10	1	-	-	-	99	Pseudomonas sp. R6(2010)	GU566346	-	-	-
W1_F14	1	-	-	-	99	Stenotrophomonas koreensis strain b87	EU434571	-	-	-
W1_P16	1	-	-	-	98	Aeromonas veronii strain IH118	EU770282	-	-	3
W2_G05	-	1	-	-	96	Tolumonas auensis DSM 9187	CP001616	-	-	-
W2_M21	-	1	-	-	98	Shewanella putrefaciens	X81623	-	-	-
Proteobacteria - Epsilonproteobacteria										
W1_J08	**9**	**10**	-	-	100	Arcobacter cryaerophilus	U34387	**13**	**63**	**52**
W2_G01	-	1	-	-	97	Arcobacter sp. D2043	FJ161215	-	1	-
W1_D06	1	-	-	-	98	Arcobacter sp. D2043	FJ161215	-	-	-
W2_C19	-	1	-	-	97	Arcobacter sp. D2043	FJ161215	-	3	-
Proteobacteria - Alphaproteobacteria										
W1_N24	1	-	-	-	100	Ochrobactrum tritici strain S117	AY972361	-	-	-
W4_I04	-	-	-	1	95	Alpha proteobacterium K6-28	EF612344	-	-	-
W4_O14	-	-	-	1	91	Azospirillum sp. BV-s	EU678791	-	-	-
Bacteroidetes										
W2_I23	**2**	**1**	**3**	**3**	97	Uncultured CFB group bacterium clone 44a-B1-14	AY082459	-	-	-
W1_B10	**4**	**3**	**1**	**1**	100	Uncultured Bacteroidetes bacterium clone RBE2CI-86	EF111174	-	-	-
W1_J04	**4**	-	**1**	-	100	Bacteroidales bacterium JN18_V15_G	EF059535	-	-	-
W2_C17	-	**3**	**2**	-	100	Uncultured bacterium clone: D85CN_B77_OTU2	AB519283	-	-	-
W4_O10	-	-	**4**	**3**	90	Cytophaga fermentans	M58766	-	-	-
W3_A07	-	-	**4**	**2**	90	Bacteria from anoxic bulk soil	AJ229237	-	-	-
W2_O07	-	1	1	1	99	Petrimonas sulfuriphila strain BN3	AY570690	-	-	-
W1_J20	4	-	-	-	99	Uncultured Bacteroidetes bacterium clone:TUT1989-B3	AB513434	-	-	-
W2_G11	1	1	-	1	99	Bacillus sp. IST-38	FM877978	-	-	2
W2_G07	-	4	-	-	92	Bacterium Oil-K-3 gene	AB081536	-	-	-
W2_A09	-	1	-	-	94	Bacteroides sp. 22C	AY554420	-	-	-
W1_J24	1	1	-	-	96	Paludibacter propionicigenes	AB078842	-	-	-
W2_K19	-	1	-	-	97	Paludibacter propionicigenes	AB078842	-	-	-
W1_F20	1	-	-	-	93	Paludibacter propionicigenes	AB078842	-	-	-
W2_M11	-	2	-	-	92	Paludibacter propionicigenes	AB078842	-	-	-
W1_P14	1	-	-	-	93	Paludibacter propionicigenes	AB078842	-	-	-
W2_E21	-	1	-	-	99	Uncultured bacterium clone EBL27	GU591518	-	-	-
W2_E03	-	1	-	-	100	Cloacibacterium normanense strain tu29	FJ544401	3	9	3

					%	Description	Accession			
W4_G04	-	-	-	1	92	Bacteria from anoxic bulk soil	AJ229237	-	-	-
W3_I15	-	-	1	-	99	Uncultured bacterium clone EUB7	AY693824	-	-	-
W4_M10	-	-	-	2	96	Uncultured *Bacteroidetes* bacterium clone QEDR3AD08	CU922148	-	-	-
W1_D18	1	-	-	-	93	Uncultured *Bacteroidetes* bacterium clone 4-191	GQ354965	-	-	-
W4_A10	-	-	-	1	99	Uncultured *Bacteroidetes* bacterium clone 4-137	GQ354954	-	-	-
W1_H06	1	-	-	-	89	*Alkaliflexus imshenetskii* strain Z-7010	AJ784993	-	-	-
W4_M16	-	-	-	2	99	Uncultured *Anaerophaga* sp. clone MDAF11	EU214540	-	-	-
W4_I20	-	-	-	2	97	Uncultured bacterium clone MS4-31	GQ354928	-	-	-
W2_M22	-	1	-	-	97	Iron-reducing enrichment clone Cl-A7	DQ676999	-	-	-
W1_D16	3	-	-	-	99	*Rikenellaceae* bacterium WN081	AB298736	-	-	-
W4_G12	-	-	-	1	99	Uncultured *Bacteroidetes* bacterium clone CAR8MG67	FJ902383	-	-	-
W3_K07	-	-	1	-	98	Uncultured bacterium clone MC1_16S_36	EU662604	-	-	-
W4_K12	-	-	-	1	97	Uncultured bacterium clone GW-20	EU407205	-	-	-
Firmicutes										
W1_H18	**3**	**3**	**1**	-	**97**	*Fusibacter* sp. SA1	AF491333	**1**	-	-
W1_H14	3	1	-	-	97	*Clostridium aminobutyricum*	X76161	-	-	-
W2_A15	-	1	2	-	99	Uncultured bacterium clone 3C7_cons	FJ810785	-	-	-
W3_A13	-	-	2	-	95	*Anaerovorax odorimutans*	AJ251215	-	-	-
W3_M09	-	-	1	-	94	*Clostridiaceae* bacterium FH042	AB298771	-	-	-
W4_M18	-	-	-	1	94	*Clostridiaceae* bacterium FH042	AB298771	-	-	-
W1_D12	1	-	-	-	98	Clostridium from anoxic bulk soil	AJ229234	-	-	-
W2_C05	-	1	-	-	100	Uncultured *Firmicutes* bacterium clone QEDN2BB09	CU926241	-	-	1
W3_O11	-	-	1	-	100	Uncultured *Firmicutes* bacterium clone QEDN2BB09	CU926241	-	-	-
W3_I03	-	-	1	1	92	Rumen bacterium R-7	AB239481	-	-	-
W3_C09	-	-	1	-	93	Rumen bacterium R-7	AB239481	-	-	-
W4_O12	-	-	-	1	98	Uncultured bacterium clone 36	FJ462043	-	-	-
W1_F06	2	-	-	-	96	Bacteria from anoxic bulk soil strain XB45	AJ229237	-	-	-
W2_M17	-	1	-	-	95	Bacteria from anoxic bulk soil strain XB45	AJ229237	-	-	-
W1_L04	1	-	-	-	97	Bacteria from anoxic bulk soil strain XB45	AJ229237	-	-	-
W2_M19	-	1	-	-	94	Bacteria from anoxic bulk soil strain XB45	AJ229237	-	-	-
W4_M04	-	-	-	1	99	*Meniscus glaucopis* strain ATCC 29398	GU269545	-	-	-
W4_I08s	-	-	-	1	98	*Faecalibacterium prausnitzii* clone 1-84	AY169429	-	-	-
W1_B20	1	-	1	-	92	Bacteria from anoxic bulk soil strain XB45	AJ229237	-	-	-
W4_M14	-	-	-	1	99	Uncultured bacterium mle1-9	AF280848	-	-	-
W1_P22	1	-	-	-	99	Rumen bacterium 38-H_9	GQ461844	-	-	-
W2_M03	-	1	-	-	99	*Streptococcus bovis* NCDO2127	AF429766	2	-	-
Synergistetes										
W3_A23	-	-	**3**	**4**	**97**	Uncultured *Aminanaerobia* bacterium clone QEDN10CG08	CU926332	-	-	-
W4_A14	-	-	-	1	92	*Synergistes* sp. NML96A088	EF551160	-	-	-
W4_C22	-	-	-	1	97	*Synergistetes* bacterium 7WAY-8-7	AB558582	-	-	-
Spirochaeta										
W1_F24	2	-	-	-	95	*Spirochaeta stenostrepta*	AB541984	-	-	-
W3_G01	-	-	1	-	96	*Spirochaeta stenostrepta*	AB541984	-	-	-
W4_K20	-	-	-	1	91	Uncultured *Spirochaeta* sp. clone: Rs-D16	AB088873	-	-	-
Tenericutes										
W1_D22	4	-	-	-	99	*Acholeplasma* sp. DM-2009 strain Lorelei	FJ590762	-	-	-
W2_G17	-	1	-	-	91	*Acholeplasma palmae*	L33734	-	-	-
Acidobacteria										
W2_G13	-	1	-	-	94	*Acidobacteria* bacterium KBS 96	FJ870384	-	-	-
W3_A21	-	-	1	-	95	*Solibacter usitatus* Ellin6076	CP000473	-	-	-
W1_B08	1	-	-	-	94	*Geothrix fermentans*	U41563	-	-	-
W1_P02	1	-	-	-	95	*Geothrix fermentans*	U41563	-	-	-
W2_M01	-	1	-	-	95	*Holophaga foetida* strain TMBS4-T	X77215	-	-	-
BRC1										
W3_G21	-	-	1	-	99	Uncultured bacterium clone: TSSUR003_L23	AB488367	-	-	-
Chloroflexi										
W4_O08	-	-	-	1	91	*Longilinea arvoryzae*	AB243673	-	-	-

Verrucomicrobia											
W3_E09	-	-	1	-	96	Opitutus sp. VeCb1	X99391	-	-	-	-
Unclassified											
W1_J02	1	-	-	-	99	Bacterium enrichment culture clone DPHB07	GQ377117	-	-	-	-
W2_C07	-	3	-	-	91	Uncultured bacterium clone AFBAB_aai01e12	EU771532	-	-	-	-
W4_I16	-	-	-	1	95	Uncultured bacterium clone BR16	GQ461624	-	-	-	-
W1_N16	1	-	-	-	98	Uncultured bacterium clone ZW-1	GU390314	-	-	-	-
W3_K01	-	-	1	-	96	Uncultured bacterium clone A1232	EU283592	-	-	-	-
W1_H12	1	-	-	-	99	Uncultured Spirochaetes bacterium clone QEEB1BC06	CU918305	-	-	-	-
W3_G07	-	-	1	-	97	Uncultured bacterium clone BS24	EU358699	-	-	-	-
W3_A11	-	-	1	-	93	Uncultured bacterium clone t30d60H49	FM956301	-	-	-	-
W4_E24	-	-	-	1	86	Uncultured Chlorobium sp. clone 4.31	GQ183439	-	-	-	-
W4_A04	-	-	-	1	93	Uncultured bacterium clone SS-36	AY945866	-	-	-	-
W1_H20	1	-	-	-	93	Uncultured bacterium clone 055B05_P_DI_P58	CT573819	-	-	-	-
W3_K03	-	-	2	-	94	Bacterium W18	DQ238245	-	-	-	-
W4_I12	-	-	-	2	94	Bacterium W18	DQ238245	-	-	-	-
W4_C24	-	-	-	1	91	Uncultured candidate division OP3 bacterium clone LiM 13F12	FN646447	-	-	-	-
W1_J06	1	-	-	-	93	Uncultured bacterium clone IR_aaa04c04	EU474642	-	-	-	-
W4_I06s	-	-	-	1	96	Uncultured bacterium clone SHA-43	AJ306766	-	-	-	-
Total	87	84	77	73				72	146	91	

Table S2. All phylotypes obtained from the primary clarifier effluents

Phylum-Class Phylotype	PC1	PC2	PC4	% match	Best matched sequence	Accession No.	W1	W2	W3	W4
Proteobacteria - Deltaproteobacteria										
PC2_C13	-	1	-	95	Desulfobulbus sp. DSM 2033	EF442993	-	-	-	-
Proteobacteria - Betaproteobacteria										
PC4_K16	6	2	2	100	Acidovorax sp. PPs-5	FJ605421	3	2	-	-
PC1_K23	1	-	-	97	Comamonadaceae bacterium CNRF14	GU300568	-	-	-	-
PC4_E02	-	-	3	99	Comamonas sp. R-25060	AM084020	-	-	-	-
PC1_C01	1	-	-	98	Comamonas denitrificans strain 110	AF233876	-	-	-	-
PC2_J21	-	1	-	98	Acidovorax caeni	AM084011	-	-	-	-
PC1_K05	1	-	-	98	Acidovorax caeni	AM084008	-	-	-	-
PC1_I01	1	-	-	99	Acidovorax sp. R-25052	AM084039	-	-	-	-
PC2_C23	-	1	-	97	Comamonas sp. PG6-1	AB277849	-	-	-	-
PC4_O24	1	1	1	99	Neisseria canis isolate VA25810gw_03	AY426973	-	-	-	-
PC1_O21	1	-	-	99	Neisseria sp. GRW59	FJ502347	-	-	-	-
PC2_C09	-	3	-	99	beta proteobacterium OcN1	AF331976	-	-	-	-
PC1_M07	1	-	-	98	Dechloromonas sp. JDS6	AY084087	1	-	-	-
PC2_A09	-	1	-	97	Dechloromonas sp. EMB 269	DQ413167	1	-	-	1
PC2_L23	-	1	-	93	Chitiniphilus shinanonensis	AB453176	-	-	-	-
PC4_I08	-	-	1	98	Zoogloea sp. EMB 357	DQ413172	-	-	-	-
PC4_C14	-	-	1	92	Vogesella sp. SK-2	AM689950	-	-	-	-
PC1_K09	1	-	-	96	Formivibrio citricus	Y17602	-	-	-	-
PC2_O21	-	1	-	99	Zoogloea oryzae	AB201044	-	-	-	-
PC1_E07	1	-	-	98	Azonexus fungiphilus	AJ630292	-	-	-	-
Proteobacteria - Gammaproteobacteria										
PC2_F23	-	4	2	100	Acinetobacter johnsonii strain CONC8	EU275352	-	-	-	-
PC4_C04	-	-	1	100	Acinetobacter sp. DNPA10	FJ404811	-	-	-	-
PC2_H19	-	3	1	100	Acinetobacter sp. DN4	AM269521	-	-	-	-
PC2_D11	-	1	-	96	Acinetobacter sp. phenon 4	AJ293690	-	-	-	-
PC2_H01	-	1	-	97	Acinetobacter sp. AOLR03	GQ916505	-	-	-	-
PC4_O08	1	2	1	100	Acinetobacter sp. Ya Gi	FJ645597	-	-	-	-
PC1_E09	4	-	1	98	Acinetobacter sp. BA17	FJ263927	-	-	-	-
PC1_E13	1	-	-	95	Moraxella sp. Everest-gws-54	EU584522	-	-	-	-

ID				%	Species	Accession				
PC2_I01	1	2	-	98	Moraxella nonliquefaciens	AF005180	-	-	-	-
PC4_K18	-	-	1	99	Enhydrobacter sp. NMC13	GU321352	-	-	-	-
PC4_O02	-	-	3	100	Aeromonas veronii strain YA090911	GU735964	1	-	-	-
PC4_G02	-	3	2	100	Aeromonas sp. CIST-WP3s2	EF428989	-	-	-	-
PC2_N17	-	1	-	99	Aeromonas veronii strain CYJ209	FJ940850	-	-	-	-
PC2_B13	-	1	-	96	Aeromonas veronii strain DSR3	EU434642	-	-	-	-
PC4_A06	-	-	2	95	Aeromonas hydrophila strain ANSE1	GU296671	-	-	-	-
PC1_M03	1	-	-	100	Escherichia coli strain AB1157	AY831405	-	-	-	-
PC4_M06	-	-	1	97	Escherichia sp. II_B13	HM028651	-	-	-	-
PC1_I21	1	-	-	98	Aeromonas punctata strain tu1	FJ544390	-	-	-	-
PC1_C03	1	-	-	99	Enterobacter sp. pp9c	GQ360072	-	-	-	-
PC2_H09	-	1	-	99	Rheinheimera sp. T3	FJ765357	-	-	-	-
PC1_G23	2	-	-	98	Actinobacillus indolicus	U65584	-	-	-	-
PC1_O05	1	-	-	99	Haemophilus parasuis strain HS82	FJ667948	-	-	-	-
PC1_O07	2	1	-	99	Actinobacillus indolicus isolate WB52/06-1	EF396307	-	-	-	-
PC1_I03	1	-	-	96	Haemophilus sp. Smarlab 3302188	AY538693	-	-	-	-
PC2_J23	-	1	-	98	Pseudomonas pseudoalcaligenes strain 23	EU780001	-	-	-	-
PC4_M20	-	-	1	98	Thiothrix sp. NKBI-C gene	AB166733	-	-	-	-
PC1_K11	1	-	-	97	Aquaspirillum sp. 411	AY904024	-	-	-	-
Proteobacteria - Epsilonproteobacteria										
PC2_N09	13	63	52	100	Arcobacter cryaerophilus	U34387	9	10	-	-
PC2_E13	-	1	-	100	Arcobacter butzleri strain ED-1	FJ968634	-	-	-	-
PC2_E17	-	3	-	97	Arcobacter sp. D2043	FJ161215	-	1	-	-
PC2_D21	-	1	-	97	Arcobacter sp. MA5 gene	AB542077	-	1	-	-
PC2_B17	1	1	-	97	Arcobacter sp. D2043	FJ161215	-	-	-	-
PC2_F21	-	1	-	98	Arcobacter cryaerophilus	U34387	-	-	-	-
PC2_J01	-	1	-	98	Dehalospirillum multivorans	X82931	-	-	-	-
PC2_G13	-	1	-	89	Sulfurospirillum sp. NO3A	AY135396	-	-	-	-
PC4_G06	-	-	2	95	Arcobacter sp. R-28314	AM084114	-	-	-	-
PC2_J17	-	1	-	94	Arcobacter cryaerophilus	U34387	-	-	-	-
Bacteroidetes										
PC2_J13	-	1	-	99	Bacteroides sp. 253c	AY082449	-	-	-	-
PC2_I03	-	2	-	98	Bacteroides uniformis	AB247146	-	-	-	-
PC1_E21	4	2	-	99	Bacteroides vulgatus ATCC 8482	CP000139	-	-	-	-
PC4_C18	-	-	1	98	Bacteroides sp. Smarlab 3302996	AY643081	-	-	-	-
PC2_D13	-	1	-	99	Bacteroides thetaiotaomicron VPI-5482	AE015928	-	-	-	-
PC2_L21	-	1	-	96	Parabacteroides distasonis strain JCM5825	EU136681	-	-	-	-
PC2_O09	-	1	-	100	Parabacteroides merdae	EU722738	-	-	-	-
PC2_N01	-	1	-	96	Porphyromonadaceae bacterium NML 060648	EF184292	-	-	-	-
PC1_A13	2	1	-	97	Prevotellaceae bacterium WR041	AB298732	-	-	-	-
PC2_G19	-	1	-	99	Bacteroides fragilis	AB542764	-	-	-	-
PC4_K02	3	9	3	100	Cloacibacterium normanense strain tu29	FJ544401	-	1	-	-
PC1_E19	2	2	-	99	Flavobacteriaceae bacterium R2A-16	EU581834	-	-	-	-
PC2_P17	-	1	-	92	Lutibacter sp. IMCC 1507	GU166749	-	-	-	-
PC2_M05	-	1	-	98	Flavobacterium sasangense strain YC6274	EU423319	-	-	-	-
PC4_M08	-	-	2	99	Uncultured Anaerophaga sp. clone MDAF17	EU214543	1	1	-	1
Firmicutes										
PC1_K01	2	1	-	99	Anaerovibrio glycerini DSM 5192(T)	AJ010960	-	-	-	-
PC1_C19	1	-	-	99	Selenomonas lacticifex strain DSM20757	AF373024	-	-	-	-
PC1_G01	1	-	-	95	Acidaminococcus fermentans DSM 20731	CP001859	-	-	-	-
PC1_G07	1	-	-	94	P.faecium (ACM3680)	X72867	-	-	-	-
PC4_G12	-	-	1	97	Veillonella sp. S101	FJ374768	-	-	-	-
PC4_C08	-	-	1	88	Anaerovibrio burkinabensis DSM 6283(T)	AJ010961	-	1	-	-
PC4_G16	-	-	1	97	Mitsuokella multacida strain C20-1	GU227153	-	-	-	-
PC4_K24	1	-	1	100	Acetoanaerobium noterae strain ATCC 35199	GU562448	-	-	-	-
PC1_K03	1	-	-	97	Fusibacter sp. SA1	AF491333	3	3	1	-
PC1_M19	1	-	-	95	Oscillospiraceae bacterium NML 061048	EU149939	-	-	-	-
PC2_G01	-	1	-	100	Clostridium orbiscindens strain KCTC5919	GU723311	-	-	-	-
PC1_K07	1	-	-	94	Clostridium orbiscindens strain AJP028.07	EU541437	-	-	-	-
PC1_C21	1	-	-	98	Clostridiaceae bacterium DJF_LS40	EU728744	-	-	-	-
PC1_E15	1	-	-	92	Acetanaerobacterium elongatum strain Z7	AY487928	-	-	-	-
PC2_A03	-	1	-	97	Clostridium sp. HFTH-1 gene	AB439724	-	-	-	-
PC2_I07	-	1	-	99	Eubacterium eligens ATCC 27750	CP001104	-	-	-	-
PC2_I21	-	1	-	97	Clostridium aminobutyricum	X76161	-	-	-	-
PC2_I15	-	1	-	99	Ruminococcus bromii L2-63	FP929051	-	-	-	-
PC1_M23	2	1	-	99	Streptococcus minor strain 29-74MPalpha	EU075082	-	-	-	-
PC1_A01	2	-	-	99	Streptococcus bovis NCDO2127	AF429766	-	1	-	-
PC2_B19	-	1	-	92	Streptococcus sp. oral clone ASCF03	AY953254	-	-	-	-
Fusobacteria										
PC2_F01	-	1	-	94	Leptotrichia sp. oral taxon 212 clone WWP_SS4_P01	GU408396	-	-	-	-
PC4_M18	-	-	1	94	Leptotrichia sp. oral taxon 212 clone WWP_SS4_P01	GU408396	-	-	-	-
PC4_A08	-	-	2	93	Leptotrichia wadei strain F0279	FJ717336	-	-	-	-
PC2_A17	-	3	-	93	Fusobacterium sp. CSL-7530	EU597748	-	-	-	-
PC2_B11	-	4	-	92	Fusobacterium ulcerans strain KCTC5932	GU723324	-	-	-	-
Total	72	146	91				87	84	77	73

ACKNOWLEDGMENTS

We thank Nancy Coglan (North City Water Reclamation Plant) and her laboratory members for providing samples of the primary clarifier effluent. We thank Greg Wanger for observing the biofilm by FE-SEM. We also thank Kelvin Li for technical assistance with executing the JCVI 16 S/18 S rRNA Pipeline, Jeff McQuaid for experimental setup, Angela Wu and Eric Son for technical assistance of clone library analysis.

AUTHOR CONTRIBUTIONS

Conceived and designed the experiments: SI SS YAG OB. Performed the experiments: SI SS. Analyzed the data: SI SS TMN-K. Contributed reagents/materials/analysis tools: SI SS TMN-K YS. Wrote the paper: SI SS KHN OB.

REFERENCES

1.   WIN (2000) Clean Safe Water for the 21st Century: Water Infrastructure Network.

2.   Pant D, Van Bogaert G, Diels L, Vanbroekhoven K (2010) A review of the substrates used in microbial fuel cells (MFCs) for sustainable energy production. Biores Technol 101: 1533–1543.

3.   Rulkens W (2008) Sewage sludge as a biomass resource for the production of energy: Overview and assessment of the various options. Energy Fuels 22: 9–15.

4.   Huang LP, Logan BE (2008) Electricity generation and treatment of paper recycling wastewater using a microbial fuel cell. Appl Microbiol Biotech 80: 349–355.

5.   Rozendal RA, Hamelers HV, Rabaey K, Keller J, Buisman CJ (2008) Towards practical implementation of bioelectrochemical wastewater treatment. Trends Biotechnol 26: 450–459.

6.   Logan BE, Hamelers B, Rozendal R, Schroder U, Keller J, et al. (2006) Microbial fuel cells: methodology and technology. Environ Sci Technol 40: 5181–5192.

7.   Rabaey K, Verstraete W (2005) Microbial fuel cells: novel biotechnology for energy generation. Trends Biotechnol 23: 291–298.

8.   Rismani-Yazdi H, Carver SM, Christy AD, Tuovinen IH (2008) Cathodic limitations in microbial fuel cells: An overview. J Power Sources 180: 683–694.

9.   Ishii S, Watanabe K, Yabuki S, Logan BE, Sekiguchi Y (2008) Comparison of electrode reduction activities of *Geobacter sulfurreducens* and an

enriched consortium in an air-cathode microbial fuel cell. Appl Environ Microbiol 74: 7348–7355.

10. Meyer S (2009) North City Water Reclamation Plant Annual Monitoring Report 2009. SDRWQCB Order No 97-03: 30.

11. Logan BE (2008) Chapter 4 Power generation. Microbial Fuel Cells. Hoboken, , New Jersey, USA: John Wiley & Sons. pp. 44–60.

12. Tsujimura S, Fujita M, Tatsumi H, Kano K, Ikeda T (2001) Bioelectrocatalysis-based dihydrogen/dioxygen fuel cell operating at physiological pH. Phys Chem Chem Phys 3: 1331–1335.

13. Di Lorenzo M, Scott K, Curtis TP, Katuri KP, Head IM (2009) Continuous feed microbial fuel cell using an air cathode and a disc anode stack for wastewater treatment. Energy Fuels 23: 5707–5716.

14. Liu H, Ramnarayanan R, Logan BE (2004) Production of electricity during wastewater treatment using a single chamber microbial fuel cell. Environl Sci Technol 38: 2281–2285.

15. Cheng S, Liu H, Logan BE (2006) Increased power generation in a continuous flow MFC with advective flow through the porous anode and reduced electrode spacing. Environ Sci Technol 40: 2426–2432.

16. Ahn Y, Logan BE (2010) Effectiveness of domestic wastewater treatment using microbial fuel cells at ambient and mesophilic temperatures. Biores Technol 101: 469–475.

17. Shimoyama T, Yamazawa A, Ueno Y, Watanabe K (2009) Phylogenetic analyses of bacterial communities developed in a cassette-electrode microbial fuel cell. Microb Environ 24: 188–192.

18. Xing D, Cheng S, Regan JM, Logan BE (2009) Change in microbial communities in acetate- and glucose-fed microbial fuel cells in the presence of light. Biosens Bioelectron 25: 105–111.

19. Liu H, Logan BE (2004) Electricity generation using an air-cathode single chamber microbial fuel cell in the presence and absence of a proton exchange membrane. Environ Sci Technol 38: 4040–4046.

20. Cheng S, Liu H, Logan BE (2006) Increased performance of single-chamber microbial fuel cells using an improved cathode structure. Electrochem Commun 8: 489–494.

21. Ishii S, Shimoyama T, Hotta Y, Watanabe K (2008) Characterization of a filamentous biofilm community established in a cellulose-fed microbial fuel cell. BMC Microbiol 8: 6.

22. Tender LM, Reimers CE, Stecher HA 3rd, Holmes DE, Bond DR, et al. (2002) Harnessing microbially generated power on the seafloor. Nat Biotechnol 20: 821–825.

23. Holmes DE, Bond DR, O›Neil RA, Reimers CE, Tender LR, et al. (2004) Microbial communities associated with electrodes harvesting electricity from a variety of aquatic sediments. Microb Ecol 48: 178–190.

24. Kiely PD, Cusick R, Call DF, Selembo PA, Regan JM, et al. (2011) Anode microbial communities produced by changing from microbial fuel cell to microbial electrolysis cell operation using two different wastewaters. Biores Technol 102: 388–394.

25. Chae KJ, Choi MJ, Lee JW, Kim KY, Kim IS (2009) Effect of different substrates on the performance, bacterial diversity, and bacterial viability in microbial fuel cells. Biores Technol 100: 3518–3525.

26. Lovley DR, Holmes DE, Nevin KP (2004) Dissimilatory Fe(III) and Mn(IV) reduction. Adv Microb Physiol 49: 219–286.

27. Bond DR, Holmes DE, Tender LM, Lovley DR (2002) Electrode-reducing microorganisms that harvest energy from marine sediments. Science 295: 483–485.

28. Bond DR, Lovley DR (2003) Electricity production by *Geobacter sulfurreducens* attached to electrodes. Appl Environ Microbiol 69: 1548–1555.

29. Holmes DE, Bond DR, Lovley DR (2004) Electron transfer by *Desulfobulbus propionicus* to Fe(III) and graphite electrodes. Appl Environ Microbiol 70: 1234–1237.

30. Jung S, Regan JM (2007) Comparison of anode bacterial communities and performance in microbial fuel cells with different electron donors. Appl Microbiol Biotechnol 77: 393–402.

31. Lovley DR, Giovannoni SJ, White DC, Champine JE, Phillips EJ, et al. (1993) *Geobacter metallireducens* gen. nov. sp. nov., a microorganism capable of coupling the complete oxidation of organic compounds to the reduction of iron and other metals. Arch Microbiol 159: 336–344.

32. Roden EE, Lovley DR (1993) Dissimilatory Fe(III) Reduction by the marine microorganism *Desulfuromonas acetoxidans*. Appl Environ Microbiol 59: 734–742.

33. Reimers CE, Girguis P, Stecher HA, Tender LM, Ryckelynck N, et al. (2006) Microbial fuel cell energy from an ocean cold seep. Geobiology 4: 123–136.

34.   Karlsson FH, Ussery DW, Nielsen J, Nookaew I (2011) A closer look at Bacteroides: phylogenetic relationship and genomic implications of a life in the human gut. Microb Ecol 61: 473–485.

35.   Gorby YA, Yanina S, McLean JS, Rosso KM, Moyles D, et al. (2006) Electrically conductive bacterial nanowires produced by *Shewanella oneidensis* strain MR-1 and other microorganisms. PNAS 103: 11358–11363.

36.   DeLong EF (1992) Archaea in coastal marine environments. PNAS 89: 5685–5689.

37.   Watanabe K, Kodama Y, Harayama S (2001) Design and evaluation of PCR primers to amplify bacterial 16 S ribosomal DNA fragments used for community fingerprinting. J Microbiol Methods 44: 253–262.

38.   Karlin S, Altschul SF (1990) Methods for assessing the statistical significance of molecular sequence features by using general scoring schemes. PNAS 87: 2264–2268.

39.   Colwell RK (2009) EstimateS: Statistical estimation of species richness and shared species from samples. Version 8.2. User›s Guide and application. Available:http://purl.oclc.org/estimates. Accessed 2011 Dec 23.

Chapter 12

APPLICATION OF ION TORRENT SEQUENCING TO THE ASSESSMENT OF THE EFFECT OF ALKALI BALLAST WATER TREATMENT ON MICROBIAL COMMUNITY DIVERSITY

Masanori Fujimoto[1], Gregory A. Moyerbrailean[1,2], Sifat Noman[1], Jason P. Gizicki[1], Michal L. Ram[1], Phyllis A. Green[3], Jeffrey L. Ram[1]

[1] Department of Physiology, School of Medicine, Wayne State University, Detroit, Michigan, United States of America

[2] Center for Molecular Medicine and Genetics, School of Medicine, Wayne State University, Detroit, Michigan, United States of America

[3] Isle Royale National Park, National Park Service, Houghton, Michigan, United States of America

ABSTRACT

The impact of NaOH as a ballast water treatment (BWT) on microbial community diversity was assessed using the 16S rRNA gene based Ion Torrent sequencing with its new 400 base chemistry. Ballast water samples from a Great Lakes ship were collected from the intake and discharge of both control and NaOH (pH 12) treated tanks and were analyzed in duplicates. One set of duplicates was treated with the membrane-impermeable DNA cross-linking reagent propidium mono-azide (PMA) prior to PCR amplification to differentiate between live and dead microorganisms. Ion Torrent sequencing generated nearly 580,000 reads for 31 bar-coded samples and revealed alterations of the microbial community structure in ballast water that had been treated with NaOH. Rarefaction analysis of the Ion Torrent sequencing data showed that BWT using NaOH significantly decreased microbial community diversity relative to control discharge (p<0.001). UniFrac distance based principal coordinate analysis (PCoA) plots and UPGMA tree analysis revealed that NaOH-treated ballast water microbial communities differed from both intake communities and control discharge communities. After NaOH treatment, bacteria from the genus *Alishewanella* became dominant in the NaOH-treated samples, accounting for <0.5% of the total reads in intake samples but more than 50% of the reads in the treated discharge samples. The only apparent difference in microbial community structure between PMA-processed and non-PMA samples occurred in intake water samples, which exhibited a significantly higher amount of PMA-sensitive cyanobacteria/chloroplast 16S rRNA than their corresponding non-PMA total DNA samples. The community assembly obtained using Ion Torrent sequencing was comparable to that obtained from a subset of samples that were also subjected to 454 pyrosequencing. This study showed the efficacy of alkali ballast water treatment in reducing ballast water microbial diversity and demonstrated the application of new Ion Torrent sequencing techniques to microbial community studies.

INTRODUCTION

Next generation sequencing techniques have been developed over the last decade such as Roche 454 pyrosequencing [1] and Solexa/Illumina sequencing [2], which enabled us to capture the diversity of microbial communities from various environments [3]–[11]. Amongst next generation sequencing techniques, Roche 454 pyrosequncing has been primarily used for 16S rRNA-based microbial community studies due to its longer average sequence length[12]–[14]. Pipelines such as Ribosomal Database Project (RDP) [15], Mothur [16], and Quantitative Insights Into Microbial Ecology (QIIME) [17] have been developed to analyze pyrosequencing output data. Until recently, the average sequence length obtained with Ion Torrent sequencing was shorter than 250 bp long [14] and was mainly used for clinical studies, such as analyzing mutations in cancer cells [18], cystic fibrosis patients [19], and viral genomes of HIV [20] that did not require longer reads. The few studies that applied the Ion Torrent platform to 16S rRNA gene-based microbial community studies [21]–[23] used 100 or 200 base chemistry and had lower taxonomic resolution than pyrosequencing. However, with the recent release of 400 base chemistry for Ion Torrent sequencing at the beginning of 2013, we were able to design primers for 16S rRNA gene sequencing of up to 410 bp including primer sequences [24], [25], a size comparable to the 454 GS Junior pyrosequencing [25].

In this study, the new Ion Torrent sequencing chemistry was applied to a study of microbial community diversity of ballast water with a focus on the assessment of alkali ballast water treatment efficacy. Ballast water of cargo ships has led to the introduction of many non-native invasive species that have negatively impacted various aquatic ecosystems [26]–[28], as well as the transportation of harmful algae [29], [30] and human pathogens [31], [32]. In order to stop the introduction of non-native species or transportation of harmful bacterial species via ballast water, treatment methods including alkali treatment have been designed to kill the majority of organisms within the ballast tanks before discharge [33].

The objective of this study was to determine the effectiveness of NaOH treatment at pH 12 in reducing microbial diversity and in altering microbial community structure in ballast water using new Ion Torrent sequencing chemistry. To assess the efficacy of the ballast water treatment (BWT), ballast water samples from a Great Lakes ship were collected from the intake and discharge of both control and NaOH (pH 12) treated tanks and were duplicated. In addition, to assess only the viable organisms among the microbial communities in these samples, microbial communities derived from the regular extraction procedure were further compared to duplicate samples that had been treated with propidium mono-azide (PMA). PMA permeates damaged membranes of dead cells and crosslinks the DNA upon exposure to light and makes them not amplifiable during the PCR reaction [34]. Since it was known that only a subset of microorganisms are able to survive in alkaline pH [33], we hypothesized that NaOH would significantly decrease the microbial diversity of the viable microorganisms in the ballast water by eliminating alkali sensitive microbial species.

METHODS

Ethics Statement

Ballast water samples were collected from the ship *M/V Indiana Harbor* (privately owned by American Steamship Company). We received permission to collect samples from the ship's ballast water from the American Steamship Company. The shipboard sample collection did not involve any endangered or protected species.

Collection of Ballast Water Samples

Ballast water samples were collected in August 2011 from the ship *M/V Indiana Harbor*, in collaboration with a study by the Great Ships Initiative (GSI). As described in the GSI report[35], the *Indiana Harbor* took on cargo near Duluth, MN and while unloading it in Gary, IN (latitude, longitude =41.6147°, −87.3252°, respectively) the ship took on ballast water which was sampled for this study from the intakes to four ballast tanks (2P, 3P, 4P, 5P). Water in two of the ballast tanks (3P and 4P) were treated with NaOH (pH 12) while the ship traveled for 3 days to Superior, WI. The pH of the ballast tanks were adjusted to pH12 by gradually adding 50% (w/v) NaOH into the ballast tanks. The other two tanks (2P and 5P) in the ship served as controls in this study. NaOH-treated tanks were neutralized using an in-tank carbonation system 18 hours prior to the discharge. The ballast water from all four ballast tanks was sampled at the port of Superior, WI (46.7430°, −92.1144°), during the discharge of the neutralized ballast water. A total of 1.0 L of ballast water collected from each tank at each sampling event was kept on ice until initial processing and preservation by methods described below. For the Gary, IN, intake samples initial processing and preservation took place 4 to 8.5 hr after collection from the ship; in Superior, WI, processing and preservation of samples was initiated for all samples within one hr (range 49 to 59 min) of collection.

Sample Processing and DNA Extraction

Except as noted in summary Table 1, four aliquots of 100 mL each were dispensed from each 1 L sample and processed with either PMA (two aliquots) or non-PMA methods (two aliquots). The 100 mL aliquot of non-PMA group was filtered using a 0.22 µm syringe filter (Thermo Nalgene cat. no. 190–2520). For PMA assay, the 100 mL aliquot of the PMA group was filtered with 0.22 µm filter membrane, and 0.8 mL of 100 µM PMA was applied to the filter associated microbes. The filter associated microbes were incubated in the dark for 15 minutes to allow PMA to bind to the DNA of dead or injured cells. Then, the microbes on the filter were exposed to an intense white LED light (Husky 180 LED Work light, Home Depot cat. no. 955–998) for 7 minutes to allow PMA to cross-link the DNA [34]. The filter associated microbes from both PMA and non-PMA groups were backwashed off the filter and preserved with 1 mL of DNAzol Direct (Molecular Research Center, cat. no. DN 131). The DNAzol Direct sample solutions were transported to Wayne State University and stored at −80°C. To purify the genomic DNA, 230 µL of DNAzol Direct sample solutions were purified using Qiagen spin column purification kits (Qiagen, cat. no. 69506), according to the manufacturer's protocol.

Table 1. Experimental design of alkali ballast water treatment

Ballast Treatment	Ballast Tank ID	Intake		Discharge	
		Non-PMA	PMA	Non-PMA	PMA
Control	2P	35N, 40N	25P, 30P	53N, 57N	45P, 49P
	5P	36N, 41N	26P, 31P	54N, 58N	46P, 50P
NaOH	3P	37N, 42N*	27P, 32P	55N, 59N	47P, 51P
	4P	38N, 43N	28P, 33P	56N, 60N	48P, 52P

Sample labels are shown for each 100 mL aliquot of ballast water samples collected at intake and discharge from both the control and NaOH treated ballast tanks. Duplicate aliquots from each ballast water sample were processed with and without propidium mono-azide (labeled ##P and ##N, respectively). *Sample 42N was spilled at the initial processing site and therefore not available for Ion Torrent sequencing.
doi:10.1371/journal.pone.0107534.t001

Ion Torrent Sequencing

The library for Ion Torrent sequencing was prepared by amplifying a 410 bp segment of the V4-V5 region of 16S rRNA gene [36], using U515F (5′-GTGCCAGCMGCCGCGGTAA-3′) [10], [37]and U926R (5′-CCGTCAATTCMTTTRAGT-3′) [8], [38] primers. The V4 and V5 regions have great sequence variability and are sensitive enough to identify diverse groups of bacterial taxa[10], [39]. This primer set is also predicted to amplify a wide range of microbial taxa by blast of public sequence databases. The primer set was also confirmed experimentally using DNA extracts of pure culture on *E. coli, Vibrio cholera* and intestinal enterococci, named in various proposed and adopted ballast water regulations [40], [41]. The forward primer was linked with the Ion adapter "A" sequence (5′-CCATCTCATCCCTGCG TGTCTCCGACTCAG-3′) and Ion-Xpress barcode sequences were inserted between the adapter "A" and the U515F primer. The reverse primer was linked to Ion adapter "p1" sequence (5′-CCTCTCTATGGGCAGTCGGTGAT-3′).

PCR amplification was performed in 50 µL reactions, using 2 U of high fidelity DNA polymerase AccuPrime Taq HiFi (Invitrogen, Grand Island, NY), 5 µL of supplied 10X buffer II, 1.0 µL of 10 µM primers, and 3 µL of template DNA. The PCR protocol was denaturation at 95°C for 5 min followed by 32 cycles of denaturation at 95°C for 30 s, annealing at 55°C for 30 s, and extension at 72°C for 2 min. The extension of 2 min was employed to minimize the formation of short fragments and chimeras [38], [42]. Thirty-two cycles was used so that the PCR amplification was stopped at the pre-saturation point (the saturation point was pre-assessed by running real time PCR with SYBR green under the same PCR condition).

After purifying PCR amplicons using Agencourt AMPure XP Beads with 1.5X concentration (Beckman Coulter, Inc., Brea, CA), the concentrations of the purified PCR products were measured using PicoGreen (Life Technology, NY), and a library mix was prepared by adding an equal mass of DNA from each barcoded sample. The amplicon size and concentration of the library mix was confirmed using TapeStation 2200 (Agilent Technologies, Inc) with high sensitivity D1K screen tape and reagents. The library mix was diluted with ultra-pure water to 26 pM. The template preparation for the PGM sequencer was prepared from 25 µL of the 26 pM solution using a One Touch 2 (OT2) system (Life Technology, Inc) according to the manufacturer›s protocol. Each PCR amplicon was hybridized on an ion sphere particle (ISP) and amplified clonally in the OT2. The clonally amplified ISPs were re-suspended and enriched manually by selecting for ISPs with sufficient amplification using Dynabeads (Life Technology, NY). Three µL of the enriched ISPs were mixed with 3 µL of PGM primers and annealed in a thermocycler with 95°C for 2 min followed by 37°C for 2 min. One µL of DNA polymerase was added to the 6 µL template ISPs-primer mix, and a total of the 7 µL mix was pipetted and added into an Ion 314 V2 chip. The mix was equally distributed across the chip by pipetting up and down, so that in principle each well in the 314 chip was filled with only a single ISP. The ISP-loaded 314 chip was mounted in the Ion PGM sequencer and DNA on the ISPs was sequenced with 800 flows after pH of the PGM system was adjusted. Two PGM sequencing runs were performed: the first run with 16 samples plus one positive control (*Vibrio cholera*JW612) and the second run with 15 samples plus two positive controls (*V. cholera* JW612 and*Enterococcus* sp.). The combined data from the two PGM runs were analyzed.

Ion Torrent Data Analysis

Ion PGM output were generated with binary alignment map format (bam) and converted to standard flowgram format (sff) using the "bam2sff" function. The resultant sff file was converted to fasta (sequence), flow (flowgram), and qual (quality score) files using Mothur version 1.30.00 (released April 2013). The fasta and qual files were processed using Ribosomal Database Project (RDP) pipeline version 10 [43] to sort the sequence data by barcodes, to trim barcode and forward primer sequences, and to filter out low quality sequences, retaining only sequences with quality scores above 20 (probability threshold of 0.01) and read lengths of >250 bp. The RDP processed (filtered/trimmed) fasta files were concatenated into a single fasta file and each sequence was transformed into valid QIIME format. Chimeric sequences were identified using the Uchime reference algorithm [44] and removed from the subsequent analysis (0.8% sequences were removed in this process). Alpha and beta diversity analyses of the ballast water samples were performed using a QIIME pipeline [17]. The QIIME formatted sequences were clustered at 97% similarity cut off, and the taxonomy of each cluster was assigned using RDP classifier at a bootstrap threshold of 70%. Rarefaction curves of the OTUs with 97% similarity cutoff were generated using the alpha rarefaction.py command in QIIME, and the effect of NaOH treatment on the microbial community diversity (number of OTUs) was assessed using a general linear model after rarefying all samples at 11,000 reads [45]. Other diversity indices including Chao1 and Shannon were also calculated using the same command. The general linear model was performed using "lm" function in R version 3.0.0 [46]. The treatment effect was also assessed using multiple comparison tests within ANOVA. P-values were controlled for multiple comparisons for *posteriori* hypotheses with Tukey›s honest significant difference (HSD). The ANOVA and multiple comparisons with Tukey›s HSD were performed using the "multicomp" package in R version 3.0.0 [46]. To assess the beta diversity between different treatment groups of ballast water samples, weighted UniFrac distances [47] between samples were calculated after rarefying all samples at 10,000 reads. The effect of alkali water treatment on ballast water microbial communities was depicted using UniFrac distance based principal coordinates analysis (PCoA) plots [48] and UPGMA tree [49]. To assess the statistical significances in microbial community dissimilarity between particular treatment groups, the Unifrac distance matrix generated in QIIME was transferred into R and ANOSIM was performed in R with the vegan package. Correlation between microbial community samples were also examined by Pearson›s correlation in R using BIOM matrix data (with 97% similarity cutoff) generated in QIIME after rarefying at 10,000 reads. Ion Torrent sequence data obtained from this project is available at the NIH Sequence Read Archive (SRA) under the project accession number SRP042367.

Genus Alishewanella Phylogenetic Tree

A genus-level phylogenetic tree for *Alishewanella* reads was constructed in MEGA version 5.2[50]. 16S rRNA gene reference sequences of all 6 known *Alishewanella* species plus some of*Alishewanella* sp. identified at the genus level were downloaded from the NCBI database and aligned with 102 randomly selected Ion Torrent reads from sample 47P (one of the NaOH treated discharge samples). The phylogenetic tree was constructed using the Neighbor-Joining algorithm [51] with the Maximum Composite Likelihood method [52]. A total of 357 positions (alter eliminating gaps with complete deletion) were used to calculate the distances for the Neighbor-Joining tree. At each node, bootstrap values with 100 replications were placed based on the frequency that the sequences appeared in the same cluster.

Pyrosequencing and Data Analysis

The Ion Torrent sequence data were compared to pyrosequencing data obtained in a pilot study on a subset of the samples (controls 26P,31P, 46P, 50P; and NaOH treatment 28P, 33P, 48P, and 52P samples) chosen from Table 1. The selected samples were PCR-amplified with barcoded universal primers targeting the V1-V2 region of the 16S rRNA gene usingAdapter/27F (5′ - GCCTTGCCAGCC CGCTC AGTCAGA GT TTGATCCT GGCTC AG - 3′) andAdapter/*Barcode*/338R (5′ -GCCTC CCTCG CGCCA TCAG *NNNN NN NN*CATGCTG CCTCCC GTAGG AGT- 3′) [53] primers. The PCR products were gel purified using a Qiagen gel purification kit (Qiagen, cat. No. 28704), concentrations of the PCR products were measured using NanoQuant (Thermo Fisher Scientific Inc, NH), and then pooled in approximately equimolar amounts. The pooled sample was sequenced by pyrosequencing on a 454 Life Science Genome Sequencer FLX (Roche) at the Environmental Genomics Core at the University of South Carolina, Columbia. Resultant sequence data in standard flowgram format (sff) format were processed using RDP pipeline version 10 [15] to sort the sequence data by barcodes, to trim barcode and forward primer sequences, and to filter out low quality sequences with minimum quality score of 20 (probability threshold of 0.01). The minimum read length was set at 150 bp. The resultant pyrosequence data were compared to those obtained using Ion Torrent sequencing by employing the RDP multiclassifier with the bootstrap threshold set at 60% [54]. Bray-Curtis dissimilarity based PCoA plot was generated using the output of the RDP multiclassifier. Bray-Curtis dissimilarity matrix was generated with the "vegan" package in R using the genus level OTUs matrix that had a relative abundance of 0.5% or more (to remove the effect of singletons and doubletons). PCoA plots were generated in R using the "labdsv" package. The correlation in microbial communities between those derived from Ion Torrent and Pyrosequensing were assessed by applying Pearson correlation analysis to the output of the RDP multiclassifier in R. Pyrosequencing data obtained from this project is available at the NIH Sequence Read Archive (SRA) under the project accession number SRP042367.

RESULTS

After filtering/trimming Ion Torrent output data, a total of 580,605 quality reads were obtained, with an average number of reads equal to 18729±4610 reads and an average sequence length of 380±3 bases per sample, as summarized in Table S1. Purified DNA from a pure culture of *Vibrio cholera* JW612, used as a control for each Ion Torrent run, generated 16270 *Vibrio* sequences out of 16280 reads (99.94%) in the first run and 28031 out of 28082 reads (99.82%) in the second run.

Rarefaction curves of samples shown in Figure 1 and Figure S1 indicate that none of the samples reached saturation, but it appears that the NaOH-treated discharge samples are closest to being saturated. The number of OTUs defined at 97% similarity cutoff were rarefied at around 11,000 reads per sample and were compared across samples (Figure 1 and Figure S1). NaOH treated discharge samples had an average 403 OTUs, which was significantly lower than the number of OTUs in NaOH intake samples (average OTUs=907, t_{27}=7.102, p<0.001) and control discharge samples (average OTUs=1302, t_{27}=13.11, p<0.001). The number of OTUs of control discharge increased slightly compared to intake water to control tanks (average OTUs=1101, t_{27}=2.933, p=0.0067). The statistical significance of the above test held after accounting for multiple comparisons using Tukey's HSD (p=0.032). Similar results were observed for other diversity indices as well (Table 2). For more detailed information, a summary of diversity measurements for each sample is available in Table S2.

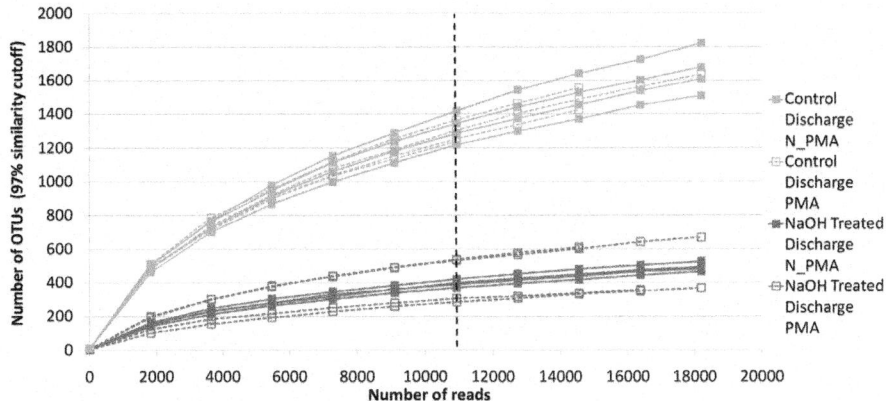

Figure 1. The rarefaction curve of Ion Torrent sequence data. OTUs were defined at 97% similarity cutoff. The figure depicts the comparison between control discharge and NaOH treated discharge samples. The breaking line was placed at around 11,000 (10,918) reads and OTUs were compared across samples when samples were rarefied at 11,000 (10,918) reads.

Table 2. Summary of averaged diversity measurements for each treatment group rarefied at 10,918 reads

	Number of OTUs	Chao1	Shannon
Control Intake	1101	1764	7.226
	(199)	(354)	(0.300)
NaOH Intake	907	1457	6.761
	(152)	(265)	(0.292)
Control Discharge	1302	2101	7.873
	(69)	(158)	(0.132)
NaOH Discharge	403	670	2.971
	(92)	(159)	(0.813)
NaOH Discharge vs. Control Discharge	p<0.001	p<0.001	p<0.001
NaOH Intake vs. NaOH Discharge	p<0.001	p<0.001	p<0.001
Control Intake vs. Control Discharge	p=0.007	p=0.011	p=0.010
Control Intake vs. Control Discharge (HSD)	p=0.032	p=0.051	p=0.045

The numbers inside the parentheses represent the standard deviation. The bottom half shows results from a statistical test comparing treatment groups of interest for each diversity index.
doi:10.1371/journal.pone.0107534.t002

16S rRNA gene amplicon sequence data analysis revealed that microbial community structure characterized at the genus level was altered by NaOH treatment (Figure 2 and Figure S2). Microbial communities of PMA-processed intake samples were similar to one another, with *Limnohabitans*, a genus of beta-proteobacteria, being particularly prominent averaging 15% of the relative abundance across PMA processed intake samples. Intake samples processed without PMA were similarly consistent across all intake samples, with *Limnohabitans* clearly present and various Cryptomonodaceae being more prominent than in the PMA-processed samples. In contrast, *Alishewanella*, a genus of gamma-proteobacteria, accounted for over 50% of the community in NaOH-treated discharge samples. A neighbor-joining tree of *Alishewanella* reads revealed that they were not homogeneous, but had variability in sequences which encompass 3 previously known *Alshewanella* species (Figure S3). The relative abundance of other gamma-proteobacteria genera such as *Rheinheimera* and *Pseudomonas* also increased as well as genera of the phylum Firmicutes such as *Bacillus* and *Exiguobacterium*.

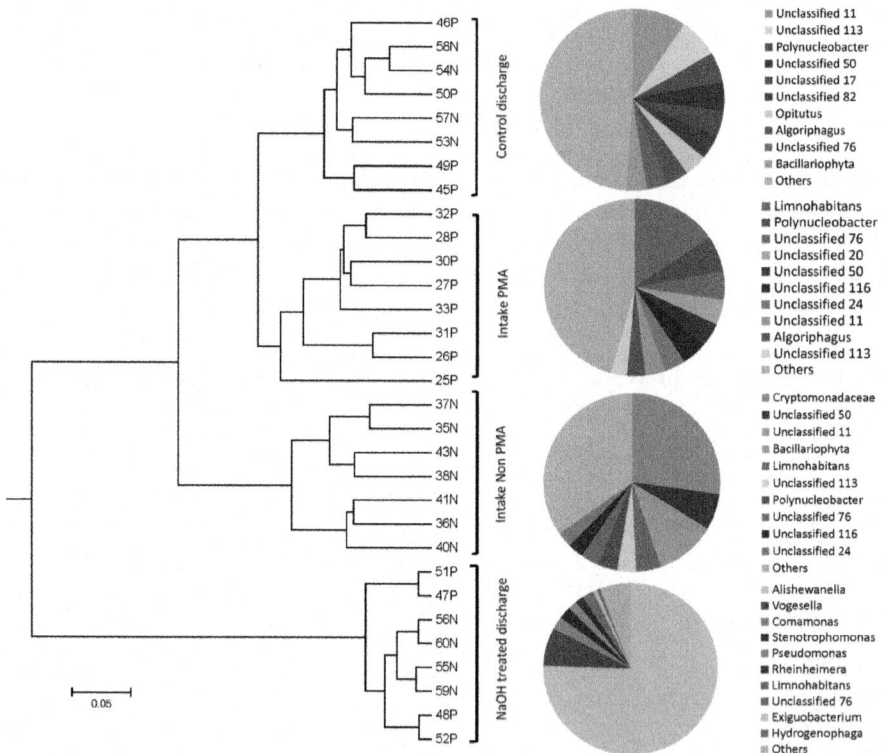

Figure 2. A UPGMA tree constructed using weighted UniFrac distance matrix among the ballast water samples. "P" and "N" in sample ID denote for PMA processed and Non-PMA processed, respectively. The Pie charts depict microbial community structure at genus level for each group. The top ten most abundant genera were shown and others were lumped in others.

Bacteria from the genera *Escherichia*, *Enterococcus*, and *Vibrio* were not detected above noise level (i.e. singleton or doubleton) in either intake or discharge samples in this study. Weighted UniFrac distance based principal coordinate analysis (PCoA) (Figure S4) and UPGMA tree (Figure 2) revealed that the microbial communities treated with NaOH were different from those of intake samples and control discharge samples, and the differences were statistically significant when assessed using Unifrac distance based ANOSIM ($p=0.001$ and $p=0.001$ for NaOH discharge vs. all intake samples, and NaOH vs. Control discharge, respectively). Correlation analysis also revealed that microbial communities of NaOH treated discharge samples did not correlate with those of other groups (Table S3). A small difference in microbial community structure between intake and discharge of control tanks was present; however, this difference was smaller than that between intake and discharge of NaOH treatment tanks (Unifrac distance in Figure 2).

The viable portion of microbial community was assessed by processing samples with PMA. Intake community structures differed between PMA-processed (i.e. live) and non-PMA processed (total DNA) duplicates as described above (Figure 2 and Figure S2). This difference between PMA and non-PMA processed intake samples was also detected in the PCoA plots (Figure S4) and the UPGMA tree (Figure 2), and the ANOSIM showed that the difference was significant ($p=0.001$). In contrast, discharge samples exhibited no significant difference in the microbial community structure between PMA and non-PMA processed samples ($p>0.10$ for both control and NaOH discharge, based on ANOSIM).

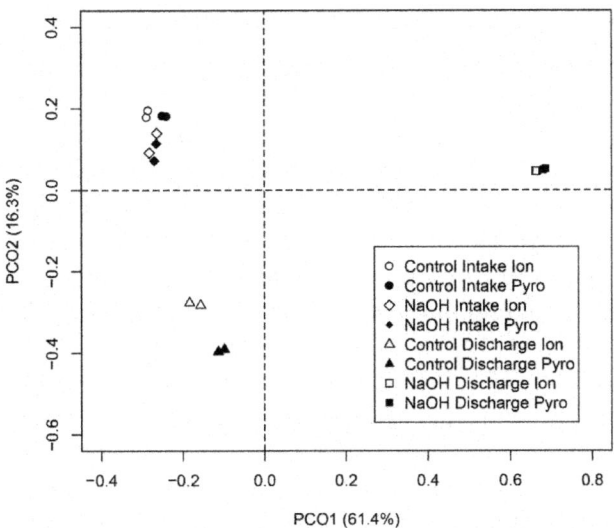

Figure 3. PCoA plot depicting a comparison of microbial communities derived from Ion Torrent and pyrosequencing. RDP multiclassifier with 60% threshold was used to

identify microbial taxa in each sample at genus level. PCoA plot was generated using Bray-Curtis dissimilarity matrix. Taxa that had the relative abundance of 0.5% or greater at least one of the samples were used to generate the matrix. The values in the parentheses indicate the percentage of eigenvalues for each axis.

The microbial community structures obtained using Ion Torrent sequencing were compared to those obtained in a pilot study using pyrosequencing on a subset of the samples (Figure 3 andFigure S5). Substantial correlation in microbial community structure occurred between Ion Torrent sequence data and pyrosequence data when compared at the genus level using Pearson correlation analysis (Table S4).

DISCUSSION

This study demonstrated the application of new Ion Torrent sequencing chemistry to the assessment of microbial community compositions and diversity. Ion Torrent sequencing generated a comparable number of sequence reads and displayed similar microbial community structure to that obtained in a pyrosequencing pilot study. Ion Torrent sequencing also proved to be accurate based on sequences generated in controls. Although the primer set used in the Ion Torrent analysis is targeted at a different region of 16S rRNA genes than the primers used for pyrosequencing, it nevertheless detected a similarly wide range and similar community assembly of microbial taxa. Many of the genera identified in these samples were the same with both methods, but at much lower cost and faster speed with the Ion Torrent.

Alkali ballast water treatment reduced microbial community diversity as shown with OTU richness and other diversity indices and altered microbial community assembly as hypothesized. The most notable change was the increase in the relative abundance of gamma-proteobacteria genera, including *Alishewanella*, *Rheinheimera*, and *Pseudomonas*, and the decrease in beta-proteobacteria including the genus *Limnohabitans* after the alkali treatment.*Alishewanella* strains were isolated from various environments including an Alkali pond in Nebraska (*Alishewanella* sp. SG13. NCBI#:HQ413096), a human fetus [55], tidal sediments[56], fermented seafood [57], landfills [58], textile dye contaminated soils [59], and industrial effluents [60]. According to a previous study by Kim et al. 2010 [58], *Alishewanella agri* BL06 can grow in the pH range between 5.5 and 12.0. Previous studies also found that genus*Alishewanella* is closely related to genera *Rheinheimera* and *Alkalimonas* [57], [61], the latter of which is known to prefer alkali conditions for the optimum growth [61]. The fact that*Alishewanella* spp. accounted for over 50% of relative abundance in ballast water microbial community after NaOH treatment suggests that some strains of *Alishewanella* are especially resistant to alkali stresses.

A previous study that tested alkali resistance among microbial species reported that 11 species of gamma-proteobacteria that they tested were all sensitive to alkali treatment [33]. However, the present study detected a significant increase in the relative abundance of gamma-proteobacteria genera of which the most prominent example was the genus *Alishewanella*. This different outcome could be due to the different experimental setting between the studies. While the previous study was performed using cultures grown in broth (thus, planktonic phase), we tested the efficacy of the alkali water treatment using real ballast water tanks where biofilm could be formed on the surfaces of the tanks [62]. Some gamma proteobacteria strains including *Alishewanella* sp. [63] and *Pseudomonas* sp. [64] are known to form biofilm. Possibly, these gamma-proteobacteria strains survived alkali treatment in the ballast tanks by forming biofilm, which is known to confer resistance to environmental stresses [65]. This suggests that studies on microbial alkali resistance with planktonic phase in laboratory setting may not apply to the real ballast water tank scenario.

Starliper et al. [33] also reported that Gram positive microorganisms, such as those belonging to the phylum Firmicutes, were more resistant to alkali stresses than other phyla they tested. The present study found a similar trend in that the relative abundance of *Bacillus* spp. increased after NaOH treatment. Although this increase in a Firmicutes genus with alkali treatment was detected, these organisms were a small fraction of the total community, accounting for less than 1% of the relative abundance.

Microbial diversity slightly increased from control intake to control discharge. These additional taxa could be derived either from sediments of the ship tank [66] or surface of the tank [62]. Since additional taxa were added during the voyage, the efficacy of the NaOH ballast water treatment should be assessed by comparing control discharge and NaOH discharge rather than NaOH intake and NaOH discharge. Despite the increase in diversity over time in the control tanks, microbial diversity nevertheless significantly decreased when the tanks were treated with NaOH.

The PMA vs. non-PMA comparisons showed that cyanobacteria/chloroplast16S rRNA gene sequences were present in intake samples, but were mostly present in PMA-permeable dead or dying cells. This suggests that the Gary, IN harbor may have had a recent algal bloom in August 2011, although no independent report has been made regarding a bloom around that time. In contrast to this apparent live-dead difference in community in environmental samples, after 3 days treatment of water in ballast tanks, no community assembly differences were present between PMA and non-PMA samples. This was possibly because the genomic materials of dead organisms degraded by the time the ballast water samples were collected.

The new Ion Torrent sequencing chemistry generated data comparable to pyrosequencing. Community structures derived from these two different sequencing techniques were congruent. Both methods have a higher likelihood of errors when homopolymer nucleotides are present since both techniques employ one nucleotide at a time without termination and are thus sensitive to insertion and deletion (in/del) errors [25], while Illumina sequencing is less affected by homopolymer issue [25]. The error rate of Ion Torrent sequencing with the new 400 base chemistry was found to be comparable to that of pyrosequencing [25].

Although our study suggests that some genera of gamma proteobacteria survived through alkali treatment, only two ballast tanks were treated in this study, which is an insufficient sample size for generalization. NaOH treated ballast tanks from other ships need to be studied before the result of this study can be generalized.

The sodium hydroxide treatment adjusted at pH 12 managed to significantly reduce the microbial species diversity of the ballast water. However, it did not manage to completely eliminate all viable microorganisms. Concerns remain with regard to the strains resistant to the alkali treatment. In our study, we focused on the effect of alkali treatment on taxa diversity; however, it is desirable to not only decrease the number of taxa but also to reduce the viable microbial quantity to minimize the impact of biota transportation via ballast water. The Great Ships Initiative (GSI) measured the microbial quantity using heterotrophic most probable number (MPN), and found that the microbial quantity of NaOH treated discharge was two orders of magnitude greater than that of the control discharge [35]. GSI concluded that this was probably due to either the sloughing off of biofilm formed on the inside of ballast tanks or sediments, and/or re-growth of bacteria that were resistant to the alkali treatment after the neutralization. The protective mechanisms of surviving bacteria in the treated tanks (i.e. such as biofilm formation) need to be analyzed and the timing of the neutralization before discharge needs to be evaluated in order design a more effective treatment. A multi stage treatment combining other chemical and ultraviolet light might be more effective at removing organisms in the ballast tanks. Alternatively, the ballast water treatment could target more specific groups of microbes that are known to be harmful and/or pathogenic.

SUPPORTING INFORMATION

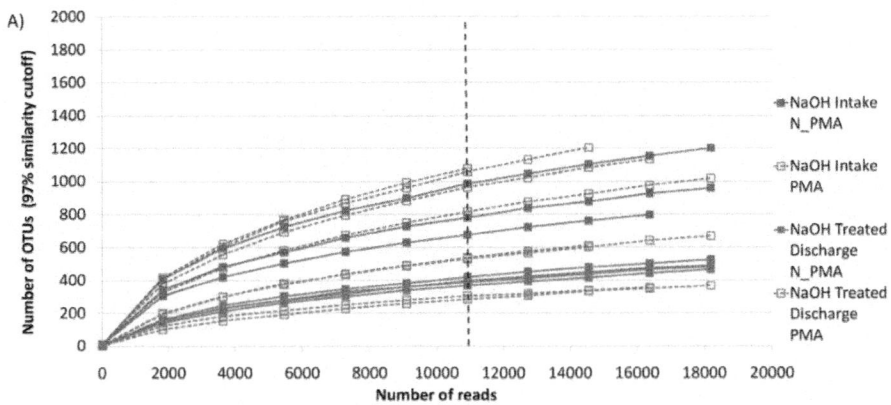

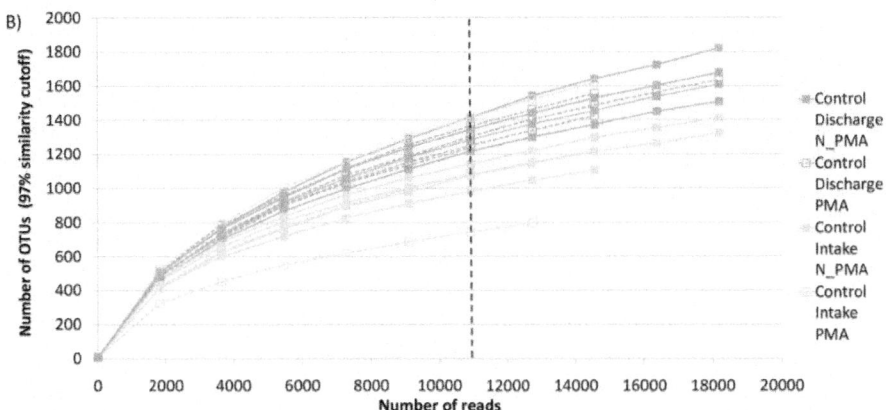

Figure S1. The rarefaction curve of Ion Torrent sequence data. OTUs were defined at 97% similarity cutoff. Panel A) displays the comparison between NaOH intake and NaOH treated discharge samples, and panel B) shows the comparison between control intake and control discharge samples. The breaking line was placed at around 11,000 (10,918) reads and OTUs were compared across samples when samples were rarefied at 11,000 (10,918) reads.

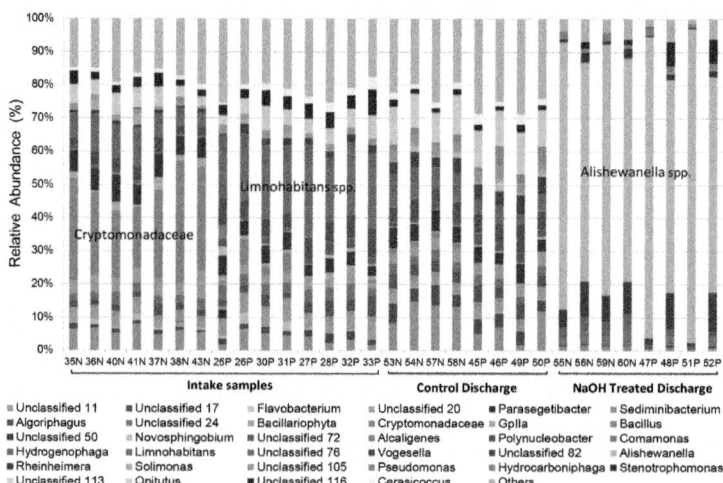

Figure S2. Microbial community assembly of ballast water samples determined at genus level using Ion Torrent. Taxa that have the relative abundance of 2.5% or greater in at least one of the samples were shown in this figure. Some dominant genera were annotated on the figure. "P" and "N" in sample ID denote for PMA processed and Non-PMA processed, respectively. Genus *Bacillus* was included although its relative abundance did not exceed 2.5% in any samples.

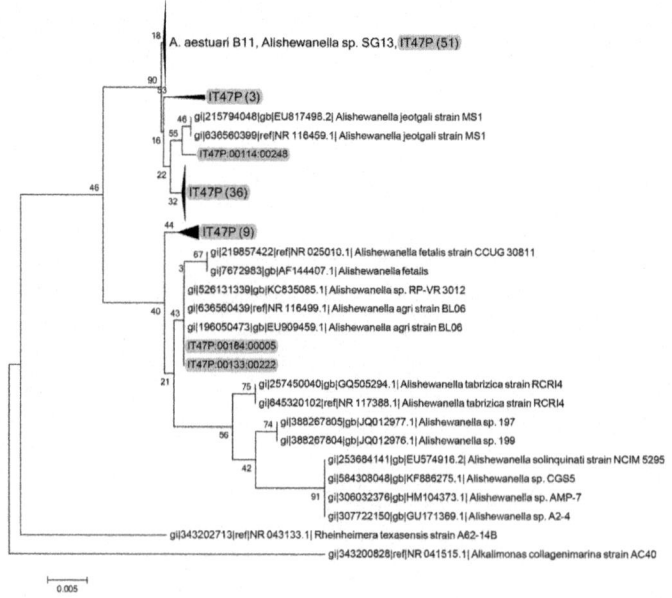

Figure S3. A genus level phylogenetic tree of *Alishewanella* **reads.** The tree was constructed using Neighbor-Joining algorithm. *Rheinheimera* and *Alkalimonas* were used as out-group. 102*Alishewanella* reads were randomly selected from sample 47P and aligned with *Alishewanella*reference sequences obtained from NCBI. The number in the parentheses represents the number of 47P *Alishewanella* reads in the respective clade.

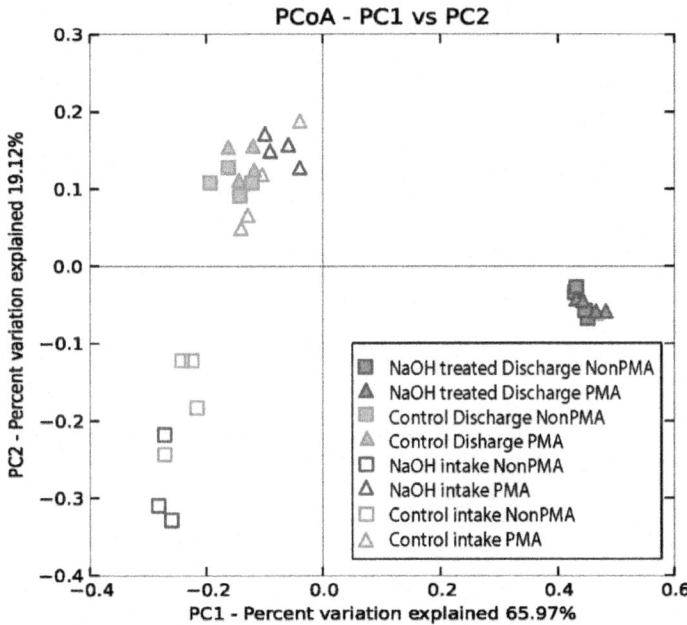

Figure S4. Principal coordinate analysis (PCoA) plots of the ballast water samples using weighted UniFrac distance.

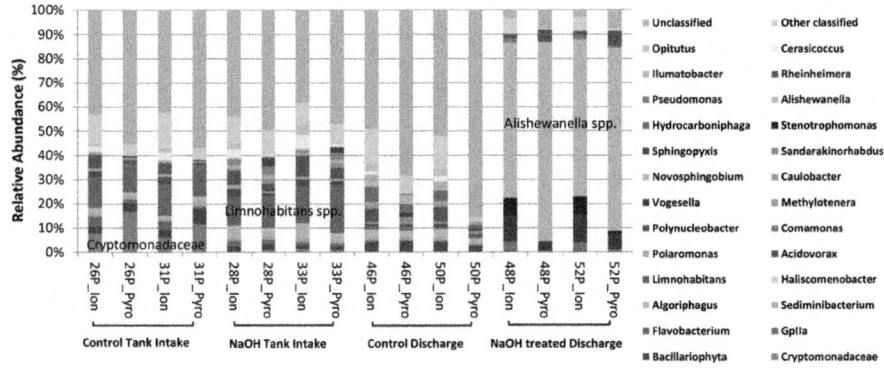

Figure S5. Comparison of microbial community assembly derived from Ion Torrent and pyrosequencing. RDP multiclassifier with 60% threshold was used to identify microbial taxa in each sample at genus level. Taxa that had the relative abundance of 1.5% or greater at least one of the samples were included in this figure.

Table S1. The number of reads and the average size obtained using Ion Torrent sequencing with the 400 chemistry for the 31 ballast water samples.

Sample ID	Ion PGM Run	Raw Reads	RDP Filtered Reads	Average Size
25P	First	19297	13136	378
26P	First	16774	11377	380
27P	First	17307	11675	379
28P	First	25342	17228	378
30P	First	18383	12228	379
31P	First	24800	16428	379
32P	First	23178	15542	378
33P	First	27276	18415	380
45P	First	27592	18427	378
46P	First	17907	11407	377
47P	First	22859	18065	386
48P	First	21938	16154	382
49P	First	24356	16466	378
50P	First	23363	15109	377
51P	First	27880	21906	386
52P	First	27872	19951	381
35N	Second	20045	15124	380
36N	Second	19930	14687	379
37N	Second	27853	20850	379
38N	Second	28432	21112	380
40N	Second	35638	26115	378
41N	Second	33958	24849	378
43N	Second	24184	18025	379
53N	Second	35933	25649	377
54N	Second	30465	22361	378
55N	Second	24442	19260	384
56N	Second	23941	19013	383
57N	Second	34711	25363	377
58N	Second	34501	24596	377
59N	Second	30604	24575	383
60N	Second	32384	25512	383
	Average	25907	18729	380
	Standard deviation	5664	4610	3

Table S2. Summary of diversity measurements for each sample rarefied at 10918 reads.

Sample_ID	Timing	Treatment	Number of OTUs	Chao1	Shannon
37N	Intake	NaOH	984.2	1531.1	6.968
38N	Intake	NaOH	778.8	1197.0	6.332
43N	Intake	NaOH	674.9	1093.8	6.498
27P	Intake	NaOH	1076.5	1743.8	6.885
28P	Intake	NaOH	962.6	1533.4	7.076
32P	Intake	NaOH	1057.7	1785.4	7.006
33P	Intake	NaOH	815.8	1312.8	6.565
35N	Intake	Control	1075.3	1715.9	6.724
36N	Intake	Control	982.7	1538.0	6.983
40N	Intake	Control	1080.4	1604.6	7.332
41N	Intake	Control	1141	1784.8	7.354
25P	Intake	Control	743.4	1204.6	7.054
26P	Intake	Control	1284.2	2228.8	7.403
30P	Intake	Control	1086.3	1754.2	7.254
31P	Intake	Control	1414.4	2285.0	7.705
55N	Discharge	NaOH	366.4	626.4	2.716
56N	Discharge	NaOH	419	688.8	3.512
59N	Discharge	NaOH	395.2	638.8	2.995
60N	Discharge	NaOH	387.7	611.5	3.405
47P	Discharge	NaOH	304	475.0	2.004
48P	Discharge	NaOH	535.8	884.8	3.800
51P	Discharge	NaOH	283.9	515.5	1.602
52P	Discharge	NaOH	530.1	921.4	3.735
53N	Discharge	Control	1417.9	2401.0	7.938
54N	Discharge	Control	1284.5	2083.5	7.713
57N	Discharge	Control	1343.7	2143.8	8.062
58N	Discharge	Control	1216.9	1926.5	7.715
45P	Discharge	Control	1303.5	2122.1	7.885
46P	Discharge	Control	1236.8	1945.7	7.947
49P	Discharge	Control	1251.8	1976.7	7.752
50P	Discharge	Control	1362.8	2210.5	7.977

Table S3. Pearson correlation analysis for Biom matrix data

	46P	49P	37N	43N	32P	58N	54N	57N	53N	41N	36N	38N	40N	50P	33P	45P	28P	51P	31P	30P	26P	35N	56N	48P	52P	25P	27P	60N	47P	55N	59N
46P	1.000	0.792	0.262	0.158	0.548	0.862	0.854	0.780	0.840	0.388	0.295	0.140	0.370	0.928	0.542	0.847	0.572	0.204	0.486	0.577	0.520	0.216	0.218	0.209	0.210	0.512	0.485	0.218	0.205	0.210	0.213
49P	0.792	1.000	0.245	0.144	0.444	0.743	0.700	0.743	0.800	0.299	0.218	0.122	0.305	0.820	0.492	0.934	0.493	0.008	0.347	0.472	0.373	0.181	0.014	0.013	0.474	0.354	0.014	0.009	0.010	0.012	
37N	0.262	0.245	1.000	0.957	0.290	0.346	0.326	0.349	0.354	0.924	0.346	0.964	0.261	0.340	0.275	0.314	0.002	0.483	0.420	0.554	0.971	0.009	0.003	0.004	0.311	0.277	0.010	0.002	0.005	0.007	
43N	0.158	0.144	0.957	1.000	0.158	0.231	0.210	0.248	0.233	0.870	0.916	0.977	0.934	0.163	0.192	0.169	0.176	0.003	0.368	0.287	0.444	0.980	0.006	0.003	0.004	0.192	0.149	0.007	0.003	0.004	0.005
32P	0.548	0.444	0.290	0.158	1.000	0.539	0.561	0.545	0.624	0.510	0.424	0.165	0.403	0.487	0.950	0.517	0.968	0.003	0.936	0.964	0.913	0.223	0.039	0.012	0.015	0.818	0.980	0.043	0.003	0.017	0.026
58N	0.862	0.743	0.346	0.231	0.539	1.000	0.978	0.858	0.918	0.480	0.383	0.224	0.468	0.934	0.533	0.839	0.559	0.133	0.503	0.593	0.557	0.291	0.147	0.137	0.138	0.506	0.478	0.148	0.133	0.139	0.142
54N	0.854	0.700	0.326	0.210	0.561	0.978	1.000	0.863	0.907	0.474	0.378	0.205	0.453	0.926	0.544	0.811	0.568	0.128	0.526	0.613	0.581	0.277	0.143	0.132	0.133	0.525	0.509	0.145	0.129	0.135	0.138
57N	0.780	0.743	0.349	0.248	0.545	0.858	0.863	1.000	0.917	0.445	0.370	0.241	0.436	0.842	0.546	0.850	0.603	0.021	0.498	0.604	0.542	0.293	0.033	0.024	0.025	0.522	0.491	0.033	0.022	0.026	0.028
53N	0.840	0.800	0.354	0.233	0.624	0.918	0.907	0.917	1.000	0.469	0.381	0.222	0.465	0.877	0.630	0.895	0.668	0.013	0.561	0.684	0.597	0.293	0.028	0.017	0.018	0.577	0.567	0.029	0.013	0.019	0.022
41N	0.388	0.299	0.924	0.870	0.510	0.480	0.474	0.445	0.469	1.000	0.946	0.469	0.469	0.261	0.359	0.263	0.283	0.001	0.702	0.624	0.762	0.907	0.019	0.006	0.007	0.480	0.504	0.021	0.001	0.009	0.013
36N	0.295	0.218	0.946	0.916	0.424	0.383	0.378	0.370	0.381	0.976	1.000	0.926	0.953	0.282	0.443	0.263	0.423	0.000	0.636	0.542	0.699	0.938	0.015	0.004	0.005	0.402	0.429	0.017	0.000	0.006	0.010
38N	0.140	0.122	0.963	0.977	0.165	0.224	0.205	0.241	0.222	0.881	0.926	1.000	0.931	0.146	0.196	0.148	0.177	0.001	0.381	0.295	0.457	0.976	0.005	0.001	0.002	0.197	0.160	0.006	0.001	0.003	0.004
40N	0.370	0.305	0.964	0.934	0.403	0.468	0.453	0.436	0.465	0.951	0.953	0.931	1.000	0.381	0.419	0.359	0.410	0.002	0.580	0.530	0.654	0.961	0.014	0.006	0.006	0.407	0.387	0.015	0.002	0.007	0.010
50P	0.928	0.820	0.261	0.163	0.487	0.934	0.926	0.842	0.877	0.385	0.282	0.146	0.381	1.000	0.469	0.899	0.503	0.088	0.422	0.522	0.476	0.220	0.093	0.093	0.476	0.407	0.099	0.088	0.092	0.095	
33P	0.542	0.492	0.340	0.192	0.950	0.533	0.544	0.546	0.630	0.530	0.443	0.196	0.419	0.469	1.000	0.537	0.959	0.001	0.909	0.954	0.879	0.261	0.034	0.010	0.013	0.799	0.946	0.038	0.001	0.015	0.023
45P	0.847	0.934	0.275	0.169	0.517	0.839	0.811	0.850	0.895	0.353	0.263	0.148	0.359	0.899	0.537	1.000	0.572	0.017	0.418	0.549	0.454	0.214	0.025	0.019	0.020	0.519	0.423	0.025	0.017	0.009	0.022
28P	0.572	0.493	0.314	0.176	0.968	0.559	0.568	0.603	0.668	0.505	0.423	0.177	0.410	0.503	0.959	0.572	1.000	0.002	0.903	0.956	0.879	0.234	0.035	0.010	0.013	0.806	0.951	0.038	0.002	0.015	0.023
51P	0.204	0.008	0.002	0.003	0.003	0.133	0.128	0.021	0.013	0.001	0.000	0.001	0.002	0.088	0.001	0.017	0.002	1.000	0.002	0.002	0.001	0.000	0.988	0.984	0.986	0.002	0.002	0.995	1.000	0.998	0.996
31P	0.486	0.347	0.483	0.368	0.936	0.503	0.526	0.498	0.561	0.702	0.636	0.381	0.580	0.422	0.909	0.418	0.903	0.002	1.000	0.953	0.984	0.433	0.037	0.011	0.014	0.761	0.948	0.043	0.002	0.016	0.025
30P	0.577	0.472	0.420	0.287	0.964	0.593	0.613	0.604	0.684	0.624	0.542	0.295	0.530	0.522	0.954	0.549	0.956	0.002	0.953	1.000	0.939	0.361	0.036	0.011	0.013	0.807	0.965	0.040	0.002	0.016	0.024
26P	0.520	0.373	0.554	0.444	0.913	0.557	0.581	0.542	0.597	0.762	0.699	0.457	0.654	0.476	0.879	0.454	0.879	0.001	0.984	0.939	1.000	0.508	0.035	0.010	0.012	0.760	0.916	0.039	0.001	0.015	0.023
35N	0.216	0.181	0.971	0.980	0.223	0.291	0.277	0.293	0.293	0.907	0.938	0.976	0.961	0.220	0.261	0.214	0.234	0.000	0.433	0.361	0.508	1.000	0.006	0.001	0.002	0.249	0.216	0.007	0.000	0.003	0.004
56N	0.218	0.014	0.009	0.006	0.039	0.147	0.143	0.033	0.028	0.019	0.015	0.005	0.014	0.099	0.034	0.025	0.035	0.988	0.037	0.036	0.035	0.006	1.000	0.994	0.995	0.032	0.041	0.993	0.988	0.994	0.996
48P	0.209	0.013	0.003	0.003	0.012	0.137	0.132	0.024	0.017	0.006	0.004	0.001	0.006	0.093	0.010	0.019	0.010	0.984	0.011	0.011	0.010	0.001	0.994	1.000	1.000	0.013	0.013	0.995	0.986	0.989	0.990
52P	0.210	0.013	0.004	0.004	0.015	0.138	0.132	0.024	0.018	0.007	0.005	0.002	0.006	0.093	0.013	0.020	0.013	0.986	0.014	0.013	0.012	0.002	0.995	1.000	1.000	0.014	0.015	0.995	0.988	0.991	0.992
25P	0.512	0.474	0.311	0.192	0.818	0.506	0.525	0.522	0.577	0.480	0.402	0.197	0.407	0.476	0.799	0.519	0.806	0.004	0.761	0.807	0.760	0.249	0.032	0.013	0.014	1.000	0.784	0.034	0.005	0.017	0.024
27P	0.485	0.354	0.277	0.149	0.980	0.478	0.509	0.491	0.567	0.504	0.429	0.160	0.387	0.407	0.946	0.423	0.951	0.002	0.948	0.965	0.916	0.216	0.041	0.013	0.015	0.784	1.000	0.044	0.003	0.018	0.027
60N	0.218	0.014	0.010	0.007	0.043	0.148	0.145	0.033	0.029	0.021	0.017	0.006	0.015	0.099	0.038	0.025	0.038	0.589	0.041	0.040	0.039	0.007	1.000	0.995	0.995	0.034	0.044	1.000	0.990	0.995	0.996
47P	0.205	0.009	0.002	0.003	0.003	0.133	0.129	0.022	0.013	0.001	0.000	0.001	0.002	0.088	0.001	0.017	0.002	1.000	0.002	0.002	0.001	0.000	0.988	0.986	0.988	0.005	0.003	0.990	1.000	0.999	0.996
55N	0.210	0.010	0.005	0.004	0.017	0.139	0.135	0.026	0.019	0.009	0.006	0.001	0.007	0.092	0.015	0.019	0.016	0.998	0.016	0.015	0.015	0.003	0.994	0.989	0.991	0.017	0.018	0.995	0.999	1.000	0.999
59N	0.213	0.012	0.007	0.005	0.026	0.142	0.138	0.028	0.022	0.013	0.010	0.004	0.010	0.095	0.023	0.022	0.023	0.996	0.025	0.024	0.023	0.004	0.996	0.990	0.992	0.024	0.027	0.996	0.997	0.999	1.000

Table S4. Pearson correlation analysis for the RDP multiclassifier output

	26P_Ion	26P_Pyro	28P_Ion	28P_Pyro	31P_Ion	31P_Pyro	33P_Ion	33P_Pyro	46P_Ion	46P_Pyro	48P_Ion	48P_Pyro	50P_Ion	50P_Pyro	52P_Ion	52P_Pyro
26P_Ion	1.000	0.851	0.827	0.810	0.986	0.899	0.841	0.831	0.406	0.311	-0.006	-0.009	0.437	0.345	-0.007	-0.009
26P_Pyro	0.851	1.000	0.493	0.586	0.797	0.957	0.520	0.583	0.189	0.237	-0.006	-0.006	0.204	0.261	-0.006	-0.006
28P_Ion	0.827	0.493	1.000	0.883	0.861	0.625	0.970	0.895	0.533	0.355	-0.007	-0.009	0.572	0.406	-0.007	-0.009
28P_Pyro	0.810	0.586	0.883	1.000	0.836	0.720	0.837	0.973	0.339	0.404	-0.008	-0.009	0.348	0.430	-0.008	-0.009
31P_Ion	0.986	0.797	0.861	0.836	1.000	0.871	0.872	0.865	0.406	0.301	-0.007	-0.009	0.435	0.330	-0.007	-0.009
31P_Pyro	0.899	0.957	0.625	0.720	0.871	1.000	0.632	0.715	0.277	0.362	-0.007	-0.006	0.292	0.410	-0.007	-0.007
33P_Ion	0.841	0.520	0.970	0.837	0.872	0.632	1.000	0.872	0.549	0.313	-0.004	-0.008	0.590	0.368	-0.005	-0.008
33P_Pyro	0.831	0.583	0.895	0.973	0.865	0.715	0.872	1.000	0.296	0.296	-0.006	-0.007	0.297	0.320	-0.006	-0.007
46P_Ion	0.406	0.189	0.533	0.339	0.406	0.277	0.549	0.296	1.000	0.753	0.251	0.249	0.966	0.751	0.249	0.248
46P_Pyro	0.311	0.237	0.355	0.404	0.301	0.362	0.313	0.296	0.753	1.000	0.460	0.466	0.684	0.932	0.460	0.465
48P_Ion	-0.006	-0.006	-0.007	-0.008	-0.007	-0.007	-0.004	-0.006	0.251	0.460	1.000	0.985	0.118	0.172	1.000	0.989
48P_Pyro	-0.009	-0.006	-0.009	-0.009	-0.009	-0.006	-0.008	-0.007	0.249	0.466	0.985	1.000	0.109	0.174	0.984	0.999
50P_Ion	0.437	0.204	0.572	0.348	0.435	0.292	0.590	0.297	0.966	0.684	0.118	0.109	1.000	0.735	0.117	0.110
50P_Pyro	0.345	0.261	0.406	0.430	0.330	0.410	0.368	0.320	0.751	0.932	0.172	0.174	0.735	1.000	0.172	0.173
52P_Ion	-0.007	-0.006	-0.007	-0.008	-0.007	-0.007	-0.005	-0.006	0.249	0.460	1.000	0.984	0.117	0.172	1.000	0.989
52P_Pyro	-0.009	-0.006	-0.009	-0.009	-0.009	-0.007	-0.008	-0.007	0.248	0.465	0.989	0.999	0.110	0.173	0.989	1.000

ACKNOWLEDGMENTS

We thank our collaborators at the National Park Service (NPS) specifically Jeffrey Henquinet, United States Geological Survey (USGS), the Great Ships Initiative (GSI), and the Great Lakes Maritime Research Institute at the University of Wisconsin, Superior campus, for assistance with implementing the shipboard sampling and subsequent on-site sample processing procedures. We also thank the laboratory of Alan Dombkowski at Wayne State University for sharing laboratory space and resources with us. We also thank two anonymous reviewers who helped to improve this manuscript.

AUTHOR CONTRIBUTIONS

Conceived and designed the experiments: JLR MF. Performed the experiments: MF JLR GAM JPG. Analyzed the data: MF. Contributed reagents/materials/analysis tools: MF JLR SN. Wrote the paper: MF JLR. Sample collection: MLR PAG.

REFERENCES

1.   Margulies M, Egholm M, Altman WE, Attiya S, Bader JS, et al. (2005) Genome sequencing in microfabricated high-density picolitre reactors. Nature 437: 376–380.

2.   Bentley DR, Balasubramanian S, Swerdlow HP, Smith GP, Milton J, et al. (2008) Accurate whole human genome sequencing using reversible terminator chemistry. Nature 456: 53–59.

3.   .Sogin ML, Morrison HG, Huber JA, Welch DM, Huse SM, et al. (2006) Microbial diversity in the deep sea and the underexplored "rare biosphere". Proc Natl Acad Sci USA 103: 12115–12120. doi: 10.1073/pnas.0605127103

4. Huber JA, Welch DBM, Morrison HG, Huse SM, Neal PR, et al. (2007) Microbial population structures in the deep marine biosphere. Science 318: 97–100. doi: 10.1126/science.1146689

5. Fierer N, Hamady M, Lauber CL, Knight R (2008) The influence of sex, handedness, and washing on the diversity of hand surface bacteria. Proc Natl Acad Sci USA 105: 17994–17999. doi: 10.1073/pnas.0807920105

6. Lauber CL, Hamady M, Knight R, Fierer N (2009) Pyrosequencing-based assessment of soil pH as a predictor of soil bacterial community structure at the continental scale. Appl Environ Microbiol 75: 5111–5120. doi: 10.1128/aem.00335-09

7. Costello EK, Lauber CL, Hamady M, Fierer N, Gordon JI, et al. (2009) Bacterial community variation in human body habitats across space and time. Science 326: 1694–1697. doi: 10.1126/science.1177486

8. Fujimoto M, Crossman JA, Scribner KT, Marsh TL (2013) Microbial community assembly and succession on Lake Sturgeon egg surfaces as a function of simulated spawning stream flow rate. Microb Ecol 66: 500–511. doi: 10.1007/s00248-013-0256-6

9. Bartram AK, Lynch MD, Stearns JC, Moreno-Hagelsieb G, Neufeld JD (2011) Generation of multimillion-sequence 16S rRNA gene libraries from complex microbial communities by assembling paired-end Illumina reads. Appl Environ Microbiol 77: 3846–3852. doi: 10.1128/aem.02772-10

10. Caporaso JG, Lauber CL, Walters WA, Berg-Lyons D, Lozupone CA, et al. (2011) Global patterns of 16S rRNA diversity at a depth of millions of sequences per sample. Proc Natl Acad Sci USA 108: 4516–4522. doi: 10.1073/pnas.1000080107

11. Caporaso JG, Lauber CL, Walters WA, Berg-Lyons D, Huntley J, et al. (2012) Ultra-high-throughput microbial community analysis on the Illumina HiSeq and MiSeq platforms. ISME J 6: 1621–1624. doi: 10.1038/ismej.2012.8

12. Metzker ML (2009) Sequencing technologies—the next generation. Nat Rev Genet 11: 31–46. doi: 10.1038/nrg2626

13. Glenn TC (2011) Field guide to next-generation DNA sequencers. Mol Ecol Resour 11: 759–769. doi: 10.1111/j.1755-0998.2011.03024.x

14. Loman NJ, Constantinidou C, Chan JZ, Halachev M, Sergeant M, et al. (2012) High-throughput bacterial genome sequencing: an embarrassment of choice, a world of opportunity. Nat Rev Microbiol 10: 599–606. doi: 10.1038/nrmicro2850

15. Cole JR, Wang Q, Cardenas E, Fish J, Chai B, et al. (2009) The Ribosomal Database Project: improved alignments and new tools for rRNA analysis. Nucleic Acids Res 37: D141–D145. doi: 10.1093/nar/gkn879

16. Schloss PD, Westcott SL, Ryabin T, Hall JR, Hartmann M, et al. (2009) Introducing mothur: open-source, platform-independent, community-supported software for describing and comparing microbial communities. Appl Environ Microbiol 75: 7537–7541. doi: 10.1128/aem.01541-09

17. Caporaso JG, Kuczynski J, Stombaugh J, Bittinger K, Bushman FD, et al. (2010) QIIME allows analysis of high-throughput community sequencing data. Nat Methods 7: 335–336. doi: 10.1038/nmeth.f.303

18. Beadling C, Neff TL, Heinrich MC, Rhodes K, Thornton M, et al. (2013) Combining highly multiplexed PCR with semiconductor-based sequencing for rapid cancer genotyping. J Mol Diagn 15: 171–176. doi: 10.1016/j.jmoldx.2012.09.003

19. Elliott AM, Radecki J, Bellal Moghis XL, Kammesheidt A (2012) Rapid detection of the ACMG/ACOG-recommended 23 CFTR disease-causing mutations using ion torrent semiconductor sequencing. J Biomol Tech 23: 24–30. doi: 10.7171/jbt.12-2301-003

20. Chang MW, Oliveira G, Yuan J, Okulicz JF, Levy S, et al. (2013) Rapid deep sequencing of patient-derived HIV with ion semiconductor technology. J Virol Methods 189: 232–234. doi: 10.1016/j.jviromet.2013.01.019

21. Whiteley AS, Jenkins S, Waite I, Kresoje N, Payne H, et al. (2012) Microbial 16S rRNA Ion Tag and community metagenome sequencing using the Ion Torrent (PGM) Platform. J Microbiol Methods 91: 80–88. doi: 10.1016/j.mimet.2012.07.008

22. Jünemann S, Prior K, Szczepanowski R, Harks I, Ehmke B, et al. (2012) Bacterial community shift in treated periodontitis patients revealed by Ion Torrent 16S rRNA gene amplicon sequencing. PLoS ONE 7: e41606. doi: 10.1371/journal.pone.0041606

23. Yergeau E, Lawrence JR, Sanschagrin S, Waiser MJ, Korber DR, et al. (2012) Next-generation sequencing of microbial communities in the Athabasca River and its tributaries in relation to oil sands mining activities. Appl Environ Microbiol 78: 7626–7637. doi: 10.1128/aem.02036-12

24. Merriman B, Torrent I, Rothberg JM (2012) Progress in Ion Torrent semiconductor chip based sequencing. Electrophoresis 33: 3397–3417. doi: 10.1002/elps.201200424

25. Jünemann S, Sedlazeck FJ, Prior K, Albersmeier A, John U, et al. (2013) Updating benchtop sequencing performance comparison. Nat Biotechnol 31: 294–296. doi: 10.1038/nbt.2522

26. Caraco NF, Cole JJ, Raymond PA, Strayer DL, Pace ML, et al. (1997) Zebra mussel invasion in a large, turbid river: phytoplankton response to increased grazing. Ecology 78: 588–602. doi: 10.1890/0012-9658(1997)078[0588:zmiial]2.0.co;2

27. Holland RE (1993) Changes in planktonic diatoms and water transparency in Hatchery Bay, Bass Island area, western Lake Erie since the establishment of the zebra mussel. J Great Lakes Res 19: 617–624. doi: 10.1016/s0380-1330(93)71245-9

28. Lodge DM (1993) Biological invasions: lessons for ecology. Trends Ecol Evol 8: 133–137. doi: 10.1016/0169-5347(93)90025-k

29. Butrón A, Orive E, Madariaga I (2011) Potential risk of harmful algae transport by ballast waters: The case of Bilbao Harbour. Mar Pollut Bull 62: 747–757. doi: 10.1016/j.marpolbul.2011.01.008

30. Doblin MA, Coyne KJ, Rinta-Kanto JM, Wilhelm SW, Dobbs FC (2007) Dynamics and short-term survival of toxic cyanobacteria species in ballast water from NOBOB vessels transiting the Great Lakes— implications for HAB invasions. Harmful Algae 6: 519–530. doi: 10.1016/j.hal.2006.05.007

31. McCarthy SA, Khambaty FM (1994) International dissemination of epidemic Vibrio cholerae by cargo ship ballast and other nonpotable waters. Appl Environ Microbiol 60: 2597–2601.

32. Ruiz GM, Rawlings TK, Dobbs FC, Drake LA, Mullady T, et al. (2000) Global spread of microorganisms by ships. Nature 408: 49–50. doi: 10.1038/35040695

33. Starliper CE, Watten BJ (2013) Bactericidal efficacy of elevated pH on fish pathogenic and environmental bacteria. J Adv Res 4: 345–353. doi: 10.1016/j.jare.2012.06.003

34. Nocker A, Cheung C-Y, Camper AK (2006) Comparison of propidium monoazide with ethidium monoazide for differentiation of live vs. dead bacteria by selective removal of DNA from dead cells. J Microbiol Methods 67: 310–320. doi: 10.1016/j.mimet.2006.04.015

35. Cangelosi A, Allinger L, Balcer M, Fanberg L, Fobbe D, et al. (2013) Final report of the shipboard testing of the sodium hydroxide (NaOH) ballast water treatment system onboard the MV Indiana Harbor. GSI/SB/F/TR/1 1–53.

36. Baker G, Smith J, Cowan DA (2003) Review and re-analysis of domain-specific 16S primers. J Microbiol Methods 55: 541–555. doi: 10.1016/j.mimet.2003.08.009

37. Turner S, Pryer KM, Miao VP, Palmer JD (1999) Investigating Deep Phylogenetic Relationships among Cyanobacteria and Plastids by Small Subunit rRNA Sequence Analysis1. J Eukaryot Microbiol 46: 327–338. doi: 10.1111/j.1550-7408.1999.tb04612.x

38. Haas BJ, Gevers D, Earl AM, Feldgarden M, Ward DV, et al. (2011) Chimeric 16S rRNA sequence formation and detection in Sanger and 454-pyrosequenced PCR amplicons. Genome Res 21: 494–504. doi: 10.1101/gr.112730.110

39. Cai L, Ye L, Tong AHY, Lok S, Zhang T (2013) Biased diversity metrics revealed by bacterial 16S pyrotags derived from different primer sets. PLoS ONE 8: e53649. doi: 10.1371/journal.pone.0053649

40. USCG (2012) Standards for Living Organisms in Ships' Ballast Water Discharged in U.S. Waters; Final Rule. In: Department of Homeland Security. pp. 17254–17320.

41. National Research Council (2011) Assessing the Relationship Between Propagule Pressure and Invasion Risk in Ballast Water: The National Academies Press.

42. Fonseca V, Nichols B, Lallias D, Quince C, Carvalho G, et al. (2012) Sample richness and genetic diversity as drivers of chimera formation in nSSU metagenetic analyses. Nucleic Acids Res 40: e66–e66. doi: 10.1093/nar/gks002

43. Cole JR, Wang Q, Cardenas E, Fish J, Chai B, et al. (2009) The Ribosomal Database Project: improved alignments and new tools for rRNA analysis. Nucleic Acids Res 37: D141–D145. doi: 10.1093/nar/gkn879

44. Edgar RC, Haas BJ, Clemente JC, Quince C, Knight R (2011) UCHIME improves sensitivity and speed of chimera detection. Bioinformatics 27: 2194–2200. doi: 10.1093/bioinformatics/btr381

45. Gihring TM, Green SJ, Schadt CW (2012) Massively parallel rRNA gene sequencing exacerbates the potential for biased community diversity comparisons due to variable library sizes. Environ Microbiol 14: 285–290. doi: 10.1111/j.1462-2920.2011.02550.x

46. R Development Core Team (2009) R: A launguage and environment for statistical computing. R foundation for Statistical Computing, Vienna, Austria.

47. Lozupone CA, Hamady M, Kelley ST, Knight R (2007) Quantitative and qualitative β diversity measures lead to different insights into factors that structure microbial communities. Appl Environ Microbiol 73: 1576–1585. doi: 10.1128/aem.01996-06

48. Borg I (2005) Modern multidimensional scaling: Theory and applications. Berlin: Springer.

49. Felsenstein J (2004) Inferring phylogenies. Sunderland, MA: Sinauer Associates.

50. Tamura K, Peterson D, Peterson N, Stecher G, Nei M, et al. (2011) MEGA5: molecular evolutionary genetics analysis using maximum likelihood, evolutionary distance, and maximum parsimony methods. Mol Biol Evol 28: 2731–2739. doi: 10.1093/molbev/msr121

51. Saitou N, Nei M (1987) The neighbor-joining method: a new method for reconstructing phylogenetic trees. Mol Biol Evol 4: 406–425.

52. Tamura K, Nei M, Kumar S (2004) Prospects for inferring very large phylogenies by using the neighbor-joining method. Proc Natl Acad Sci USA 101: 11030–11035. doi: 10.1073/pnas.0404206101

53. Hamady M, Walker JJ, Harris JK, Gold NJ, Knight R (2008) Error-correcting barcoded primers for pyrosequencing hundreds of samples in multiplex. Nat Methods 5: 235–237. doi: 10.1038/nmeth.1184

54. Wang Q, Garrity GM, Tiedje JM, Cole JR (2007) Naïve Bayesian Classifier for Rapid Assignment of rRNA Sequences into the New Bacterial Taxonomy. Appl Environ Microbiol 73: 5261–5267. doi: 10.1128/aem.00062-07

55. Vogel BF, Venkateswaran K, Christensen H, Falsen E, Christiansen G, et al. (2000) Polyphasic taxonomic approach in the description of Alishewanella fetalis gen. nov., sp. nov., isolated from a human foetus. Int J Syst Evol Microbiol 50: 1133–1142. doi: 10.1099/00207713-50-3-1133

56. Roh SW, Nam Y-D, Chang H-W, Kim K-H, Kim M-S, et al. (2009) Alishewanella aestuarii sp. nov., isolated from tidal flat sediment, and emended description of the genus Alishewanella. Int J Syst Evol Microbiol 59: 421–424. doi: 10.1099/ijs.0.65643-0

57. Kim M-S, Roh SW, Nam Y-D, Chang H-W, Kim K-H, et al. (2009) Alishewanella jeotgali sp. nov., isolated from traditional fermented food, and emended description of the genus Alishewanella. Int J Syst Evol Microbiol 59: 2313–2316. doi: 10.1099/ijs.0.007260-0

58. Kim M-S, Jo SK, Roh SW, Bae J-W (2010) Alishewanella agri sp. nov., isolated from landfill soil. Int J Syst Evol Microbiol 60: 2199–2203. doi: 10.1099/ijs.0.011684-0

59. Kolekar YM, Pawar SP, Adav SS, Zheng L-Q, Li W-J, et al. (2013) Alishewanella solinquinati sp. nov., isolated from soil contaminated with

textile dyes. Curr Microbiol 67: 454–459. doi: 10.1007/s00284-013-0385-7

60. Jain R, Jha S, Adhikary H, Kumar P, Parekh V, et al. (2014) Isolation and Molecular Characterization of Arsenite-Tolerant Alishewanella sp. GIDC-5 Originated from Industrial Effluents. Geomicrobiol J 31: 82–90. doi: 10.1080/01490451.2013.811317

61. Ma Y, Xue Y, Grant WD, Collins NC, Duckworth AW, et al. (2004) Alkalimonas amylolytica gen. nov., sp. nov., and Alkalimonas delamerensis gen. nov., sp. nov., novel alkaliphilic bacteria from soda lakes in China and East Africa. Extremophiles 8: 193–200. doi: 10.1007/s00792-004-0377-4

62. Drake LA, Meyer AE, Forsberg RL, Baier RE, Doblin MA, et al. (2005) Potential invasion of microorganisms and pathogens via 'interior hull fouling': biofilms inside ballast water tanks. Biol Invasions 7: 969–982. doi: 10.1007/s10530-004-3001-8

63. Jung J, Chun J, Park W (2012) Genome Sequence of Extracellular-Protease-Producing Alishewanella jeotgali Isolated from Traditional Korean Fermented Seafood. J Bacteriol 194: 2097. doi: 10.1128/jb.00153-12

64. Palleroni NJ (2010) Pseudomonas. Topley & Wilson's Microbiology and Microbial Infections. Hoboken, NJ: John Wiley & Sons, Ltd.

65. Landini P (2009) Cross-talk mechanisms in biofilm formation and responses to environmental and physiological stress in Escherichia coli. Res Microbiol 160: 259–266. doi: 10.1016/j.resmic.2009.03.001

66. Drake LA, Doblin MA, Dobbs FC (2007) Potential microbial bioinvasions via ships' ballast water, sediment, and biofilm. Mar Pollut Bull 55: 333–341. doi: 10.1016/j.marpolbul.2006.11.007

CITATION

CHAPTER 1

Wang X, Gai Z, Yu B, et al. Degradation of Carbazole by Microbial Cells Immobilized in Magnetic Gellan Gum Gel Beads. *Applied and Environmental Microbiology.* 2007;73(20):6421-6428. doi:10.1128/AEM.01051-07.

CHAPTER 2

Tadahiro Suzuki and Yumiko Iwahashi, "Low Toxicity of Deoxynivalenol-3-Glucoside in Microbial Cells," Toxins 2015, 7(1), 187-200; doi:10.3390/toxins7010187.

CHAPTER 3

More VS, Tallur PN, Niyonzima FN, More SS. Enhanced degradation of pendimethalin by immobilized cells of *Bacillus lehensis* XJU. *3 Biotech.* 2015;5(6):967-974. doi:10.1007/s13205-015-0299-0.

CHAPTER 4

Berlowska J, Kregiel D, Ambroziak W. Physiological tests for yeast brewery cells immobilized on modified chamotte carrier. *Antonie Van Leeuwenhoek.* 2013;104(5):703-714. doi:10.1007/s10482-013-9978-1.

CHAPTER 5

Andreas Blecha, Kristof Zarschler, Klaas A Sjollema, Marten Veenhuis and Gerhard Rödel, "Expression and cytosolic assembly of the S-layer fusion protein mSbsC-EGFP in eukaryotic cells," Microbial Cell Factories20054:28, DOI: 10.1186/1475-2859-4-28.

CHAPTER 6

Mozzetti et al.: New method for selection of hydrogen peroxide adapted bifidobacteria cells using continuous culture and immobilized cell technology. Microbial Cell Factories 2010 9:60.

CHAPTER 7

Meerza Abdul Razak and Buddolla Viswanath, "Comparative studies for the biotechnological production of l-Lysine by immobilized cells of wild-type *Corynebacterium glutamicum* ATCC 13032 and mutant MH 20-22 B," Volume 5, Issue 5, pp 765-774, doi: 10.1007/s13205-015-0275-8.

CHAPTER 8

Rachel C Wagner, Sikandar Porter-Gill and Bruce E Logan, "Immobilization of anode-attached microbes in a microbial fuel cell," AMB Express2012 2:2, DOI: 10.1186/2191-0855-2-2.

CHAPTER 9

Fehlbaum S, Chassard C, Haug MC, Fourmestraux C, Derrien M, Lacroix C (2015) Design and Investigation of PolyFermS *In Vitro* Continuous Fermentation Models Inoculated with Immobilized Fecal Microbiota Mimicking the Elderly Colon. PLoS ONE 10(11): e0142793. doi:10.1371/journal.pone.0142793

CHAPTER 10

Orellana MV, Pang WL, Durand PM, Whitehead K, Baliga NS (2013) A Role for Programmed Cell Death in the Microbial Loop. PLoS ONE 8(5): e62595. doi:10.1371/journal.pone.0062595.

CHAPTER 11

Ishii S, Suzuki S, Norden-Krichmar TM, Nealson KH, Sekiguchi Y, Gorby YA, et al. (2012) Functionally Stable and Phylogenetically Diverse Microbial Enrichments from Microbial Fuel Cells during Wastewater Treatment. PLoS ONE 7(2): e30495. doi:10.1371/journal.pone.0030495.

CHAPTER 12

Fujimoto M, Moyerbrailean GA, Noman S, Gizicki JP, Ram ML, Green PA, et al. (2014) Application of Ion Torrent Sequencing to the Assessment of the Effect of Alkali Ballast Water Treatment on Microbial Community Diversity. PLoS ONE 9(9): e107534. doi:10.1371/journal.pone.0107534.

INDEX